Control
of
Embryonic
Gene Expression

Editor

M. A. Q. Siddiqui, Ph.D.
Associate Member
Biochemistry Department
Roche Institute of Molecular Biology
Nutley, New Jersey

CRC Press, Inc.
Boca Raton, Florida

Library of Congress Cataloging in Publication Data
Main entry under title:

Control of embryonic gene expression.

Bibliography: p.
Includes index.
1. Embryology. 2. Gene expression. 3. Cell
differentiation. 4. Cellular control mechanisms.
I. Siddiqui, M. A. Q. [DNLM: 1. Gene expression
regulation. 2. Embryo. 3. Genes, Regulator.
4. Morphogenesis. 5. Genetic code. QH 453 C764]

QL971.C64   1983   591.3'3   82-17840
ISBN 0-8493-5756-X

Direct all inquiries to CRC Press, Inc., 2000 Corporate Blvd., N.W., Boca Raton, Florida, 33431.

© 1983 by CRC Press, Inc.

International Standard Book Number 0-8493-5756-X

Library of Congress Card Number 82-17840
Printed in the United States

PREFACE

Recent studies in well-defined systems have disclosed many new facts concerning biochemical and physiological aspects of fertilization, morphogenesis, and development of various forms of life. The crucial area of ignorance in our understanding of embryology, however, is the mechanism that controls these events. Modern embryologists are faced with the task of defining and interpreting the classical embryological concepts on induction, determination, pattern formation, etc. in molecular terms. The goal is to provide a satisfactory basis for understanding the mechanisms whereby an individual cell develops and maintains the diversity of cell types characteristic of the adult state out of the original single cell, the fertilized ovum.

The present state of knowledge strongly supports the view that development and differentiation is a function of variable gene activity. Although we are still on the threshold of knowledge about genes and their functions, it is becoming clear that a coherent view, at least for a few fundamental processes involved in regulation of gene expression, might emerge. It was desirable, therefore, to assemble the available information on molecular biology of early embryonic development and present it in this book as a comprehensive collection of existing data with the pertinent literature review and the evaluation. The book also serves as a useful guide for understanding the basic embryological concepts, which are discussed in a manner intelligible to newcomers to molecular embryology. In addition, it conveys effectively the substance of experimental approaches for future investigations. The subject matter of each chapter naturally reflects the interest of the author engaged in the particular domain of research. The first two chapters discuss the complex regulatory mechanisms underlying fertilization in multicellular organisms. The following chapters attempt to present the fundamental issues concerning the regulation of development. The potential of RNA as a developmental signal is the focus of several articles.

I am greatly indebted to all the authors for making a conscientious effort to provide sufficient background information and present in-depth analyses of the work with a balanced presentation of facts, concepts, and projections. I am also indebted to Ms. Katherine Tighe for secretarial assistance.

M. A. Q. Siddiqui
Editor

THE EDITOR

M. A. Q. Siddiqui, Ph.D., is an Associate Member in the Department of Biochemistry, Roche Institute of Molecular Biology, Nutley, N. J. Dr. Siddiqui is also a Visiting Professor of Biochemistry, Rutgers State University, New Brunswick, N. J. and Professor Extraordinary, Universidad Austral de Chile, Valdivia, Chile. Dr. Siddiqui received his Ph.D. in 1967 from the University of Houston, Texas. He received the postdoctoral training at the University of California, Berkeley, in 1967 to 1969, and served as a Research Associate at the Roche Institute of Molecular Biology from 1969 to 1972. He was appointed an Assistant Member in the Department of Biochemistry at the Institute in the year 1972. He assumed his present position in 1979.

Dr. Siddiqui was elected to memberships of the American Society of Biological Chemists and the International Cell Research Organization. He is also a member of the International Society of Differentiation and Society for Developmental Biology. He has lectured as an invited speaker at many international meetings and symposia and has organized and conducted workshops on molecular cloning at the Institute of Developmental Biology, Peking, China, and at the Institute of Biochemistry, Universidad Austral de Chile, Valdivia.

Dr. Siddiqui is the author of about 50 scientific publications and review articles and co-edited a book. His current major research interest involves control of gene expression in early embryonic development.

CONTRIBUTORS

Fiorenza De Bernardi, Ph.D.
Associate Professor
Zoological Laboratory
Department of Biology
State University
Milano, Italy

Rosaria De Santis, Ph.D.
Investigator
Stazione Zoologica
Napoli, Italy

M. R. Dohmen, Ph.D.
Scientist
Zoological Laboratory
University of Utrecht
Utrecht, The Netherlands

William R. Jeffery, Ph.D.
Associate Professor
Department of Zoology
The University of Texas at Austin
Austin, Texas

Alberto Monroy, Ph.D.
Director
Stazione Zoologica
Napoli, Italy

Tamito Noto, Ph.D.
Lecturer
Embryology Laboratory
Department of Anatomy
Faculty of Medicine
and
Assistant Professor
Faculty of Dentistry
Kagoshima University
Kagoshima, Japan

Silvio Ranzi, Ph.D.
Professor
Zoological Laboratory
Department of Biology
State University
Milano, Italy

Floriana Rosati, Ph.D.
Associate Professor
University of Siena
Siena, Italy

Joel Schindler, Ph.D.
Assistant Professor
Department of Anatomy and Cell Biology
University of Cincinnati College of
  Medicine
Cincinnati, Ohio

Michael I. Sherman, Ph.D.
Full Member
Department of Cell Biology
Roche Institute of Molecular Biology
Nutley, New Jersey

M. A. Q. Siddiqui, Ph.D.
Associate Member
Biochemistry Department
Roche Institute of Molecular Biology
Nutley, New Jersey

Harold C. Slavkin, D.D.S.
Professor of Biochemistry and Nutrition
Department of Basic Sciences/Section of
  Biochemistry
                  and
Laboratory Chief
Laboratory for Developmental Biology
Andrus Gerontology Center
University of Southern California
Los Angeles, California

Margarita Zeichner-David, Ph.D.
Research Assistant Professor
Department of Basic Sciences/Section of
  Basic Sciences
Laboratory for Developmental Biology
Andrus Gerontology Center
University of Southern California
Los Angeles, California

TABLE OF CONTENTS

Chapter 1

THE ROLE OF THE EGG SURFACE IN DEVELOPMENT

M. R. Dohmen

TABLE OF CONTENTS

I. INTRODUCTION

In a number of recent models the cell surface functions as the main site of regulatory activity in differentiation and pattern formation. Brunner[1] has proposed a model in which the interaction between the cell membrane and the genome explains how differential transcription is directed. He assumes that differentiation is triggered by signals from the microenvironment of the cell rather than by the expression of an endogenous "master" program in the genome. These signals are perceived by receptors in the membrane and relayed to the genome. The genome is programmed to produce a certain sequence of receptors in the plasma membrane. At each stage of differentiation receptors are expressed for the perception of that signal which induces the progression to the next differentiation stage. Membrane-mediated control of differentiation is also assumed in the model of Sumper,[2] which explains certain aspects of the differentiation and pattern formation in *Volvox*.[3] This model is based on the assumption that specific recognition proteins, able to mediate cell-cell contacts, are sequentially expressed at successive cleavages and thus provide a mechanism for counting cell divisions and triggering the differentiations in specific cells. McMahon[4] has proposed a model for position determination in which cells are assumed to interact via contact sensing molecules on their plasma membranes. These respond to cell contact by regulating the intracellular concentrations of a second messenger such as cyclic AMP. The model attempts to explain the development of the pseudoplasmodium of the slime mold *Dictyostelium discoideum* where cells become committed to differentiate into either spore or stalk cells on the basis of their position in the slug.

The emphasis in these models is mostly on cellular interactions and how they bring about differentiation and pattern formation. In this review the focus is mainly on the uncleaved egg, whose surface is thought to be involved in a number of processes that are of vital importance for normal development. These processes include (1) establishing of polarity and regulation of ooplasmic segregation; (2) establishing the conditions for correct fertilization, viz., ensuring species-specific sperm penetration, restricting possible sperm entrance points, and providing blocks to polyspermy; (3) mediating egg activation, the continuation of meiosis, and the onset of cleavage; and (4) determining the cleavage pattern. Of course, not all of these processes occur in every species. In particular, the events associated with fertilization are very variable between different species. The latter events will not be discussed in this review as they are treated elsewhere in this book.

An important aspect of the organization of an egg is its capacity for compartmentalization. It is the basis for differentiation as it provides the initial inhomogeneities that will finally result in cells with different potentials. This mechanism may operate in a direct way by depositing specific determinant factors in certain blastomeres, or it may operate in an indirect way by creating morphogenetic fields or other systems that allow cells to interact in such a way that the proper determinations occur in the right place at the right time. A combination of both processes probably occurs in every species, with the possible exception of mammals and triclad turbellarians, where there are no indications of cytoplasmic localizations in the egg that operate in determination.

Experiments described in the second section of this chapter indicate that the egg surface plays an important role in bringing about or maintaining cytoplasmic inhomogeneities in the egg. This implies that the egg surface itself is not homogeneous and the emphasis in this review is on this aspect.

II. EXPERIMENTS INVOLVING THE EGG CORTEX

A. Centrifugation Experiments

Subjecting eggs to moderate centrifugal forces results in stratification of the cytoplasmic inclusions. Higher forces will separate the eggs in a heavy and a light fragment. It is not known to what extent endoplasmic cytoskeletal structures or other components that are responsible for endoplasmic organization are resistant to the disruptive centrifugal force. Reports by Conklin[5] on eggs of the mollusc *Crepidula*, and by Lillie[6] on eggs of the annelid *Chaetopterus* indicate that a cytoplasmic "spongioplasm" structure is unaffected by centrifugation. These studies have not been repeated with modern techniques to visualize this spongioplasm, e.g., high voltage electron microscopy or immunofluorescence. It is generally assumed that the egg surface is not displaced by centrifugation, although it may be stretched by the elongation of the egg. The existence of a cortical cytoplasm as a separate entity displaying some rigidity has been doubted in some species, as in centrifuged eggs cytoplasmic inclusions may be found touching the plasma membrane.[7] However, in eggs possessing cortical granules some sort of cortical structure clearly exists, as the cortical granules are not displaced by centrifugation, e.g., in eggs of the sea urchin *Arbacia*,[8] the annelid *Chaetopterus*,[6] and the scaphopod *Dentalium*. A more detailed account of cortical structure will be given in the next section of this chapter.

A large number of reports exists on the effects of centrifugation of eggs on development. It appears that in many species the eggs develop into abnormal organisms after centrifugation, but in many other species the stratification of the cytoplasmic inclusions does not impair normal development. In the latter category, normal development may ensue only under certain conditions. In ascidians, for instance, centrifugation is harmless only when it is carried out on unfertilized eggs.[9,10] In the molluscs *Pholas* and *Spisula* normal veliger larvae develop from eggs in which the first cleavage has not been equalized by centrifugation.[11] In the snail *Lymnaea*, the effect of centrifugation is much more harmful immediately prior to third cleavage than at earlier or later stages.[12]

Abnormal development is generally explained by arguing that the normal position of cytoplasmic determinants in the egg is disturbed by the stratification of the cytoplasm. Ascidian eggs demonstrate this quite convincingly: eggs centrifuged after fertilization give rise to embryos in which all organ primordia are present, but their relative positions are abnormal.[9] Such a clear example can only be provided by organisms showing a highly mosaic type of development, where the determinants for different organ primordia each have a distinct localization in the egg and are independent of each other. In eggs whose development is not purely mosaic, centrifugation may have effects whose causal analysis is far more complicated. To explain the fact that in many species the stratification of the cytoplasm does not preclude normal development, the existence of a cortical scaffold is generally invoked. This scaffold is not supposed to suffer from centrifugation. In Raven's words, the cortex "carries a morphogenetic field, based on local differences in its structure and properties, which has in part a mosaic character. It has been established during oogenesis by the interaction of the oocyte with the surrounding elements of the gonad."[13] If at the time of centrifugation the cytoplasmic determinants are already segregated and rigidly bound to the cortical scaffold, centrifugation will have little effect on further development. This mechanism apparently operates in polar lobe-forming molluscan eggs.

The morphogenetic determinants present in the polar lobe exert a dominant influence on development.[14] This can be demonstrated by removing the polar lobe at first cleavage and studying the resulting development. The larva reared from a lobeless egg invariably shows a large number of characteristic defects. The same defects are observed when the polar lobe of a *Dentalium* egg is removed after centrifugation.[15] This demonstrates that the lobe-specific determinants have not been removed from the polar lobe although the cytoplasm of the lobe may have a composition which is completely different from the normal one. Similarly, in *Ilyanassa* eggs Clement[16] has shown that vegetal egg fragments from which the normal cytoplasm had been removed by centrifugation were able to differentiate lobe-dependent structures. These results clearly demonstrate that morphogenetic determinants may be rigidly fixed to a localized region of the egg surface.

In eggs of the snail *Bithynia* the effects of centrifugation on morphogenetic determinants can be observed directly and the results fully confirm the conclusions drawn from the experiments on *Dentalium* and *Ilyanassa*. In *Bithynia* eggs the polar lobe contains a characteristic aggregate of vesicles called the vegetal body.[17] This aggregate cannot be displaced unless centrifugal forces are employed which disrupt part of the eggs. If these high forces are used, the vegetal body is displaced in about 40% of the eggs. When the eggs are centrifuged before cleavage, they cleave normally afterwards and a polar lobe is normally formed, but this lobe does not contain the vegetal body in a number of eggs. If such a lobe, which lacks the vegetal body, is removed, the egg may develop normally.[14] This suggests strongly that the vegetal body contains the necessary determinants and demonstrates at the same time that there is indeed a cortical mechanism that is able to bind cytoplasmic determinants very strongly.

The difference between the effects of centrifugation before and after fertilization in ascidian eggs can be explained by arguing that localization of determinants in these eggs takes place after fertilization. In fact this can be clearly seen by the appearance of differently colored cytoplasmic regions in ascidian eggs.[18] Centrifuging eggs before fertilization has no effect on morphogenesis as the cortical scaffold exerts its positioning effect afterwards. Once localized, however, the determinants appear not to be as rigidly fixed to the cortex as the polar lobe determinants in molluscan eggs. They can be displaced by centrifugation and consequently disorder results. However, the assumption that there is a rigidly fixed preformed cortical pattern, established during oogenesis and triggered into action sooner or later, is untenable in many cases. A few examples will be discussed. In many species the site of sperm entrance has a definite relationship with the cleavage pattern. It can be argued that both the sperm entrance point and the cleavage pattern are determined by a cortical prepattern. Apart from eggs possessing a micropyle in the surrounding chorions, e.g., in insects, cephalopods, and fish,[19] there are a few reports that the sperm enters the egg at a predetermined site. For example, in *Rana* sperm entrance is limited to the pigmented animal hemisphere[20] and in *Discoglossus* the sperm can only enter through the animal dimple.[21] In ascidian eggs the sperm seems to enter the egg at the vegetal pole,[22] but animal halves can also be fertilized so there is no strict determination in this respect. In the mollusc *Cumingia* and the annelid *Chaetopterus* the first cleavage plane has a tendency to pass through the sperm entrance point. This relationship can be uncoupled by centrifugation. After centrifugation the first cleavage plane tends to parallel the axis of stratification.[23] In the molluscs *Pholas* and *Spisula* the dorsal blastomere D is normally located opposite to the sperm entrance point. After centrifugation the dorsal blastomere can occupy any position respective to the sperm entrance point.[11] In the ctenophore *Pleurobrachia* Freeman[24] suspects that the site where the first cleavage is initiated corresponds to the sperm entrance point. This site indicates the future oral pole of the embryo and the plane of the first cleavage corresponds to the sagittal plane of the embryo. The first cleavage plane can be shifted by centrifugation and the newly established initiation site will be the future oral pole.

Other examples are provided by reversals of animal-vegetative polarity. This axis has been shown in many species to originate from the position of the oocyte in the ovary. The site of attachment of the oocyte generally becomes the vegetal pole of the egg.[25,26] Guerrier[27,28] has shown that in *Parascaris* and *Limax* eggs this initial polarity can be reversed by centrifugation or compression before the extrusion of the second polar body. Before going into a discussion of these results, a survey of other types of experiments will be presented.

B. Micromanipulation and Other Experiments

Probably the best known among these experiments are the graftings of grey crescent cortex in eggs of *Xenopus*.[29,30] Removal of the cortex in the grey crescent region completely prevented morphogenesis, although the removed layer was only 1-μm thick. Grafting the grey crescent cortex into the ventral side of another fertilized egg induced the formation of a double embryo. In sea urchin eggs more than half of the cytoplasm may be sucked out with a micropipette and still the eggs may give rise to normal plutei.[31] Apparently the endoplasm is relatively unimportant: the cortex contains the vital determinants and this structure is supposed to remain intact in an egg from which cytoplasm is removed. In *Dentalium* eggs 60% of the ooplasm has been removed via a pipette inserted in the vegetal hemisphere of the uncleaved egg. Nevertheless, the eggs developed normally.[32] Similarly, partial removal of the ooplasm from eggs of the ascidian *Ciona* did not impair normal development, though only if carried out on unfertilized eggs.[33] In insect eggs, localized damage to the cortical region or to the blastoderm can produce localized lesions in later stages and this has been used to establish fate maps. The extent to which cortical patterns controlling differentiation can be reestablished following experimental intervention varies; in some species there is considerable flexibility while in the higher diptera and some coleoptera there is little or none.[34] Whether these results should be explained by assuming a mosaic of different and independent determinants or by the action of one or more gradients is still an unresolved question.[35] In *Smittia* eggs, Kalthoff[36] has demonstrated that the anterior determinants are not exclusively localized in the yolk-free cortex. In *Loligo* eggs, regional treatment with small blocks of agar soaked in a solution of cytochalasin B resulted in positionally related effects during organogenesis.[37] In the sea urchin *Paracentrotus* treatment of eggs immediately after fertilization with a detergent resulted in a large percentage of larvae lacking an archenteron, spicules, and mesenchyme cells. These results suggest a cortical influence on the determinants of entomesodermic structures.[38]

All these observations support the view that the cortex is the repository of developmental determinants. However, a number of experiments are difficult to reconcile with the assumption that there is a cortical prepattern. In unfertilized eggs of *Ciona*, partial removal of either cortex or ooplasm has no effect on development.[33] After fertilization, the cortex may still be partially removed without harmful consequences, but removal of ooplasm results in abnormal development. Ortolani[39] has shown that every fragment of the unfertilized or fertilized egg of *Ascidia malaca* and *Phallusia*, if sufficiently large, will give rise to perfectly normal tadpoles, irrespective of the plane of section. Later experiments on *Ascidia malaca* showed that when unfertilized eggs were divided into equal parts by equatorial, meridional, or oblique sections, both parts could be fertilized and developed into complete larvae.[40] In the nemertine *Cerebratulus* unfertilized eggs can be cut equatorially into animal and vegetal halves and then fertilized. Each fragment frequently develops into a normal larva.[41] Thus, the putative cortical pattern may be severely disturbed in some species without impairing normal development.

C. Discussion

It is evident that the egg surface has the capacity to position and bind cytoplasmic components. This is most convincingly demonstrated by the experiments involving the vegetal body of the snail *Bithynia*. The spatial coordinates of the egg of *Bithynia* are predetermined during oogenesis. This can be inferred from the observation that the vegetal body originates during vitellogenesis at the basal pole of the oocyte. This pole is apparently determined to become the future vegetal pole of the egg. There are various other eggs in which the basal pole of the oocyte can be unambiguously identified as the future vegetal pole of the egg.[26] In some eggs this animal-vegetative polarity can be reversed by experimental intervention during a certain period. This means that the polarity, though present, is not rigidly imprinted in the cortex during that period.

The most extensive series of studies on this problem has been done with zygotes of brown algae. In the absence of any other stimulus the apical-basal polarity is determined by the sperm entrance point, but a large array of external stimuli can overrule this initial determination, e.g., the direction of illumination, electrical currents, centrifugation in combination with pH treatment, and the presence of other eggs.[26] In *Pelvetia* zygotes a local influx of ions and secretion of wall material occur at the presumptive rhizoid site before the polar axis is definitely fixed.[42] If the polar axis is shifted by reversing the orienting light, the local inward current changes accordingly to the new dark side of the zygote within 40 min, followed in about an hour by the deposition of wall material at the new site of inward current. These observations suggest that polar axis fixation results from stabilization of current-producing membrane components. Quatrano et al.[43] have proposed a model in which cytoskeletal filaments act in the transport of membrane components, involved in transporting ions, to the dark side of the light gradient. After having accumulated, they can shift again to another site if the orientation of the light is changed. This process can go on until the membrane components are definitely anchored by cytoskeletal elements.

In animal eggs a similar process may occur. The microenvironment of the oocyte in the ovary serves to initiate a labile apical-basal polarity. This initial polarity can be changed by a variety of stimuli, probably by acting on the cortical cytoskeleton. The timing of the definite fixation of polarity probably differs between species. In *Parascaris* and *Limax* the positioning of the second maturation spindle seems to trigger the definite polar axis fixation. The determination of other spatial coordinates, for example, the plane of bilateral symmetry, is often triggered by the entry of the sperm. A clear example of the initial lability of the spatial pattern induced by the sperm entry is provided by the *Xenopus* egg. The position of the grey crescent is dependent both on the distribution of yolk within the egg and on the point of sperm entry.[44] Its location corresponds to the future blastopore site. Rotation experiments show that the determining effect of the sperm can be overruled.[45] In eggs orientated with the sperm entrance point on top, a blastopore will form in the sperm entrance region, contrary to the rule that the blastopore appears opposite to the sperm entrance point. This reversal occurs even when the experiment is started after formation of the grey crescent as well as shortly before first cleavage. These results imply that the grey crescent may not be the ultimate determinant for the dorsal side and hence Curtis' experiments[29,30] may have to be reevaluated.

The sperm entrance often provokes a visible reaction of the egg surface. Such a reaction may be wave-like propagated from the site of sperm entrance. In *Xenopus* a series of waves has been observed between fertilization and first cleavage. Immediately following fertilization a wave-like propagation of dark-light-dark zones from the site of sperm entrance called the "activation wave"[46] presumably reflects the movement of the front of cortical granule breakdown. Sometime later two "postfertilization waves"[47] proceed from the sperm entrance point to the opposite side of the eggs. In axolotl eggs, just before first cleavage two "surface contraction waves"[48] proceed from the animal pole to the vegetal pole. In *Rana* the pigmented area of the cortex contracts towards the animal pole after fertilization or activation. The function of this contraction might be to move the sperm nucleus towards the animal pole.[20] Nondirectional stimuli may also provoke a cortical response. For instance, a heat shock can induce the appearance of the grey crescent in unfertilized axolotl eggs.[49] These examples demonstrate that the egg surface is a dynamic system that is capable of reacting on stimuli. Such a reaction might unmask a preformed but cryptic cortical pattern, but it is also possible that the stimulus triggers some cortical process that establishes a pattern whose spatial orientation is not preestablished but is dependent on the position of the stimulus, e.g., the sperm entrance point. In view of the experiments described above the latter mechanism seems to be the most plausible one. Still the microsurgical experiments carried out on eggs of *Cerebratulus*[41] and of several ascidians[39,40] remain hard to explain. In these eggs an initial labile polarity is clearly present in the unfertilized egg. In *Cerebratulus* eggs, for instance, a small peduncle frequently marks the vegetal pole of the egg. Fragments of these eggs are each capable of establishing the necessary set of spatial coordinates. A simple shifting of provisionally localized polar determinants to another site is impossible, because if they were initially localized, they must be absent in one of the two egg fragments. Possibly the initial pattern is not a mosaic of distinct factors. It may be formed by a gradient in which each segment has regulative capacity. Freeman's[24] statement that "eggs from some groups of animals have essentially no promorphological organization at the end of oogenesis" is probably true as far as the localization of cytoplasmic morphogenetic determinants is concerned. However, the eggs are clearly biased to develop along spatial coordinates that are evident already in the unfertilized egg. This implies the existence of a promorphological organization at the end of oogenesis, however labile this organization may be.

III. THE STRUCTURE OF THE CELL SURFACE

A. The Cortical Cytoplasm

In all cells, eggs as well as differentiated cells, a distinct cortical cytoplasm probably exists. In murine fibroblasts[50] the cytocortex is characterized by a high concentration of ground substance and a complete exclusion of all cellular organelles. It contains very short, interconnected threads and electrondense globules, forming a network. In erythrocytes a membrane cytoskeleton, composed of a specific protein (spectrin), actin, and other elements has been demonstrated.[51] Moore et al.[52] have proposed a model in which α-actinin associates directly with the membrane, possibly through interaction with an integral membrane protein. Short actin filaments interact with the α-actinin at one end and are tied together to form a submembrane matrix by cross bridges of actin-binding protein and myosin. Actin-binding protein is thus able to provide stability for the submembrane matrix while the myosin-actin interaction exerts a contractile force that can be used to power membrane movements.

Detailed knowledge of cortical structure has been provided by the study of ciliated protozoa. These organisms have a highly complex and easily observable cell surface organization that might serve as a model system for other organisms in which cortical patterns are often hard to detect. In *Tetrahymena* a distinct cortical layer, the epiplasm, is characterized by two major proteins, probably in association with actin.[53] One of the most spectacular properties of the ciliate cortex is its potential for pattern maintenance. A reversed ciliary pattern, caused by grafting a small piece of cortex into a host after 180° rotation, has been preserved through 1500 fissions in *Tetrahymena*.[54] The mechanism responsible for this cortical inheritance has been termed "cytotaxis" by Sonneborn[55] and "structural guidance" by Frankel.[56] The local cortical environment created by the position and orientation of preexisting ciliary units within a ciliary row determines the position and orientation of new units within the same row. The old structures serve as scaffolds for the new ones.[57] Apart from this mechanism for short-range positioning, there is also a mechanism for long-range positioning in ciliates. The latter mechanism operates through gradients of positional value. These gradients should probably not be thought of as gradients of a diffusible morphogen, as this concept is difficult to apply to a unicellular system. What else we should think of is not clear at present.

The existence of a distinct cortical cytoplasm in eggs was postulated from the observation that in sea urchin eggs the peripheral region shows birefingence.[58,59] A variety of biophysical and micromanipulation studies of the rigidity of the surface have confirmed the existence of a distinct cortical layer in sea urchin eggs. Its thickness has been measured by various methods and the values obtained range between 2 and 6 μm. In a recent study of sea urchin eggs Higashi[60] found that the cortical layer in the fertilized egg is not a gel but consists of a hard matrix. It exists also in the unfertilized egg, but in a more fluid form. Schatten and Mazia[61] have shown by scanning electron microscopy that the cortex consists of an overlapping network of fibers measuring between 50 and 200 nm. The thickness of the cortex, as measured on SEM micrographs, ranges between 0.2 and 0.5 μm, which is far less than the values obtained with other methods. The cortex contains myosin and actin and seems to cause a spreading surface deformation at fertilization which may be involved in the secretion of the cortical granules, in the detachment of the vitelline sheet from the egg surface, and perhaps in the rapid block to polyspermy.[61] A cortical protein with spectrin-like properties has been isolated from sea urchin cortices by Kane.[62] Harris[63] has observed a spiral cortical fiber system, visible with phase-optics, in *Strongylocentrotus* eggs between 30 and 70 min after fertilization. Franke et al.[64] have studied the cortex of various amphibian oocytes. It appears to consist of a mechanically stable meshwork of microfilament webs, dense aggregates of coiled filaments, paracrystalline arrays of microfilaments, microtubules, and vesicles. Actin is located in the microvilli and in small aggregates in the cortex. The authors suggest that the oocyte cortex contains a considerable amount of actin and micro-filaments in storage forms. Long bundles of parallel microfilaments as they have been observed in the cleavage furrows of axolotl eggs[65] are only rarely found in the oocyte cortex.

The mechanism that brings about unequal cleavage is particularly intriguing. Extreme examples are the meiotic divisions which give rise to the polar bodies. The mechanism responsible for the formation of polar bodies does not reside in any particular characteristic of the meiotic apparatus. If a meiotic apparatus is displaced by centrifugation, it may induce the formation of normally sized blastomeres, the so-called giant polar bodies.[5] The meiotic apparatus becomes rigidly anchored to the cortex during metaphase and then the peripheral aster disassembles prior to the internal aster, thus allowing the spindle to move closer to the egg surface.[75] Czihak[76] has suggested that a similar phenomenon possibly occurs during the formation of the micromeres in sea urchin eggs. The vegetal cortex might differentially depolymerize microtubules so that the peripheral asters become smaller and the spindle moves closer to the vegetal pole. In ciliated protozoa the role of the cell surface in positioning the nucleus and determining its behavior has been thoroughly investigated.[77] A particularly striking example is provided by the macronucleus of *Stentor*.[78] This nucleus is located beneath and parallel to the narrowest cortical striping of the cell's right side. When the direction of the striping is changed by microsurgery, the position of the macronucleus is shifted accordingly. In stentors with reversed asymmetry of the cortical pattern, the position of the macronucleus is also reversed. It is concluded, therefore, that the cortical striping determines the orientation of the macronucleus. Franke et al.[64] have shown that there is no prepattern of contractile rings of filaments in amphibian oocytes. The formation of these structures is induced immediately before actual cleavage occurs. The nature of the stimulus is obscure.[79] Calcium appears to play an important role in activating the cortical contractile system, as microinjection of calcium or treatment with ionophores can provoke the appearance of various kinds of cortical contractions.[80-83] The waves of cortical granule breakdown also depend on the presence of calcium. Microinjection of calcium into amphibian oocytes induces dehiscence of cortical granules.[84] In eggs of the medaka a wave of release of free calcium can be shown to precede the wave of cortical granule breakdown.[85]

Apart from the phenomena already described above, there are many other contractile events to be observed in eggs. A prominent example is provided by the constriction of polar lobes in eggs of a number of molluscs and annelids. A contractile ring composed of cyto-chalasin B-sensitive filaments has been shown to cause this constriction[86] which is independent of any nuclear activity as it occurs also in enucleated vegetal halves.[87,88] In many eggs the maturation divisions are accompanied by extensive deformations. The most conspicuous deformations have been observed in eggs of the annelid *Tubifex*. Shimizu[89] has described two successive deformations during maturation: grooves caused by contracting microfilaments appear first on the equatorial surface, then on the animal hemisphere. In eggs of the barnacle, *Pollicipes polymerus*, a series of constriction rings moves slowly from the animal to the vegetal pole.[90] The constrictions are inhibited by cytochalasin B, but not by colchicine. Pulsations have been described in the eggs of *Nereis*.[91] The pulsations start immediately before the breakdown of the germinal vesicle, about 15 min after insemination, and proceed continuously with the exception of a short rest period of 3 to 5 min which occurs at about 25 min after insemination. In eggs of the bivalve molluscs *Barnea candida* Pasteels[92] has observed that occasionally undulating movements of the vitelline membrane occur. They may be provoked with greater intensity by cold treatment. These undulations result from successive waves of elongation of the microvilli. This elongation does not involve any undulation of the plasma membrane. Rhythmic movements of the fertilization membrane have also been observed in the eggs of *Chaetopterus, Mactra,* and *Nereis.*[93] The function of this surface activity is generally obscure. A role in cytoplasmic localization or in egg activation seems to be plausible. However, when the peristaltic constrictions in *Pollicipes* eggs are inhibited by cytochalasin B, ooplasmic segregation continues, and when ooplasmic segregation is inhibited by antimycin A, the constriction continues.[90]

B. The Plasma Membrane

The prevailing model of plasma membrane structure[94] depicts the membrane as a fluid lipid matrix in which integral proteins are embedded, whereas peripheral proteins are bound weakly to the surface. The lipid matrix is responsible for the basic membrane structure and the proteins are considered to be responsible for specific functions. A single phospholipid such as phosphatidylcholine could satisfy the structural requirements of the basic structure.[95] The observation that the membrane of a single cell may contain as many as one hundred or more different lipids suggests that lipids play other functional roles. The ability of lipids to adopt nonbilayer configurations in addition to the bilayer phase provides a variety of possibilities for the direct involvement of lipids in many functional abilities of biological membranes.[95] However, as far as specific functions are concerned, most data refer to membrane proteins. In accordance with the fluid nature of the membrane, most membrane proteins are randomly distributed over the cell surface, as far as can be judged from experiments in which the exposed proteins or carbohydrate groups were labeled with antibodies, lectins, anionic or cationic probes, etc.[96] The mobility and distribution of membrane proteins are thought to be affected by cortical microfilaments and microtubules, and this may result in nonrandom patterns. Well-known examples of stable patterns of membrane components are found in the intercellular junctions. The different types of junctions are characterized in freeze-fracture preparations by organized arrays of inner-membrane particles (IMP).[97] Other examples are found in the ciliated protozoa *Tetrahymena* and *Paramecium* which display two types of stable repetitive IMP arrays, viz., ciliary ''necklaces'' and trichocyst or mycocyst insertion sites, regularly alternating over the cell surface.[98] Examples of nonrandom distribution of antigens, surface immunoglobulins, and anionic sites have also been reported.[96,99]

Redistribution of membrane components can be brought about in many ways: by temperature changes, alterations in membrane lipid composition, pH changes, enzymes, membrane-active drugs, labeling of surface antigens, lectin receptors and anionic sites with appropriate ligands, etc.[96] The involvement of cytoskeletal elements is suggested by several observations of striking correlations between patterns of membrane components and patterns of underlying cytoskeletal elements. For example, in fibroblasts the concanavalin A receptors can be induced to form linear arrays of clusters which appear to be aligned with intracellular stress fibers.[100] Singer[101] observed a correlation between linear or fusiform aggregates of pinocytotic vesicles on fibroblasts and cortical microfilament bundles. Elgsaeter et al.[102] found that the aggregation of spectrin in erythrocyte ghosts correlated with aggregation of IMP. Even peripheral proteins such as fibronectin, which is loosely associated with the external membrane surface,[103] seem to maintain transmembrane connections with the cortical cytoskeleton. Double-label immunofluorescence has shown a definite correspondence between the fibrillar arrays of fibronectin and intracellular actin.[104] Analogous investigations have failed to detect any relationship between fibronectin and microtubules or intermediate filaments. As neither actin nor fibronectin are integral membrane proteins, there must be intervening proteins connecting the two.

The role of cytoskeletal elements can be experimentally demonstrated by disrupting them with colchicin, cytochalasin B, local anesthetics,[105] or other agents. After disruption ligand-induced redistribution of surface receptors is facilitated.[95,105] Disassembly of microfilaments by cytochalasin B results in release of fibronectin from the cell surface.[104] The current view is that microtubules and microfilaments play opposing roles in the distribution of membrane components. Microtubules, in this concept, serve to anchor the components and microfilaments tend to redistribute them by their contractile activity. The interplay between these two opposing but coordinated systems determines the topography of the membrane.[96,99]

Eggs are highly specialized cells and we may expect that this is also manifested in the properties of the plasma membrane and in its modulations during development. One of the properties that can be investigated is the lateral mobility of membrane components. This can be accurately measured by means of the fluorescence photobleaching recovery (FPR or FRAP) technique.[106-108] This technique involves labeling a specific cell surface receptor with a fluorescent ligand followed by laser-induced bleaching of the fluorescence in a small region. The time required for the recovery of the original amount of fluorescence in this region is a measure of the rate of lateral diffusion of neighboring unbleached fluorophores into the bleached region. In mouse eggs Johnson and Edidin[109] measured the diffusion of antigens in this way and found a very high diffusion coefficient in the unfertilized egg. After fertilization a dramatic reduction was observed. On the other hand, in sea urchin eggs an increase in membrane fluidity was observed after fertilization.[110] The latter measurements were done by using a spin label fatty acid, so the fluidity of the lipid matrix is measured. In FPR measurements of the lateral mobility of proteins the fluidity of the lipid matrix may be of minor importance if the mobility is restricted by cytoskeletal elements. The results obtained by Johnson and Edidin[109] may thus reflect a sudden restriction of lateral mobility by anchoring of the antigens to cortical elements following fertilization, independent of any change in lipid matrix fluidity.

Evidence for a high degree of stability of membrane components is provided by Bluemink et al.[111] who showed that in *Xenopus* eggs the preexisting and the nascent membrane are continuous but nevertheless maintain their highly different IMP densities. Bluemink and Tertoolen[112] suggest that IMP-associated filaments probably are instrumental in creating long-range stability in the plane of the membrane. These filaments may control cell surface topography and as such they may function in establishing polarity and bilateral symmetry in the amphibian egg. Other examples of stable patterns have been found in sea urchin eggs. After fertilization the membrane of the cortical granules is incorporated in the plasma membrane. The mosaic membrane thus produced is composed of smooth patches, derived from the cortical granule membranes, and microvillous patches, derived from the original plasma membrane. This mosaic surface is then reorganized by the elongation of microvilli and by the reduction in size of the smooth patches. The mosaic character of the surface is still recognizable after microvillar elongation, indicating that the different membranous components do not intermix.[113] However, recent measurements of surface area indicate that the mosaic membrane is not a long-lasting composite structure. A large area of egg surface is resorbed within 10 min after fertilization.[114] A considerable degree of stability in sea urchin and mouse eggs is demonstrated by the persistence of the sperm plasma membrane as a discrete patch in the egg membrane.[115] This shows that there is a limited lateral mobility and a low turnover of surface components in these eggs.

If the spatial organization of the egg is primarily under the control of the egg surface, one might expect to find patterns of membrane components that correlate with animal-vegetative polarity and cytoplasmic localizations. These patterns have only rarely been found. In *Xenopus* eggs the overall densities of IMP in the animal and the vegetal hemisphere do not differ significantly.[116] However, for IMP sizes smaller than 81 Å, a significant difference was found, more small IMPs being present in the animal hemisphere. Labeling of externally exposed carbohydrate groups with lectins is a widely employed technique for studying the distribution of glycoproteins or glycolipids on the cell surface.[117,118] In this way a polar organization of the egg membrane has been demonstrated in a few egg types. In mouse eggs the animal pole region does not bind concanavalin A prior to the formation of the second polar body.[119] This negative area is sequestered with the second polar body, leaving an entirely positive egg surface after maturation. The presence or absence of concanavalin A binding correlates with the presence or absence of microvilli,[120] so the mosaicism should be explained by structural variation of the surface architecture instead of unequal molecular distribution of concanavalin A receptors. In hamster eggs lectin binding is localized in small spots over the entire egg surface except for the surface of the first polar body.[121] This also suggests a polar organization of the egg membrane. In *Ascidia* eggs *Dolichos* lectin shows a patchy distribution on unfertilized eggs. Concurrently with the cytoplasmic redistribution that follows fertilization, the surface staining becomes strictly localized to the vegetal pole of the egg. In the course of cleavage it concentrates on the B-4,1 blastomeres from which the muscles of the tail arise.[122] In *Discoglossus* eggs, a specific pattern of lectin binding transiently appears in the animal dimple when this structure is completely formed. The binding disappears immediately after fertilization.[123] Other studies using lectins show either the complete absence of binding or a uniform binding on the entire surface, depending on stage and species used.

The spatial organization of an egg may manifest itself in the generation of electrical currents. Local accumulations of ion-transporting channels in the plasma membrane probably bring about this phenomenon. The zygotes of brown algae have been most thoroughly studied in this respect. Before the apical-basal polarity of the zygotes is irreversibly established, a current of calcium ions enters the presumptive apical role and leaves the basal pole.[124] In several other developing or regenerating systems electrical currents have been measured, as reviewed by Borgens et al.[125] Jaffe et al.[126] have suggested that electrical currents may serve to self-electrophorese cytoplasmic components, e.g., negatively charged vesicles, and thus control development by bringing about cytoplasmic localizations. In *Fucus* zygotes localization of fucoidin, a sulfated polysaccharide, is brought about by the directional transport of Golgi vesicles containing fucoidin to the rhizoid pole. This directional transport may be caused by electrophoresis, but as it can be prevented by treatment with cytochalasin B and D[127] it is more likely to be brought about by microfilament activity. The fact that only a few examples of membrane mosaicism in eggs have been found is not surprising if we take into account that finding suitable probes for detecting specific membrane components is not an easy task. Many efforts will be needed to make some progress in this field.

C. The Surface Architecture

The interaction between the cortical cytoskeleton and the plasma membrane can lead to a variety of surface structures: microvilli, blebs, ridges, folds, ruffles, spikes, spines, etc. Microvilli are found on almost every egg species. These structures are among the best studied surface phenomena.[128] The first stage in the formation of a microvillus is the attachment of a densely staining material to the cytoplasmic face of the plasma membrane. This material then appears to induce the formation of a microfilament bundle and subsequently the microvillus grows out.[129] The plaque of dense material anchors the bundle of microfilaments end-on to the plasma membrane, while they are braced to the membrane along the length of the microvillus by α-actinin.[128] This solid construction does not preclude dynamic behavior and rapid reorganization of the cell surface. A few examples illustrate this.

In sea urchin eggs two bursts of microvillar elongation occur after fertilization.[130] In *Spisula* eggs a structural reorganization of microvilli takes place upon insemination.[131] The bases of the microvilli no longer anastomose with one another to form the complex reticular pattern seen in the unfertilized egg. At the same time the number of IMP in the microvillar membranes increases approximately twofold. In the mollusk *Bankia*, the pattern of the microvilli changes from an irregular array before fertilization into a hexagonal pattern after fertilization.[132] In *Tubifex*[133] and *Nereis*[134] most microvilli disappear after fertilization. In eggs of the molluscs *Barnea* waves of elongation of microvilli have been observed by Pasteels.[92] Experimental intervention can provoke considerable changes in number and distribution of microvilli. A spectacular example can be seen in the reaction of mouse oocytes on cytochalasin B.[135,136] The oocytes undergo "pseudocleavage" into two compartments, one containing the nucleus, the other anucleate. The surface of the nucleate compartment is completely devoid of microvilli, whereas the anucleate part is densely covered with them. The microvilli are initially withdrawn from the surface of the oocyte. After the pseudo-cleavage the microvilli reappear on the anucleate compartment. When lymphocytes are incubated in hypertonic medium, the microvilli on the cell surface are first transformed into lamellae. These are translocated to one pole of the cell and then transformed again into microvilli, thus forming a cap of accumulated microvilli at one pole and leaving a smooth surface on the remaining part of the cell.[137]

These phenomena illustrate the dynamic potential of the surface architecture. However, stable mosaic patterns have also been observed. The absence of microvilli in a neatly defined region at the animal pole of mouse eggs has already been mentioned.[71,119,120] In eggs of sea anemone the area adjacent to the pronucleus displays short villi, 2 to 3 μm in length, whereas the remaining surface is covered with 10- to 20-μm long villous structures.[138] In the unfertilized egg of *Xenopus*[139] the density of microvilli is greater on the vegetal hemisphere than on the animal hemisphere and the vegetal microvilli are more slender than those on the animal half of the egg. After fertilization the microvilli are almost completely withdrawn. This suggests a relationship with the electrical currents observed in the same species. Robinson[140] found that a current enters the animal hemisphere and leaves the vegetal hemisphere in fully grown oocytes. The current falls to nearly zero in response to maturation-producing agents such as progesterone. A number of molluscan eggs exhibit striking examples of mosaicism in their surface structure. In the egg of *Nassarius reticulatus* the pattern of microvilli differs between the animal and vegetal hemisphere (Figures 1 and 2). On the vegetal hemisphere the microvilli are arranged in parallel rows, whereas on the animal hemisphere no such pattern can be distinguished. On polar lobe-forming eggs a number of local surface differentiations associated with the presence of morphogenetic determinants have been described.[141] The eggs of *Nassarius reticulatus* display an area at the vegetal pole whose surface architecture differs from the remaining egg surface. The two domains are separated by a sharply defined boundary line (Figure 3). The surface architecture of the vegetal pole area may vary considerably. This variation is probably not due to stage-dependent modulations but to different fixation procedures, as illustrated by Figures 4 to 6. This vegetal area corresponds exactly to the area with which the oocyte was attached to the ovarian wall and thus lends some support to Raven's[13] view that a mosaic structure may be imprinted on the oocyte surface by the surrounding structures in the ovary.

In eggs of *Buccinum* the vegetal pole area is characterized by the extensive outgrowth and branching of microvilli, resulting in a typical pattern of surface ridges (Figures 7 and 8). A dense cortical cytoplasmic layer is present exclusively in this area.[141] Its detailed structure and composition have not yet been elucidated. The eggs of *Crepidula* show a symmetrical pattern of folds at the vegetal pole.[141] These folds probably arise at the end of the maturation and are preceded by an elongated streak of blebs and villi. The intriguing suggestion that this streak and the symmetry axis of the subsequently arising pattern of folds may indicate a prepattern of the direction of the first cleavage plane is presently being investigated. Unpublished observations indicate that similar mosaics in surface architecture also occur on other polar lobe-forming eggs, e.g., in the molluscs *Littorina* and *Bithynia*, and in the annelid *Sabellaria*. The vegetal surface differentiations in all these polar lobe-

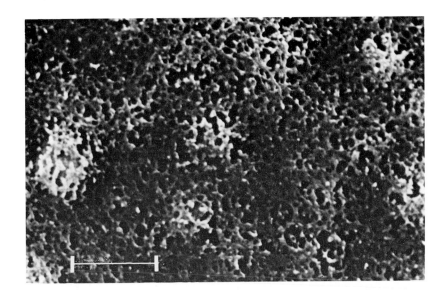

FIGURE 1. *Nassarius reticulatus:* pattern of microvilli on the animal hemisphere. Remnants of extracellular material partially cover the micovilli. Bar indicates 4 μm.

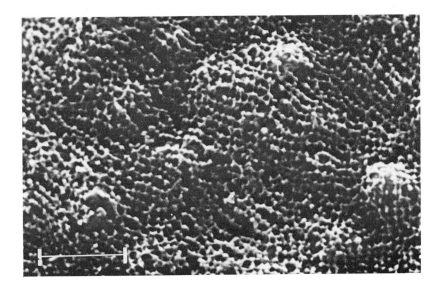

FIGURE 2. *Nassarius reticulatus:* pattern of microvilli on the vegetal hemisphere. The microvilli are arranged in parallel rows. Remnants of extracellular material partially cover the microvilli. Bar indicates 4 μm.

forming eggs are shunted to the D-quadrant macromere, together with the morphogenetic determinants of the polar lobe.[14] The most conspicuous example is provided by *Crepidula* (Figure 9). This observation supports the view that the particular surface characteristics are somehow associated with the localization of polar lobe determinants. Additional arguments are provided by the finding that in the annelids *Clepsine* and *Tubifex* similar mosaic surface patterns are associated with the location of the pole plasms which are very prominent in these species and whose morphogenetic role has been experimentally demonstrated.[142]

The functional significance of these local differentiations is not clear. Their most obvious

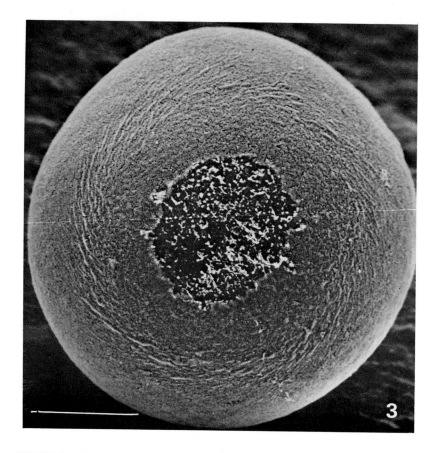

FIGURE 3. *Nassarius reticulatus:* a locally differentiated area on the vegetal pole of the uncleaved egg is characterized by the absence of the microvilli that otherwise cover the entire egg surface. The surface architecture of the vegetal pole patch may differ considerably, as shown in Figures 4 to 6. Bar indicates 50 μm.

common feature is an increase in surface area, suggesting an increased transport. Maybe an increased influx of ions results in an electrical current through the egg. A local increase in surface activity, manifested by the appearance of folds, villi, or blebs, has been observed many times, particularly in association with the formation of cleavage furrows.[143] These surface protrusions may provide the extra membrane needed for furrow formation. They may also constitute a general way of reacting on the stimuli that give rise to cytokinetic events. The type of reaction may depend on the animal species. This can be illustrated by comparing the reactions occurring at the animal and vegetal pole in the eggs of *Bithynia* and *Crepidula*. In eggs of *Bithynia*, immediately before the extrusion of the first polar body an array of long villi appears both at the animal and the vegetal pole. These areas are structurally almost indistinguishable. Both are a prelude to a cytokinetic event, viz., the formation of a polar body and the protrusion of a polar lobe. The main difference is that the villi at the animal pole disappear as soon as the polar body has been formed, whereas at the vegetal pole the villi increase in number and eventually cover the polar lobe entirely. They remain unchanged after the resorption of the lobe, like the vegetal folds in the eggs of *Crepidula* (Figure 9). In *Crepidula* eggs the cleavage furrows are associated with extensive arrays of folds (Figure 10) that resemble the folds on the vegetal pole. Thus in both egg types surface activity accompanies cytokinesis and the type of activity, folds or villi, is dependent on the species. In *Nassarius*, however, surface structures like those on the vegetal pole are never found in association with cytokinetic events.

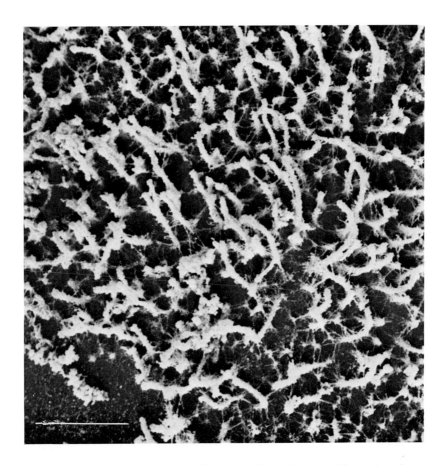

FIGURE 4. *Nassarius reticulatus*: vegetal pole area of the uncleaved egg. Fixation in a mixture of glutaraldehyde and osmiumtetroxide at 0°. The whole area is densely populated with long, erect villi which are interconnected by thin strands of extracellular material, probably carbohydrate. Bar indicates 10 μm.

Polar lobe-like protrusions specifically covered with microvilli have also been observed on thyroid follicular epithelial cells of young rats.[144] The so-called mitosis-associated apical protrusions (MAAP), extending into the follicular lumen, display a high concentration of microvilli and strikingly resemble a small polar lobe, including its characteristic surface architecture. The function of these MAAPs is unknown. A satisfactory explanation for the occurrence of these surface phenomena should take into account the diversity in architecture and their relative stability. Experiments in which the effect of suppressing these surface phenomena are studied will have to provide part of the answer.

D. The External Surface

Eggs may have very elaborate external layers, cellular and noncellular. An important function of these layers consists of ensuring correct fertilization. Apart from this aspect almost nothing is known. External layers may function to keep the blastomeres from falling apart. The planarian *Dendrocoelum* constitutes a peculiar case. In this species the blastomeres dissociate from the two-cell stage onwards. They form a mass of free blastomeres which is kept together by the external egg layers.[145]

Whether peripheral components contribute to the mechanical stability of eukaryotic membranes is not known. The major components conferring shape and rigidity to animal cells appear to be internal — the cytoskeleton or the spectrin network of erythrocytes. The possibility of a peripheral component acting as a shape-maintaining element is illustrated

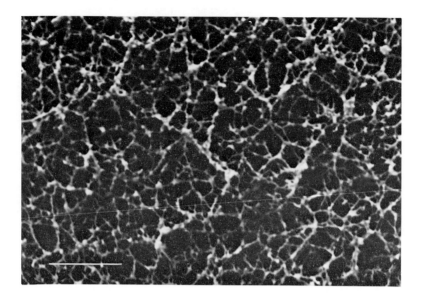

FIGURE 5.   *Nassarius reticulatus:* vegetal pole area of the uncleaved egg. Fixation in
glutaraldehyde, followed by osmiumtetroxide at room temperature. It is impossible to tell
whether the anastomosing network covering the vegetal pole area consists of villous structures
or represents extracellular carbohydrate strands. Bar indicates 5 μm.

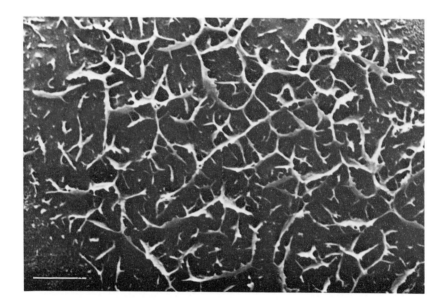

FIGURE 6.   *Nassarius reticulatus:* vegetal pole area of the uncleaved egg. Fixation in the
presence of Alcian Blue, according to the procedure of Behnke and Zelander in *J. Ultrastruct.
Res.,* 31, 424, 1970. The ruffles shown in this picture have only been observed once. Normally,
the surface is covered with villi after the Alcian Blue fixation. Bar indicates 4 μm.

by a prokaryote, *Halobacterium salinarium*, where a cell surface glycoprotein exerts this
function.[146] In eggs, an indication for a shape-maintaining role of an extracellular component,
viz., the vitelline layer of sea urchin eggs, is provided by the observation that after fertilization
a disarrayal of microvilli occurs.[61] The vitelline layer might hold the microvilli of the

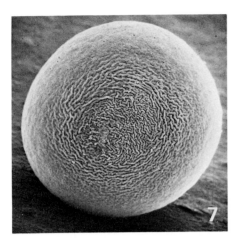

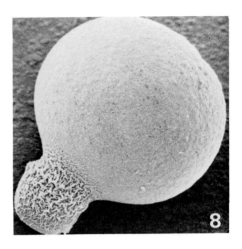

FIGURE 7. *Buccinum undatum:* the surface in the vegetal pole area forms an extensive system of branching microvilli, resulting in a typical array of ridges.

FIGURE 8. *Buccinum undatum.* The polar lobe is formed.

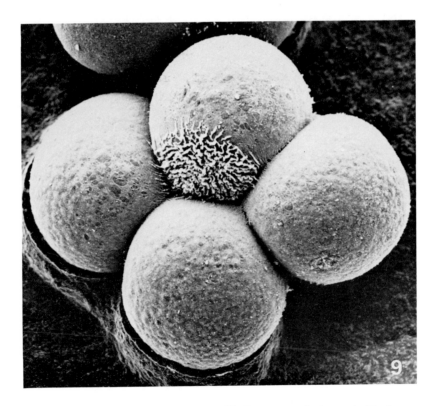

FIGURE 9. *Crepidula fornicata:* 4-cell stage. The D-macromere is characterized by the presence of an array of folds at the vegetal pole. At this site the second polar lobe has been resorbed.

unfertilized egg in their characteristic array. Fertilization results in the elevation of the vitelline layer, thus loosening its contact with the microvilli.

The potential of peripheral components to exert control on cellular structure and function is clearly demonstrated by the effects of fibronectin and hyaluronic acid. Fibronectin added

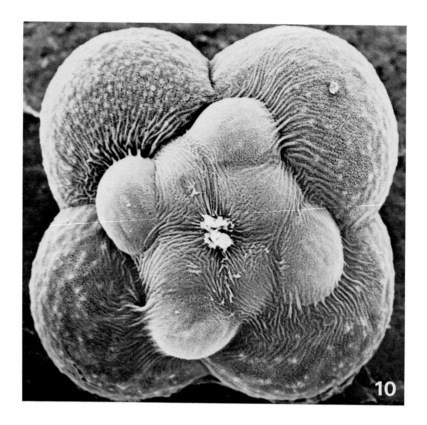

FIGURE 10.   *Crepidula fornicata:* 12-cell stage. Extensive folding of the surface accompanies cleavage in this species. The folds largely disappear during later stages of the cleavage cycle.

to chondroblasts alters their phenotype — from typically epithelioid cells they become fibroblastic.[147] Hyaluronic acid added to chondroblasts inhibits both the synthesis and deposition of sulfated proteoglycans by these cells.[148] In eggs, external components are known to have important effects as soon as cleavage has started. Intercellular communication probably proceeds to a large extent through membrane glycoproteins and glycolipids. Their carbohydrate residues, which are exposed externally, have an enormous potential for structural heterogeneity. They act as sensors for all kinds of signals and serve as specific cellular recognition determinants.[149,150] The potential of surface-associated material in eggs has been demonstrated by Brachet.[151] Surface material was obtained by disaggregating cleaving sea urchin eggs in sea water lacking Ca^{2+} and Mg^{2+} and collecting the medium. When this medium was added to fertilized eggs, development was arrested either during early cleavage or at the blastula stage, depending on the species of the recipient eggs.

The role of proteoglycan in early sea urchin development has been investigated by inhibition of its synthesis at various stages.[152] It appears that the embryos cannot develop beyond the blastula stage when the synthesis of proteoglycan is blocked. Developmental arrest at the blastula stage could be cancelled by the administration of exogenous proteoglycan from postgastrulae, but not by proteoglycan from blastulae.[153] On the other hand, when embryos were administered proteoglycan prepared from different embryonic stages, they ceased to develop at a stage corresponding to the stage from which the proteoglycan was extracted. The authors explain these results by assuming that proteoglycan is transferred into the nucleus and affects nuclear activity directly. Another possibility is that extracellular proteoglycan acts indirectly on nuclear activity by means of transmembrane control mechanisms. In mouse eggs an effect of glycoproteins on compaction and trophoblast adhesion has been demonstrated by inhibiting glycosylation with tunicamycin.[154] However, the intracellular or extracellular location of these glycoproteins remains to be established.

In later development extracellular matrix components are known to play important roles in the control and regulation of developmental processes, e.g., in somite chondrogenesis[155] and corneal development.[156,157] Several different mechanisms may operate in the interaction between extracellular matrix and cells. An autocatalytic process makes collagen stimulate its own production by corneal epithelial cells.[156] Prevention of precocious differentiation and stimulation of migration and proliferation is thought to result from the interaction between hyaluronate and corneal fibroblasts.[157] It appears that each stage in development is characterized by a unique extracellular matrix, whose components are important both structurally and as environmental signals influencing cell behavior.

E. The Egg Surface during Cleavage

The changes in surface structure and composition during cleavage have been investigated in several organisms. A few clear examples of early differentiations as expressed by surface features of particular cells have been reported. In sea urchin embryos, concanavalin A induces a clustered or capped distribution of receptors on the micromeres but not on the mesomeres or macromeres.[158] In *Lymnaea palustris* a unique patch of microvilli appears at the 8-cell stage on the macromeres in the vicinity of the vegetal cross-furrow. Their appearance and then disappearance by the 24-cell stage coincides with the migration of a particular type of cytoplasmic localizations, the subcortical accumulations (SCA), to the center of the egg. At later stages only the cells 4d and $2d^{22}$ are densely populated by microvilli.[159] In mouse embryos, fibronectin is not detectable in early preimplantation stages. It is first found between the cells of the inner cell mass of late blastocysts.[160] The varying distribution of membrane-associated enzymes, lectin receptors, and antigens during early development of mammalian embryos has been reviewed by Sherman.[161] The specific role of the original egg surface during the cleavage stage is largely unknown. The original plasma membrane of the uncleaved egg probably remains on the periphery during cleavage. The extra membrane needed for the formation of blastomeres seems to be produced in the region of the furrow.[143] The resulting blastomeres thus contain two distinct membrane domains, the preexisting membrane at the periphery and the new membrane internally. The persistence of local surface differentiations during cleavage, as observed in molluscan eggs,[141] supports the view that the original egg membrane persists at the periphery of the cleaving egg. As the determination of polarity, cleavage pattern, and cytoplasmic localization are assumed to be functions of the original egg surface, it is likely that these aspects of development remain under its control during cleavage and that other functions such as intercellular communication and the determination of the microenvironment of the blastomeres are controlled by the newly formed membrane.

IV. CONCLUDING REMARKS

The role of the egg surface is highly diverse. Its properties affect almost every aspect of development — polarity, cytoplasmic localization, fertilization, activation, position of nuclei, asters and cleavage planes, nuclear activity, etc. The mechanisms operating in these regulatory activities remain to be established in most cases. The interplay between cortical cytoplasm and plasma membrane seems to be the dominant mechanism that regulates the majority of cellular functions. Whether a pattern initially established by the egg surface serves as a primer for the organization of the endoplasm is still an open question. The degree of organizational independence of the endoplasm is hard to measure, as the elimination of the influence of the surface is hardly feasible.

REFERENCES

1. **Brunner, G.,** Membrane impression and gene expression. Towards a theory of cytodifferentiation, *Differentiation,* 8, 123, 1977.
2. **Sumper, M.,** Control of differentiation in *Volvox carteri.* A model explaining pattern formation during embryogenesis, *FEBS Lett.,* 107, 241, 1979.
3. **Wenzl, S. and Sumper, M.,** Evidence for membrane-mediated control of differentiation during embryogenesis of *Volvox carteri, FEBS Lett.,* 107, 247, 1979.
4. **McMahon, D.,** A cell contact model for position determination in development, *Proc. Natl. Acad. Sci. U.S.A.,* 70, 2396, 1973.
5. **Conklin, E. G.,** Effects of centrifugal force on the structure and development of the eggs of *Crepidula, J. Exp. Zool.,* 22, 311, 1917.
6. **Lillie, F. R.,** Polarity and bilaterality of the annelid egg. Experiments with centrifugal force, *Biol. Bull.,* 16, 54, 1909.
7. **Elbers, P. F.,** Over de beginoorzaak van het Li-effect in de morphogenese, een electronenmicroscopisch onderzoek aan eieren van *Limnaea stagnalis* en *Paracentrotus lividus,* Thesis, University of Utrecht, Netherlands, 1959.
8. **Moser, F.,** Studies on a cortical layer response to stimulating agents in the *Arbacia* egg. I. Response to insemination, *J. Exp. Zool.,* 80, 423, 1939.
9. **Conklin, E. G.,** The development of centrifuged eggs of ascidians, *J. Exp. Zool.,* 60, 1, 1931.
10. **La Spina, R.,** Lo spostamento dei plasmi mediante centrifugazione nell'uovo vergine di ascidie e il consequente sviluppo, *Acta Embryol. Morphol. Exp.,* 2, 66, 1958.
11. **Guerrier, P.,** Les caractères de la segmentation et la détermination de la polarité dorsoventrale dans le développement de quelques Spiralia. III. *Pholas dactylus* et *Spisula subtruncata* (Mollusques Lamellibranches), *J. Embryol. Exp. Morphol.,* 23, 667, 1970.
12. **Raven, Chr. P. and Tates, A. D.,** Centrifugation of *Limnaea* eggs at stages immediately preceding third cleavage, *Proc. K. Ned. Akad. Wet. Ser. C,* 64, 129, 1961.
13. **Raven, Chr. P.,** Mechanisms of determination in the development of gastropods, in *Advances in Morphogenesis,* Vol. 3, Academic Press, New York, 1964, 1.
14. **Dohmen, M. R. and Verdonk, N. H.,** The ultrastructure and role of the polar lobe in development of molluscs, in *Determinants of Spatial Organization,* Subtelny, S. and Konigsberg, I. R., Eds., Academic Press, New York, 1979.
15. **Verdonk, N. H.,** The effect of removing the polar lobe in centrifuged eggs of *Dentalium, J. Embryol. Exp. Morphol.,* 19, 33, 1968.
16. **Clement, A. C.,** Development of the vegetal half of the *Ilyanassa* egg after removal of most of the yolk by centrifugal force, compared with the development of animal halves of similar visible composition, *Dev. Biol.,* 17, 165, 1968.
17. **Dohmen, M. R. and Verdonk, N. H.,** The structure of a morphogenetic cytoplasm, present in the polar lobe of *Bithynia tentaculata* (Gastropoda, Prosobranchia), *J. Embryol. Exp. Morphol.,* 31, 423, 1974.
18. **Conklin, E. G.,** The organization and cell-lineage of the ascidian egg, *J. Acad. Natl. Sci. (Philadelphia),* 13, 1, 1905.
19. **Austin, C. R.,** *Fertilization,* Prentice-Hall, Englewood Cliffs, N.J., 1965.
20. **Elinson, R. P.,** Site of sperm entry and a cortical contraction associated with egg activation in the frog, *Rana pipiens, Dev. Biol.,* 47, 257, 1975.
21. **Campanella, C.,** The site of spermatozoon entrance in the unfertilized egg of *Discoglossus pictus* (Anura): an electron microscopic study, *Biol. Reprod.,* 12, 439, 1975.
22. **Reverberi, G.,** Ascidians, in *Experimental Embryology of Marine and Fresh-Water Invertebrates,* Reverberi, G., Ed., North-Holland, Amsterdam, 1971, 507.
23. **Pease, D. C.,** The influence of centrifugal force on the bilateral determination and the polar axis of *Cumingia* and *Chaetopterus* eggs, *J. Exp. Zool.,* 84, 387, 1940.
24. **Freeman, G.,** The establishment of the oral-aboral axis in the ctenophore embryo, *J. Embryol. Exp. Morphol.,* 42, 237, 1977.
25. **Raven, Chr. P.,** *Morphogenesis: The Analysis of Molluscan Development,* 2nd ed., Macmillan, New York, 1966.
26. **Kühn, A.,** *Entwicklungsphysiologie,* 2nd ed., Springer-Verlag, Berlin, 1965.
27. **Guerrier, P.,** Les facteurs de polarisation dans les premiers stades du développement chez *Parascaris equorum, J. Embryol. Exp. Morphol.,* 18, 121, 1967.
28. **Guerrier, P.,** Origine et stabilité de la polarité animale végétative chez quelques spiralia, *Annal. Embryol. Morphol.,* 1, 119, 1968.
29. **Curtis, A. S. G.,** Cortical grafting in *Xenopus laevis, J. Embryol. Exp. Morphol.,* 8, 163, 1960.
30. **Curtis, A. S. G.,** Morphogenetic interactions before gastrulation in the amphibian *Xenopus laevis* — the cortical field, *J. Embryol. Exp. Morphol.,* 10, 410, 1962.

31. **Hörstadius, S., Lorch, J. J., and Danielli, J. F.,** Differentiation of the sea urchin egg following reduction of the interior cytoplasm in relation to the cortex, *Exp. Cell Res.,* 1, 188, 1950.
32. **Van den Biggelaar, J. A. M.,** The determinative significance of the geometry of the cell contacts in early molluscan development, *Biol. Cell.,* 32, 155, 1978.
33. **Abbate, C. and Ortolani, G.,** The development of *Ciona* eggs after partial removal of cortex or ooplasm, *Acta Embryol. Morphol. Exp.,* 4, 56, 1961.
34. **Counce, S. J.,** The causal analysis of insect embryogenesis, in *Developmental Systems: Insects,* Vol. 2, Counce, S. J. and Waddington, C. H., Eds., Academic Press, London, 1973, 1.
35. **Sander, K.,** Specification of the basic body pattern in insect embryogenesis, *Adv. Insect Physiol.,* 12, 125, 1976.
36. **Kalthoff, K.,** Analysis of a morphogenetic determinant in an insect embryo (*Smittia sp.,* Chironomidae, Diptera), in *Determinants of Spatial Organization,* Subtelny, S. and Konigsberg, I. R., Eds., Academic Press, New York, 1979, 97.
37. **Arnold, J. M. and Williams-Arnold, L. D.,** Cortical-nuclear interactions in cephalopod development: cytochalasin B effects on the informational pattern in the cell surface, *J. Embryol. Exp. Morphol.,* 31, 1, 1974.
38. **Lallier, R.,** Recherches sur la polarite de l'oeuf de l'oursin *Paracentrotus lividus, Acta Embryol. Exp.,* 1, 47, 1978.
39. **Ortolani, G.,** Cleavage and development of egg fragments in ascidians, *Acta Embryol. Morphol. Exp.,* 1, 247, 1958.
40. **Reverberi, G. and Ortolani, G.,** Twin larvae from halves of the same egg in ascidians, *Dev. Biol.,* 5, 84, 1962.
41. **Freeman, G.,** The role of asters in the localization of the factors that specify the apical tuft and the gut in the nemertine *Cerebratulus lacteus, J. Exp. Zool.,* 206, 81, 1978.
42. **Nuccitelli, R.,** Oöplasmic segregation and secretion in the *Pelvetia* egg is accompanied by a membrane-generated electrical current, *Dev. Biol.,* 62, 13, 1978.
43. **Quatrano, R. S., Brawley, S. H., and Hogsett, W. E.,** The control of the polar deposition of a sulfated polysaccharide in *Fucus* zygotes, in *Determinants of Spatial Organization,* Subtelny, S. and Konigsberg, I. R., Eds., Academic Press, New York, 1979, 77.
44. **Brachet, J.,** An old enigma: the gray crescent of amphibian eggs, *Curr. Top. Dev. Biol.,* 11, 133, 1977.
45. **Ubbels, G. A.,** Determination of dorso/ventral polarity in the anuran egg: reversal experiments in *Xenopus laevis, 3e Colloq. Annu. Soc. Franc. Biol. Dev. Toulouse,* 1979.
46. **Hara, K. and Tydeman, P.,** Cinematographic observation of an "activation wave" (AW) on the locally inseminated egg of *Xenopus laevis, Wilhelm Roux' Arch. Entwicklungsmech. Org.,* 186, 91, 1979.
47. **Hara, K., Tydeman, P., and Hengst, R. T. M.,** Cinematographic observation of "post-fertilization waves" (PFW) on the zygote of *Xenopus laevis, Wilhelm Roux' Arch. Entwicklungsmech. Org.,* 181, 189, 1977.
48. **Hara, K.,** Cinematographic observation of "surface contraction waves" (SCW) during the early cleavage of axolotl eggs, *Wilhelm Roux' Arch. Entwicklungsmech. Org.,* 167, 183, 1971.
49. **Beetschen, J.-C.,** Récherches experimentales sur la symétrisation de l'oocyte et de l'oeuf d'axolotl: facteurs conditionnant l'apparition précoce du croissant gris à la suite d'un choc thermique, *C. R. Acad. Sci. Paris Ser. D,* 288, 643, 1979.
50. **Temmink, J. H. M. and Spiele, H.,** Different cytoskeletal domains in murine fibroblasts, *J. Cell Sci.,* 41, 19, 1980.
51. **Marchesi, V. T.,** Spectrin: present status of a putative cytoskeletal protein of the red cell membrane, *J. Membr. Biol.,* 51, 101, 1979.
52. **Moore, P. B., Ownby, Ch. L., and Carraway, K. L.,** Interactions of cytoskeletal elements with the plasma membrane of sarcoma 180 ascites tumor cells, *Exp. Cell. Res.,* 115, 331, 1978.
53. **Williams, N. E., Vaudaux, P. E., and Skrives, L.,** Cytoskeletal proteins of the cell surface in *Tetrahymena.* I. Identification and localization of major proteins, *Exp. Cell Res.,* 123, 311, 1979.
54. **Ng, S. F. and Frankel, J.,** 180° rotation of ciliary rows and its morphogenetic implications in *Tetrahymena pyriformis, Proc. Natl. Acad. Sci. U.S.A.,* 74, 1115, 1977.
55. **Sonneborn, T. M.,** Does preformed cell structure play an essential role in cell heredity?, in *The Nature of Biological Diversity,* Allen, J. M., Ed., McGraw-Hill, New York, 1963, 165.
56. **Frankel, J.,** Positional information in unicellular organisms, *J. Theor. Biol.,* 47, 439, 1974.
57. **Frankel, J.,** An analysis of cell-surface patterning in *Tetrahymena,* in *Determinants of Spatial Organization,* Subtelny, S. and Konigsberg, I. R., Eds., Academic Press, New York, 1979, 215.
58. **Mitchison, J. M. and Swann, M. M.,** Opical changes in the membranes of the sea-urchin egg at fertilization, mitosis and cleavage, *J. Exp. Biol.,* 29, 357, 1952.
59. **Monroy, A. and Montalenti, G.,** Variations of the submicroscopic structure of the cortical layers of fertilized and parthenogenetic sea-urchin eggs, *Biol. Bull.,* 92, 151, 1947.

60. **Higashi, A.,** The thickness of the cortex and some analytical experiments of thixotropy in sea urchin eggs, *Annot. Zool. Jpn.,* 45, 119, 1972.

61. **Schatten, G. and Mazia, D.,** The surface events at fertilization: the movements of the spermatozoon through the sea urchin egg surface and the roles of the surface layers, *J. Supramol. Struct.,* 5, 343, 1976.

62. **Kane, R. E.,** Preparation and purification of polymerized actin from sea urchin extracts, *J. Cell Biol.,* 66, 305, 1975.

63. **Harris, P.,** A spiral cortical fiber system in fertilized sea urchin eggs, *Dev. Biol.,* 68, 525, 1979.

64. **Franke, W. W., Rathke, P. C., Seib, E., Trendelenburg, M. F., Osborn, M., and Weber, K.,** Distribution and mode of arrangement of microfilamentous structures and actin in the cortex of the amphibian oocyte, *Cytobiologie,* 14, 111, 1976.

65. **Bluemink, J. G.,** The first cleavage of the amphibian egg: an electron microscope study of the onset of cytokinesis in the egg of *Ambystoma mexicanum, J. Ultrastruct. Res.,* 32, 142, 1970.

66. **Virtanen, I. and Miettinen, A.,** The role of actin in the surface integrity of cultured rat liver parenchymal cells, *Cell Biol. Int. Rep.,* 4, 29, 1980.

67. **Oppenheimer, S. B., Bales, B., Brenneman, G., Knapp, L., Lesin, E., Neri, A., and Pollock, E. G.,** Modulation of agglutinability by alteration of the surface topography in mouse ascites tumor cells, *Exp. Cell Res.,* 105, 291, 1977.

68. **Anderson, E.,** Oocyte differentiation in the sea urchin, *Arbacia punctulata,* with particular reference to the origin of cortical granules and their participation in the cortical reaction, *J. Cell Biol.,* 37, 514, 1968.

69. **Monné, L. and Härde, S.,** On the cortical granules of the sea urchin egg, *Arch. Zool. Stockholm Ser. II,* 1, 127, 1951.

70. **Vacquier, V. D.,** The isolation of intact cortical granules from sea urchin eggs: calcium ions trigger granule discharge, *Dev. Biol.,* 43, 62, 1975.

71. **Nicosia, S. V., Wolf, D. P., and Inoue, M.,** Cortical granule distribution and cell surface characteristics in mouse eggs, *Dev. Biol.,* 57, 56, 1977.

72. **Reverberi, G.,** The ultrastructure of *Dentalium* egg at the trefoil stage, *Acta Embryol. Morphol. Exp.,* 1, 31, 1970.

73. **Rappaport, R.,** Cytokinesis in animal cells, *Int. Rev. Cytol.,* 31, 169, 1971.

74. **Rappaport, R. and Rappaport, B. N.,** Establishment of cleavage furrows by the mitotic spindle, *J. Exp. Zool.,* 189, 189, 1974.

75. **Burgess, D. R.,** Ultrastructure of meiosis and polar body formation in the egg of the mud snail, *Ilyanassa obsoleta,* in *Cell Shape and Surface Architecture,* Revel, J. P., Henning, U., and Fox, C. F., Eds., Alan R. Liss, Inc., New York, 1977, 569.

76. **Czihak, G.,** The role of astral rays in early cleavage of sea urchin eggs, *Exp. Cell Res.,* 83, 424, 1974.

77. **de Terra, N.,** Some regulatory interactions between cell structures at the supramolecular level, *Biol. Rev.,* 53, 427, 1978.

78. **Tartar, V.,** *The Biology of Stentor,* Pergamon Press, New York, 1961.

79. **Rappaport, R.,** Effects of continual mechanical agitation prior to cleavage in echinoderm eggs, *J. Exp. Zool.,* 206, 1, 1978.

80. **Conrad, G. W. and Davis, S. E.,** Microiontophoretic injection of calcium ions or of cyclic AMP causes rapid shape changes in fertilized eggs of *Ilyanassa obsoleta, Dev. Biol.,* 61, 184, 1977.

81. **Conrad, G. W. and Davis, S. E.,** Polar lobe formation and cytokinesis in fertilized eggs of *Ilyanassa obsoleta.* III. Large bleb formation caused by Sr^{2+}, ionophores X 537A and A 23187, and compound 48/80, *Dev. Biol.,* 74, 152, 1980.

82. **Gingell, D.,** Contractile responses at the surface of an amphibian egg, *J. Embryol. Exp. Morphol.,* 23, 583, 1970.

83. **Schroeder, T. E. and Strickland, D. L.,** Ionophore A 23187, calcium and contractility in frog eggs, *Exp. Cell Res.,* 83, 139, 1974.

84. **Hollinger, T. G., Dumont, J. N., and Wallace, R. A.,** Calcium-induced dehiscence of cortical granules in *Xenopus laevis* oocytes, *J. Exp. Zool.,* 210, 107, 1979.

85. **Gilkey, J. C., Jaffe, L. F., Ridgway, E. B., and Reynolds, G. T.,** A free calcium wave traverses the activating egg of the medaka, *Oryzias latipes, J. Cell Biol.,* 76, 448, 1978.

86. **Conrad, G. W., Williams, D. C., Turner, F. R., Newrock, K. M., and Raff, R. A.,** Microfilaments in the polar lobe constriction of fertilized eggs of *Ilyanassa obsoleta, J. Cell Biol.,* 59, 228, 1973.

87. **Morgan, T. H.,** The separation of the egg of *Ilyanassa* into two parts by centrifuging, *Biol. Bull.,* 68, 280, 1935.

88. **Verdonk, N. H., Geilenkirchen, W. L. M., and Timmermans, L. P. M.,** The localization of morphogenetic factors in uncleaved eggs of *Dentalium, J. Embryol. Exp. Morphol.,* 25, 57, 1971.

89. **Shimizu, T.,** Surface contractile activity of the *Tubifex* egg: its relationship to the meiotic apparatus functions, *J. Exp. Zool.,* 208, 361, 1979.

90. **Arey Lewis, C., Chia, F.-S., and Schroeder, T. E.,** Peristaltic constrictions in fertilized barnacle eggs (*Pollicipes polymerus*), *Experientia,* 29, 1533, 1973.

91. **Hoadley, L.,** Pulsations in the *Nereis* egg, *Biol. Bull.,* 67, 484, 1934.

92. **Pasteels, J. J.,** Les mouvements corticaux de l'oeuf de *Barnea candida Exp.* (Mollusque Bivalve) étudiés au microscope électronique, *J. Embryol. Exp. Morphol.,* 16, 311, 1966.

93. **Pasteels, J. J.,** Mouvements localisés et rythmiques de la membrane de fécondation chez les oeufs fécondés et activés *(Chaetopterus, Mactra, Nereis), Arch. Biol. Liege,* 61, 199, 1949.

94. **Singer, S. J. and Nicolson, G. L.,** The fluid mosaic model of the structure of cell membranes, *Science,* 175, 720, 1972.

95. **Cullis, P. R. and De Kruyff, B.,** Lipid polymorphism and the functional roles of lipids in biological membranes, *Biochim. Biophys. Acta Rev. Biomembr.,* 559, 399, 1979.

96. **Nicolson, G. L., Poste, G., and Ji, T. H.,** The dynamics of cell membrane organization, in *Dynamic Aspects of Cell Surface Organization,* Poste, G. and Nicolson, G. L., Eds., North-Holland, Amsterdam, 1977, 1.

97. **Staehelin, L. A.,** Structure and function of intercellular junctions, *Int. Rev. Cytol.,* 39, 191, 1974.

98. **Beisson, J.,** Non-nucleic acid inheritance and epigenetic phenomena, in *Cell Biology,* Vol. 1, Goldstein, L. and Prescott, D. M., Eds., Academic Press, New York, 1977, 375.

99. **Stackpole, C. W.,** Topological differentiation of the cell surface, in *Progress in Surface and Membrane Science,* Vol. 12, Cadenhead, D. A. and Danielli, J. F., Eds., Academic Press, New York, 1978, 1.

100. **Ash, J. F. and Singer, S. J.,** Concanavalin A induced transmembrane linkage of concanavalin A surface receptors to intracellular myosin-containing filaments, *Proc. Natl. Acad. Sci. U.S.A.,* 73, 4575, 1976.

101. **Singer, I. I.,** Microfilaments bundles and the control of pinocytotic vesicle distribution at the surfaces of normal and transformed fibroblasts, *Exp. Cell Res.,* 122, 251, 1979.

102. **Elgsaeter, A., Shotton, D. M., and Branton, D.,** Intramembrane particle aggregation in erythrocyte ghosts. II. The influence of spectrin aggregation, *Biochim. Biophys. Acta,* 426, 101, 1976.

103. **Yamada, K. M. and Olden, K.,** Fibronectins — adhesive glycoproteins of cell surface and blood, *Nature (London),* 275, 179, 1978.

104. **Hynes, R. O., Destree, A. T., Perkins, M. E., and Wagner, D. D.,** Cell surface fibronectin and oncogenic transformation, *J. Supramol. Struct.,* 11, 95, 1979.

105. **Nicolson, G. L. and Poste, G.,** Cell shape changes and transmembrane receptor uncoupling induced by tertiary amine local anesthetics, *J. Supramol. Struct.,* 5, 65, 1976.

106. **Peters, R., Peters, J., Tews, K. H., and Bähr, W.,** A microfluorimetric study of translational diffusion in erythrocyte membranes, *Biochim. Biophys. Acta,* 367, 282, 1974.

107. **Jacobson, K., Derzko, Z., Wu, E.-S., Hou, Y., and Poste, G.,** Measurement of the lateral mobility of cell surface components in single, living cells by fluorescence recovery after photobleaching, *J. Supramol. Struct.,* 5, 565, 1976.

108. **Cherry, R. J.,** Rotational and lateral diffusion of membrane proteins, *Biochim. Biophys. Acta Rev. Biomembr.,* 559, 289, 1979.

109. **Johnson, M. and Edidin, M.,** Lateral diffusion in plasma membrane of mouse egg is restricted after fertilization, *Nature (London),* 272, 448, 1978.

110. **Campisi, J. and Scandella, C. J.,** Fertilization-induced changes in membrane fluidity of sea urchin eggs, *Science,* 199, 1336, 1978.

111. **Bluemink, J. G., Tertoolen, L. G. J., Ververgaert, P. H. J. Th., and Verkley, A. J.,** Freeze-fracture electron microscopy of preexisting and nascent cell membrane in cleaving eggs of *Xenopus laevis, Biochim. Biophys. Acta,* 443, 143, 1976.

112. **Bluemink, J. G. and Tertoolen, L. G. J.,** Freeze fracture electron microscopy of the plasma membrane of the *Xenopus* egg: evidence for particle-associated filaments, *Cytobiologie,* 16, 358, 1978.

113. **Eddy, E. M. and Shapiro, B. M.,** Changes in the topography of the sea urchin egg after fertilization, *J. Cell Biol.,* 71, 35, 1976.

114. **Schroeder, Th. E.,** Surface area change at fertilization: resorption of the mosaic membrane, *Dev. Biol.,* 70, 306, 1979.

115. **Gabel, Ch. A., Eddy, E. M., and Shapiro, B. M.,** After fertilization, sperm surface components remain as a patch in sea urchin and mouse embryos, *Cell,* 18, 207, 1979.

116. **Bluemink, J. G. and Tertoolen, L. G. J.,** The plasma-membrane IMP pattern as related to animal/vegetal polarity in the amphibian egg, *Dev. Biol.,* 62, 334, 1978.

117. **Nicolson, G. L.,** The interaction of lectins with animal surfaces, *Int. Rev. Cytol.,* 39, 89, 1974.

118. **Oppenheimer, S. B.,** Interactions of lectins with embryonic cell surfaces, in *Current Topics in Developmental Biology,* Vol. 11, Moscona, A. A. and Monroy, A., Eds., Academic Press, New York, 1977, 1.

119. **Johnson, M. H., Eager, D., Muggleton-Harris, A., and Grave, H. M.,** Mosaicism in the organization of concanavalin A receptors on surface membrane of mouse egg, *Nature (London),* 257, 321, 1975.

120. **Eager, D. D., Johnson, M. H., and Thurley, K. W.,** Ultrastructural studies on the surface membrane of the mouse egg, *J. Cell Sci.,* 22, 345, 1976.

121. **Yanagimachi, R. and Nicolson, G. L.,** Lectin-binding properties of hamster egg zona pellucida and plasma membrane during maturation and preimplantation development, *Exp. Cell Res.,* 100, 249, 1976.

122. **Ortolani, G., O'Dell, D. S., and Monroy, A.,** Localized binding of *Dolichos* lectin to the early *Ascidia* embryo, *Exp. Cell Res.,* 106, 402, 1977.

123. **Denis-Donini, S. and Campanella, Ch.,** Ultrastructural and lectin binding changes during the formation of the animal dimple in oocytes of *Discoglossus pictus* (Anura), *Dev. Biol.,* 61, 140, 1977.

124. **Robinson, K. R. and Jaffe, L. F.,** Polarizing fucoid eggs drive a calcium current through themselves, *Science,* 187, 70, 1975.

125. **Borgens, R. B., Vanable, J. W., and Jaffe, L. F.,** Bioelectricity and regeneration, *BioScience,* 29, 468, 1979.

126. **Jaffe, L. F., Robinson, K. R., and Nuccitelli, R.,** Local cation entry and self-electrophoresis as an intracellular localization mechanism, *Ann. N.Y. Acad. Sci.,* 238, 372, 1974.

127. **Brawley, S. H. and Quatrano, R. S.,** Sulfation of fucoidin in *Fucus* embryos. IV. Autoradiographic investigations of fucoidin sulfation and secretion during differentiation and the effect of cytochalasin treatment, *Dev. Biol.,* 73, 193, 1979.

128. **Kenny, A. J. and Booth, A. G.,** Microvilli: their ultrastructure, enzymology and molecular organization, *Essays Biochem.,* 14, 1, 1978.

129. **Tilney, L. G. and Cardell, R. R.,** Factors controlling the reassembly of the microvillous border of the small intestine of the salamander, *J. Cell Biol.,* 47, 408, 1970.

130. **Schroeder, T. E.,** Microvilli on sea urchin eggs: a second burst of elongation, *Dev. Biol.,* 64, 342, 1978.

131. **Longo, F. J.,** Cortical changes in *Spisula* eggs upon insemination, *J. Ultrastruct. Res.,* 56, 226, 1976.

132. **Popham, J. D.,** The fine structure of the oocyte of *Bankia australis* (Teredinidae, Bivalvia) before and after fertilization, *Cell Tiss. Res.,* 157, 521, 1975.

133. **Shimizu, T.,** The fine structure of the *Tubifex* egg before and after fertilization, *J. Fac. Sci. Hokkaido Univ. Ser. VI Zool.,* 20, 253, 1976.

134. **Pasteels, J. J.,** La réaction corticale de fécondation de l'oeuf de *Nereis diversicolor,* étudiée au microscope electronique, *Acta Embryol. Morphol. Exp.,* 6, 155, 1966.

135. **Wassarman, P. M., Albertini, D. F., Josefowicz, W. J., and Letourneau, G. E.,** Cytochalasin B-induced pseudo-cleavage of mouse oocytes *in vitro:* asymmetric localization of mitochondria and microvilli associated with a stage-specific response, *J. Cell Sci.,* 21, 523, 1976.

136. **Wassarman, P. M., Ukena, T. E., Josefowicz, W. J., Letourneau, G. E., and Karnovsky, M. J.,** Cytochalasin B-induced pseudo-cleavage of mouse oocytes *in vitro.* II. Studies of the mechanism and morphological consequences of pseudo-cleavage, *J. Cell Sci.,* 26, 323, 1977.

137. **Yahara, I. and Kakimoto-Sameshima, F.,** Mechanism of translocation of microvilli accompanying cap formation of surface receptors, *Cell Struct. Function,* 4, 143, 1979.

138. **Spaulding, J. G.,** The structure and distribution of cytospines on the surface of the egg of the sea anemone *Peachia quinquecapitata, Am. Zool.,* 19, 956, 1979.

139. **Monroy, A. and Baccetti, B.,** Morphological changes of the surface of the egg of *Xenopus laevis* in the course of development. I. Fertilization and early cleavage, *J. Ultrastruct. Res.,* 50, 131, 1975.

140. **Robinson, K. R.,** Electrical currents through full-grown and maturing *Xenopus* oocytes, *Proc. Natl. Acad. Sci. U.S.A.,* 76, 837, 1979.

141. **Dohmen, M. R. and van der Mey, J. C. A.,** Local surface differentiations at the vegetal pole of the eggs of *Nassarius reticulatus, Buccinum undatum* and *Crepidula fornicata* (Gastropoda, Prosobranchia), *Dev. Biol.,* 61, 104, 1977.

142. **Dohmen, M. R. and Verdonk, N. H.,** Cytoplasmic localizations in mosaic eggs, in *Maternal Effects in Development,* Newth, D. R. and Balls, M., Eds., Cambridge University Press, London, 1979, 127.

143. **Bluemink, J. G. and de Laat, S. W.,** Plasma membrane assembly as related to cell division, in *The Synthesis, Assembly and Turnover of Cell Surface Components,* Poste, G. and Nicolson, G. L., Eds., North-Holland, Amsterdam, 1977, 403.

144. **Zeligs, J. D. and Wollman, S. H.,** Mitosis in thyroid epithelial cells *in vivo.* V. Characteristic apical protrusions are associated with mitosis in unstimulated glands, *J. Ultrastruct. Res.,* 68, 308, 1979.

145. **Gallien, L.,** *L'oeuf et les Stades Initiaux de l'Ontogenese,* Presses Universitaires de France, Paris, 1979, 180.

146. **Mescher, M. F. and Strominger, J. L.,** The shape-maintaining component of *Halobacterium salinarium*: a cell surface glycoprotein, in *Cell Shape and Surface Architecture,* Revel, J. P., Henning, U., and Fox, C. F., Eds., Alan R. Liss, Inc., New York, 1977, 459.

147. **West, Ch. M., Lanza, R., Rosenbloom, J., Lowe, M., and Holtzer, H.,** Fibronectin alters the phenotypic properties of cultured chick embryo chondroblasts, *Cell,* 17, 491, 1979.

148. **Handley, C. J. and Lowther, D. A.,** Inhibition of proteoglycan biosynthesis by hyaluronic acid in chondrocytes in cell culture, *Biochim. Biophys. Acta,* 444, 69, 1976.

149. **Burger, M. M. and Jumblatt, J.,** Membrane involvement in cell-cell interactions: a two-component model system for cellular recognition that does not require live cells, in *Cell and Tissue Interactions,* Lash, J. W. and Burger, M. M., Eds., Raven Press, New York, 1977, 155.

150. **Moscona, A. A. and Hausman, R. E.,** Biological and biochemical studies on embryonic cell-cell recognition, in *Cell and Tissue Interactions,* Lash, J. W. and Burger, M. M., Eds., Raven Press, New York, 1977, 173.

151. **Brachet, J. L.,** Nucleocytoplasmic interactions in morphogenesis, *Proc. R. Soc. London B,* 178, 227, 1971.

152. **Kinoshita, S. and Saiga, H.,** The role of proteoglycan in the development of sea urchins. I. Abnormal development of sea urchin embryos caused by the disturbance of proteoglycan synthesis, *Exp. Cell Res.,* 123, 229, 1979.

153. **Kinoshita, S. and Yoshii, K.,** The role of proteoglycan synthesis in the development of sea urchins. II. The effect of administration of exogenous proteoglycan, *Exp. Cell Res.,* 124, 361, 1979.

154. **Surani, M. A. H.,** Glycoprotein synthesis and inhibition of glycosylation by tunicamycin in preimplantation mouse embryos: compaction and trophoblast adhesion, *Cell,* 18, 217, 1979.

155. **Lash, J. W. and Vasan, N. S.,** Tissue interactions and extracellular matrix components, in *Cell and Tissue Interactions,* Lash, J. W. and Burger, M. M., Eds., Raven Press, New York, 1977, 101.

156. **Hey, E. D.,** Interaction between the cell surface and extracellular matrix in corneal development, in *Cell and Tissue Interactions,* Lash, J. W. and Burger, M. M., Eds., Raven Press, New York, 1977, 115.

157. **Toole, B. P., Okayama, M., Orkin, R. W., Yoshimura, M., Muto, M., and Kaji, A.,** Developmental roles of hyaluronate and chondroitin sulfate proteoglycans, in *Cell and Tissue Interactions,* Lash, J. W. and Burger, M. M., Eds., Raven Press, New York, 1977, 139.

158. **Roberson, M., Neri, A., and Oppenheimer, S. B.,** Distribution of concanavalin A receptor sites on specific populations of embryonic cells, *Science,* 189, 639, 1975.

159. **Morrill, J. B. and Macey, L. E.,** Cell surface changes during the early cleavages of the pulmonate snail, *Lymnaea palustris, Am. Zool.,* 19, 902, 1979.

160. **Wartiovaara, J., Leivo, I., and Vaheri, A.,** Expression of the cell surface-associated glycoprotein, fibronectin, in the early mouse embryo, *Dev. Biol.,* 69, 247, 1979.

161. **Sherman, M. I.,** Developmental biochemistry of preimplantation mammalian embryos, *Ann. Rev. Biochem.,* 48, 443, 1979.

Chapter 2

THE PROGRAM OF FERTILIZATION AND ITS CONTROL*

Floriana Rosati, ** **Rosaria De Santis, and Alberto Monroy**

TABLE OF CONTENTS

* This article is dedicated to the memory of Valerio Monesi.

** On leave from the University of Siena.

I. INTRODUCTION

Sexual reproduction, one of the major advances in evolution, is the process whereby two individuals belonging to the same species exchange their genetic material thereby giving rise to organisms genetically different from their "parents". We do not wish to enter into a discussion, which by necessity would be largely speculative, on the evolutionary history of the gametes (see Maynard Smith[1]). However, a brief discussion on the mechanisms that are of general relevance for the interpretation of the complex systems underlying fertilization in multicellular organisms is perhaps not out of place.

Despite the various differences, the mating process has at least two general characteristics shared by all organisms:

1. It begins with a process of "recognition" between the cells of the opposite sexes, and in fact only between opposite sexes, whose specificity depends on the molecular organization of the receptors on the cell surfaces.
2. As a result of this, the cell membranes are dissolved, at the point of contact, and this allows the transfer of the genetic material accompanied or not by cell fusion.

In bacteria, in which the sexual events involve the transfer of part or all of the chromosome, the difference between the donor ($+$, also called male) and the recipient ($-$, also called female) cells appears to be determined by a sexuality factor (F) which endows the cells (F^+) with the ability to transfer the chromosome to the recipient cells (F^-). Bacteria are a model in which sex determination and recognition mechanisms are closely associated and they are in fact controlled by the same factor. The interaction depends on the presence of the F antigen at the cell surface of the F^+ or Hfr (the bacteria into which the F factor has become integrated in the chromosome) males. The F antigen allows the F^+ or Hfr cells to recognize those cells which do not carry the factor and which lack the surface structures specified by F.[2] Kennedy et al.[3] have shown that the proteins coded for by the *tra* operon of F (the transcriptional unit containing the genes required for the synthesis of the F-pili and for the stabilization of the mating aggregates, for DNA transfer, and for the *surface exclusion* which prevents the interaction of donor cells with one another) are associated with the cell envelope. Thus, the appearance of the sexuality factors has resulted in the encoding of certain proteins in the cell membrane; one of these proteins became responsible for the recognition mechanisms.[4] In the blue-green alga, *Chlamydomonas*, high molecular weight particles containing glycoproteins have been recognized as the mating-type specific recognition substances.[5]

In eukaryotes, rather elaborate interacting systems have become associated with sexual reproduction. Although receptor molecules have not yet been identified in many systems, specific cell recognition appears to depend on glycoproteins on the gamete surface. In systems in which these molecules have been chemically characterized a combined genetic analysis to find out how their synthesis is controlled by the mating type is still lacking. The most detailed analysis of the recognition factors has been conducted in the yeast, *Hansenula wingei* (reviewed by Crandall).[6] Two factors have been isolated and their complementarity has been established. Complementarity is believed to depend on the formation of weak chemical bonds, the most important being hydrogen bonds.[6] The two recognition factors, designated 5f and 21f, are both mannoproteins. The 5f agglutinin has six combining sites[7] while 21f is univalent; when mixed together the complex is inactive. Since differentiation of sexually reactive cells is *genetically determined* but induced during the cell cycle and one of the most important parts of this process depends on the presence of certain glycoproteins in the membrane, the question is whether or not the sexual factors are present in the vegetative cells in a "masked" form or whether they are synthesized as a result of induction. The former alternative is suggested by the work of Goudenough and Jurivich.[8] Antisera against gametic flagella are bound by the membrane both of the vegetative cells and the gametes, as well as by the membranes of both mating types. The most relevant difference is that while the flagella of the gametes are agglutinated by their tips, those of the vegetative cells are agglutinated along their whole length. It is suggested that during gametogenesis the cells acquire the ability to rearrange some molecules already present in the membrane of the vegetative cells into patches which can mediate agglutination. In this connection, it is worth recalling the recent findings of Romrell and O'Rand,[9] (see also Section II.C.2) that cell surface isoantigens of the primary spermatocytes and early spermatids are able to cap and form patches, while those of later stages are not. Although more information is needed, the rearrangement of molecules in restricted areas of the gamete plasma membrane may be a general mechanism whereby the sexually reactive cells acquire the ability to recognize each other.

II. THE PREPARATORY EVENTS OF FERTILIZATION

Until their interaction and fusion at fertilization, both the oocyte and the spermatozoon are in a sense "dormant" cells. After the gametes reach a certain stage of the maturation process (which in the case of the spermatozoon, but not necessarily of the oocyte, implies the completion of meiosis), they are kept in a metabolic condition in the gonads, which ensures their extended survival *within the gonad* while their life expectancy *outside* of the gonad is very short. As a result of fertilization, the egg is triggered into embryogenesis. Since activation of the egg can be accomplished by a number of chemical and physical agents (artificial parthenogenesis) it is assumed that apart from its genetic contribution, the role of the spermatozoon in fertilization is to activate a machinery that is "ready-to-go". Fusion with the egg also activates the spermatozoon. In the case of the spermatozoon, however, only its nucleus undergoes activation; and in fact it is only when it is *within the egg cytoplasm* that its DNA reacquires its replicative and transcriptional activities. At a variance with the egg, no "parthenogenetic" conditions for the spermatozoon have thus far been found.

The study of the "ready-to-go" machinery of the egg is central to the understanding not only of the process of the "activation" of the egg upon fertilization, but also of the sequence of events of early development. Hence, we will discuss the processes that occur in the oocyte during oogenesis and maturation and which may be held responsible for both the metabolic repression and acquisition of the properties that enable the mature egg to respond to the activating trigger.

In recent years a combination of genetic and immunological analyses has led to important discoveries on the differentiation of surface antigens of the spermatozoon and these may be relevant to the understanding of the sperm-egg interaction. Thus, we will also briefly discuss this topic.

A few words of presentation of the "dramatis personae" are appropriate in introducing the subject.

A. The Egg and the Spermatozoon

The gonad is a composite structure resulting from the germ cells and a somatic component. The differentiation of the gonad has been most intensively studied in vertebrates in which it is now well established that the germ cells (gonocytes) originate in an extragonadal territory from which they migrate into the somatic rudiment (see review in Monroy and Moscona).[10] Differentiation of the gametes depends on the interaction between the germ cells and the cells of the somatic rudiment of the gonad.

The genetic sex of the gonocytes invading the genital ridge has been thought to induce the differentiation of the gonad into the male or female direction. However, it has recently been shown that in chick embryos completely deprived of their germ cells the differentiation

of the somatic component of the gonad proceeded almost normally in spite of the absence
of germ cells.[10a] A key role in the differentiation of the gonad has been attributed to the H-
Y antigen which is expressed at the surface of the genetically male gonocytes (review by
Ohno).[11] Support of this view came from experiments showing that the genital ridge cocultured with testicular cells stripped of the H-Y antigen differentiates into an ovary.[12,13]
Although there are still many unanswered questions, especially about its mode of action,
the role of this antigen in sex differentiation appears at least "highly plausible".[13a]

The outcome of spermatogenesis is a very sophisticated cell containing all the components
required for the interaction and fusion with the oocyte. The spermatozoon may thus be
considered a functionally self-contained unit. On the other hand, the egg is a functionally
and morphologically composite structure which results from the oocyte — the germ cell —
and a number of investments, all of which play an important role in fertilization. For example,
the receptors on which the specific recognition and binding of the spermatozoon and the
egg depend are integral components of the sperm plasma membrane, while in the case of
the egg they appear to be located in the vitelline coat, a glycoprotein envelope external to
the oocyte plasma membrane. The oocyte with its vitelline coat is also surrounded by a
glycoprotein jelly layer (or layers), by follicle cells which are derivatives of the somatic
component of the gonad, or by a tough chorion.

Sperm-egg interaction thus requires a complex series of steps which involve not only the
plasma membrane of the gametes, but also components of the spermatozoon which are
exposed in the course of its interaction with the egg (in the acrosome reaction; Section II.B)
and structures external to the oocyte itself.

B. The Interacting Surfaces of the Gametes and the Acrosome Reaction

The question is, how does the species-specific recognition of the gametes occur? This is
equivalent to asking, what gamete structures are involved in the recognition process? As for
the egg, in principle recognition may occur at the level of the jelly coat(s) surrounding the
egg, of the vitelline coat (and its equivalent in mammals, the *zona pellucida*) or of the egg
plasma membrane. Studies of the last few years point to the vitelline coat as the structure
endowed with the property to species-specifically recognize and bind the spermatozoa[14-23]
(Figure 1).

As for the spermatozoon, the situation appears to be more complex. The crux of the
problem concerns the condition of the spermatozoon when it reaches the vitelline coat; i.e.,
whether or not by this time the acrosome reaction has already occurred. If the acrosome
reaction takes place *prior* to the interaction of the spermatozoon with the vitelline coat, then
the "receptors" should be associated with some of the acrosome components that become
exposed as a result of the reaction, i.e., the acrosomal content or the surface of the acrosomal
process. (In this connection a detailed study of the sperm-egg interaction in starfish, which
to our knowledge is the only well-documented case of the acrosome reaction taking place
before the spermatozoon has reached the vitelline coat, and in fact at the surface of the jelly
coat, may give interesting clues.) On the other hand, if the acrosome reaction takes place
upon the interaction of the spermatozoon with the vitelline coat, this implies that the acrosome
is still intact at the time of the interaction; in this case the receptors should be located on
the plasma membrane of the sperm head. It is a widespread assumption that in the sea urchin
the acrosome reaction is triggered by some component of the egg jelly *before* the spermatozoon reaches the vitelline coat. In fact, solutions of jelly coat are powerful inducers of
the acrosome reaction[24,25] and transmission electron micrographs show that many of the
spermatozoa embedded in the jelly coat as well as spermatozoa attached to the vitelline coat
have undergone the acrosome reaction.[19,21] Hence, the conclusion that binding of the spermatozoon to the vitelline coat must be preceded by the dehiscence of the acrosome. The
recent discovery, at least in some animals, of the protein "bindin" which is released from

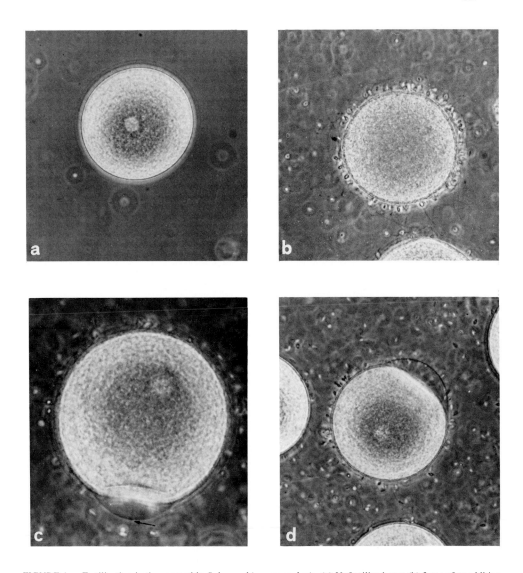

FIGURE 1. Fertilization in the sea urchin *Sphaerechinus granularis*. (a) Unfertilized egg; (b) 2 sec after addition of sperm: spermatozoa attached to the egg surface; (c) beginning of fertilization membrane elevation: the flagellum of a fertilizing spermatozoon is still visible (arrow); and (d) on the upper right corner the fertilization membrane has elevated; note that the spermatozoa have already detached from the elevated fertilization membrane, while they still adhere to the rest of the egg surface. (e) Fertilization of *Ciona intestinalis* shows an egg after removal of the follicle cells: spermatozoa attached to the vitelline coat; (f) two eggs recovered from a female hamster 5 hr after artificial insemination and incubated for 5 min with capacitated spermatozoa.[147] One of the eggs (A) has many spermatozoa attached to the *zona pellucida* (negative zona reaction), while the other egg (B) has none (positive zona reaction).

the acrosome vesicle upon its dehiscence[26-28] and which specifically "glues" the acrosome process to the egg surface indicates this protein as being the sperm component responsible for the specificity of the binding of the spermatozoon to the egg (Figure 2). In view of the very short life span of the spermatozoa after having undergone the acrosome reaction,[29] one would assume that for the acrosome reaction to lead to a successful sperm-egg fusion it should occur very close to the vitelline coat. The view that the interaction of the spermatozoon with the vitelline coat is the event that in the sea urchin triggers the acrosome reaction is supported by the observation of Aketa and Ohta[30] showing that in *Pseudocentrotus depressus* the spermatozoa bind to the vitelline coat in an unreacted condition. (The difficulty here is

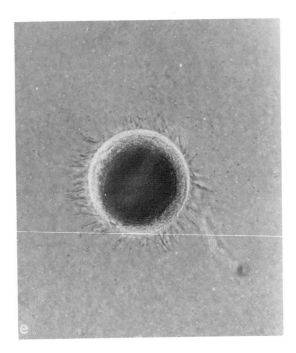

FIGURE 1e.

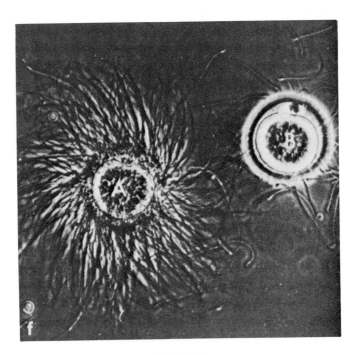

FIGURE 1f.

of being sure of the complete removal of the jelly coat, and in fact, of how to draw a sharp line between jelly coat and vitelline coat.)[31] The observations of Chandler and Heuser[32] using the rapid freeze-fracture method also support the same view. It has also been shown that *Arbacia* jelly coat solutions at physiological pH fail to trigger the acrosome reaction.[33]

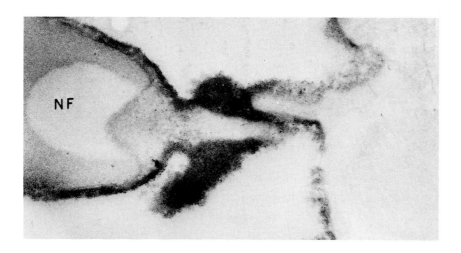

FIGURE 2. Localization of "bindin" in the sperm-to-egg binding site as shown by its reaction with 3,3-diaminobenzidine-hydrochloride. Bindin also appears on the surface of the egg microvilli. NF, nuclear fossa. (Magnification × 79,500.) (From **Moy, G. W. and Vacquier, V. D.,** in *Current Topics in Developmental Biology,* Vol. 13, Moscona, A. A. and Monroy, A., Eds., Academic Press, New York, 1979, 31. With permission.)

In our view, the only way to satisfactorily answer these questions is to try and design experiments reproducing as closely as possible the conditions under which the encounter between spermatozoa and eggs occur in the sea, which are very different from laboratory conditions. An alternative to be considered is that spermatozoa carry two sets of binding sites, one on the plasma membrane and the other on the membrane of the acrosomal process, the latter being involved in the fusion of the spermatozoon with the egg plasma membrane.

The precise sequence of events of sperm-egg interaction in mammals is still a debated question. According to some authors[22] the inner acrosomal membranes exposed as a result of the acrosome reaction is where species-specific recognition occurs. However, recent experiments[34] using a fluorescent probe for the acrosome reaction point to the sperm plasma membrane as being responsible for the recognition process. The acrosome reaction follows the binding to the zona pellucida and in fact only unreacted spermatozoa can bind to the zona. Thus, it is necessary to study the individual steps independently of each other to understand the molecular basis of this process. In this connection, the eggs of the Ascidians have been proven to be an interesting model. In the Ascidians the processes of sperm-egg recognition and binding can be studied independently of those of the acrosome reaction and sperm penetration through the vitelline coat (the so-called chorion).[17,35] These studies have been made possible by the discovery that glycerol treatment of the unfertilized eggs results in the complete removal of the follicle cells (the outermost egg envelope) and in the cytolysis of the test cells and of the oocyte. The morphology of the vitelline coat remains unchanged and more importantly, it retains the ability to species-specifically bind spermatozoa. On the other hand, it loses the ability to elicit an acrosome reaction; i.e., the binding of the spermatozoa to the vitelline coat is not accompanied by any obvious alteration of the plasma membrane of the sperm head such as occurs in the acrosome reaction.[35] These observations show that, at least in *Ciona*, the receptors for the specific recognition and binding are associated with the sperm plasma membrane; also, binding is not necessarily followed by an acrosome reaction. The lack of acrosome reaction in the glycerol-treated eggs is not due to the removal of the follicle cells; spermatozoa attached to the vitelline coat of eggs from which the follicle cells have been removed mechanically (by shaking) are able to undergo an acrosome reaction.[17]

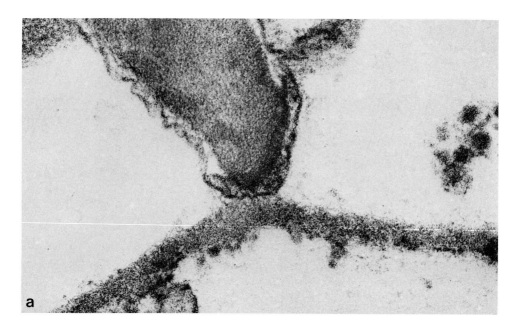

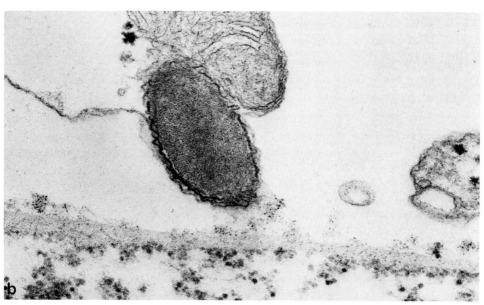

FIGURE 3.   (a) A spermatozoon of *Ciona intestinalis* is attached to an isolated vitelline coat. Thin fibrils connect the plasma membrane of the tip of the head of the spermatozoon to the outer surface of the vitelline coat. The acrosome vesicle is still intact. (Magnification × 119,000.) (b) A spermatozoon bound to the vitelline coat of a glycerol-treated egg; staining with ferritin-conjugated Con-A. The tufts of fibrils at the outer surface of the vitelline coat and the fibrils connecting the outer surface of the vitelline coat to the plasma membrane of the spermatozoon selectively bind the ferritin-conjugated Con-A. (Magnification × 63,000.) (From **De Santis, R., Jamunno, G., and Roasti, F.,** *Dev. Biol.,* 74, 490, 1980. With permission.)

The spermatozoa bind to tufts of very thin fibrils which protrude from the outer surface of the vitelline coat (Figure 3). The tufts react with Ruthenium Red and bind Concanavalin A (Con A) and fucose binding protein (FBP).[35] Similar tufts also emerge from the inner side of the vitelline coat; however, while they stain with Ruthenium Red and bind Con A, they fail to bind FBP. The asymmetry of the vitelline coat with respect to the presence of fucosyl residues parallels the difference in the ability to bind spermatozoa. Isolated vitelline coats were challenged with spermatozoa; binding occurred only at the outer surface. Hence, fucosyl residues appear to be the most important structural components of the sperm receptors on the vitelline coat.

Direct evidence has come from experiments showing that L-fucose (but not D-fucose or other tested monosaccharides) competitively interferes with binding of the spermatozoa to the vitelline coat and with fertilization.[18]

About ten protein bands are resolved by gel electrophoresis from *Ciona* vitelline coats; only three of them react with ^{125}I-FBP.[36] This is interesting in connection with the recent findings of Bleil and Wassarman,[20] that the *zona pellucida* of the mouse oocyte is made up of three glycoproteins, only one which is endowed with sperm-receptor activity.

The study of fertilization of *Ciona* eggs deprived of their follicle cells by shaking has shown that of the many spermatozoa that become attached to the vitelline coat, only a very few undergo the acrosome reaction.[35] The same is true in the mammals.[34] We have postulated[35] a diversity of the molecular organization of the receptors, both on the vitelline coat and on the sperm plasma membrane, whereby a specific molecular fit between the interacting receptors is required for the bound spermatozoa to be triggered into the acrosome reaction. This raises the interesting possibility of "clones" of spermatozoa, each expressing at their surface one or several classes of receptors, each with a different steric configuration.[23]

Let us now turn to the acrosome reaction. The acrosome reaction is the first event that is switched on after the sperm-egg interaction and is considered a prerequisite for the fusion of the spermatozoon with the egg. Typically, it involves the fusion of the plasma membrane of the tip of the sperm head with the acrosome membrane thus resulting in the exocytosis of the acrosomal vacuole and, in a number of spermatozoa, in the polymerization of actin[37] (Figure 4). Using two ionophores, Tilney et al.[37a] have recently shown that the two processes are controlled by different ions. The first event is controlled by Ca^{2+} influx into the spermatozoon, while actin polymerization is controlled by a rise in intracellular pH. The two events are probably connected. In this regard, it is interesting that in the sea urchin spermatozoa the acrosome reaction (induced in vitro by treatment with a solution of jelly coat) is accompanied by a large uptake of Ca^{2+}.[38] It has been computed that the localized Ca^{2+} concentration in the sperm head reaches as much as 10 mM.

In 1954, J. C. Dan[39] showed that Ca^{2+} is required for the induction of the acrosome reaction; more recently, Takahashi and Sugyiama[40] have presented evidence that calcium is required only for the acrosome reaction. In mammals Ca^{2+} is required both for the acrosome reaction[41] and for the sperm-egg fusion.[42]

At this point it is of interest to mention a recent study[43] on the localization of calmodulin (the Ca^{2+}-binding protein; see Section III.D) in the mammalian spermatozoon. In the non-reacted spermatozoa, calmodulin is localized around the acrosome, in a band around the lower third of the head and the base of the flagellum. In the reacted spermatozoa calmodulin is no longer visible on the anterior part of the head.

The mechanism whereby Ca^{2+} triggers the fusion between the plasma membrane and the acrosomal membrane is one aspect of the membrane fusion mechanism in general. Indeed, it can be considered a process of exocytosis involving the exposed surfaces of the cytoplasm-facing part (PS) of the plasma membrane of the spermatozoon tip with that of the acrosomal membrane (i.e., as a PS-PS fusion).[44] We do not intend here to enter into the intricacies of the problem of exocytosis; the recent discussion of Satir[44] and of Plattner[45] may be consulted. Following the description of Satir[44] the essential steps may be summarized as follows: the first step is membrane recognition which results in the close apposition of the membrane lipid bilayers. This is followed by a formation of a well-organized array of particles, probably a concentration of transmembrane proteins which function as ion channels and in particular as Ca^{2+} channels. The outer surface of this array can directly or indirectly be connected to

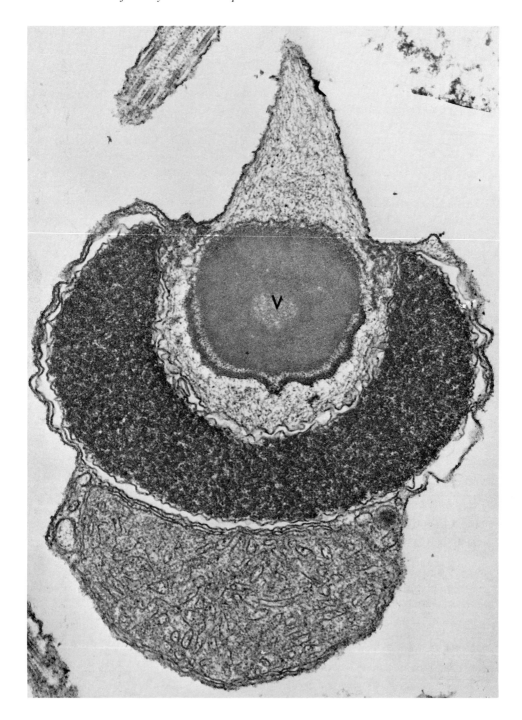

FIGURE 4.   Thin section through a *Pisaster* sperm which had been washed twice in 0.5 *M* NaCl containing 1 m*M* EGTA, then fixed 200 sec after the addition of X537A. It is of particular interest that the profilactin has been converted into filaments, but the acrosome vacuole (V) remains intact. The rest of the sperm remains unaltered. (Magnification × 74,000.) (Reproduced from *The Journal of Cell Biology*, 1978, Vol. 77, page 536. By copyright permission of The Rockefeller University Press, New York.)

receptor proteins which respond to an appropriate membrane stimulation that results in the opening of the calcium channels. Fusion of the lipid bilayer then follows. This scheme fits some observations on the exocytosis of the acrosome vesicle. The site at which the acrosomal membrane fuses with the plasma membrane is defined by a ring of particles on the P face of both membranes, while the point at which fusion occurs within the ring is particle-free.[36] Whether or not the particles, which are expressions of an aggregation of proteins, are involved in the fusion is still unclear.[36,45] It has been suggested that during acrosome exocytosis[44] they may destabilize the lipid bilayer within the ring so that membrane fusion occurs in this limited area following Ca^{2+} intake.

A detailed description of the acrosome reaction would be outside the scope of this chapter. The reader is referred to the reviews of Colwin and Colwin,[46] J. C. Dan,[39] Monroy,[47] and Monroy and Moscona.[10]

Finally, a few words about the interaction of the reacted spermatozoa with the egg plasma membrane, which is the process leading to the incorporation of the spermatozoon into the egg. The fusion actually occurs — except in the animals whose spermatozoa lack an acrosome — through the fusion of the egg plasma membrane with the inner acrosomal membrane. Whether or not specific receptors exist on the egg plasma membrane and on the acrosomal membrane is not known. FBP binding sites have been found on the plasma membrane of the *Ciona* egg.[48] However, we cannot say whether or not they are involved in the sperm-egg fusion. We do know (unpublished observations) that spermatozoa are able to attach to the surface of the vitelline coat — free eggs and enter the egg; the details of the process, however, have not been worked out (see also Section III.D for a discussion on the mechanism of fusion).

C. Gene Expression during Gametogenesis

The synthesis of macromolecules, in particular RNA and proteins during oogenesis, has been intensely studied in the last 2 decades. The information obtained has advanced our understanding of the molecular basis of the "programming" of early development. From genetic studies it was known that many events taking place during early development are controlled by genes which express themselves during oogenesis (so-called "maternal effect"). It is now clear that oogenesis is a time of intense transcriptional activity which results in the synthesis of a large variety of gene products, the majority of which are not meant for immediate translation but are rather stored in a nontranslatable condition until some time after fertilization.

The questions we shall discuss in this section deal with the part of the program of early development that is relevant to the interaction of the gametes and to the egg response to fertilization.

1. Oocyte Maturation: The Acquisition by the Oocyte of the Ability to Interact with the Spermatozoon

Cytoplasmic maturation was originally described by Delage[49] as the process whereby the oocyte acquires the ability to interact and fuse with the spermatozoon, thus giving rise to the zygote which is able to divide and in turn give rise to an embryo. The two processes are relatively independent of each other. In fact, under laboratory conditions immature oocytes can fuse with spermatozoa; however, in such a case no activation, i.e., no cell division follows.

In the majority of animal eggs meiosis is not completed until after fertilization: in some animals it is arrested at the prophase of the first division; in others it proceeds to the metaphase of the second division. Fertilization triggers the resumption of the meiotic process. On the other hand, activation followed by cell division is strictly dependent on completion of meiosis. Hence, the oocyte acquires the ability to interact with the spermatozoon independently of the stage of meiosis attained in the course of maturation. Delage[49] postulated that cytoplasmic maturation was dependent on the admixture of the contents of the germinal vesicle (GV) with the oocyte cytoplasm that occurs at the onset of maturation, i.e., at the time of the breakdown of the GV (GVBD). However, in a number of eggs, such as for example those of the annelids and of the mollusks, the egg is fertilized as a primary oocyte, i.e., while the nucleus (GV) is in the prophase of the first meiotic division and its nuclear envelope is still intact. Whether or not a transfer of GV contents to the cytoplasm occurs also in this case is not known.

Let us start by considering the changes of the oocyte surface which lead to the acquisition of the ability to interact with the spermatozoon. First of all, they involve the differentiation of the sperm receptors on the vitelline coat. A particular molecular organization of the oocyte plasma membrane is also required for it to fuse with the acrosome membrane and to respond to fusion with a "propagated response", i.e., with the setting in motion of the chain of reactions leading to the activation of the egg.

As discussed in Section II.B, the sperm receptors are structure components of the vitelline coat (whose equivalent in, e.g., reptiles and mammals, is the *zona pellucida*). The origin of the vitelline coat — whether from the egg or from the follicle cells — has been an open question until recently. Observations by Bleil and Wassarman[50] and from our laboratory (Rosati et al.[50a]) have now provided evidence that, at least in mammals and ascidians, the vitelline coat is synthesized by the oocyte. Having shown that fucose is the key component of the sperm receptors,[18,35,36,48] we have followed incorporation of [^{3}H]L-fucose injected into the ovary of *Ciona*. The autoradiographs do not show any radioactivity in the follicle cells at any stage of oocyte differentiation. The tracer appears first in the cytoplasm of the oocyte and then accumulates in the vitelline coat (Figure 5). No incorporation has ever been observed in the vitelline coat of mature, i.e., meiotic oocytes; thus showing that fucose is incorporated in the vitelline coat during the period of the oocyte growth. This we take as evidence that the fucose-containing structures of the vitelline coat, and we assume these to be the building blocks of the sperm receptors, are synthesized by the oocyte. We propose that their assembly is directed by the oocyte plasma membrane and that the assembly of the sperm receptors is part of the oocyte differentiation program. Hence, the differentiation of the sperm receptors appears to occur early in the course of oogenesis and in fact long before the onset of oocyte maturation. This is in agreement with the fact that spermatozoa can attach to and fuse with immature oocytes (see in particular Longo[51] for the sea urchin and Usui and Yanagimachi[52] for mammals). In fact, in the sea urchin sperm-oocyte fusion may occur in previtellogenic oocytes.[51] In particular, the electron micrographs of the latter author show that the attached spermatozoa have undergone acrosome reaction, which again supports the notion that the sperm receptors are fully functional even in the youngest oocytes.

Cytoplasmic maturation involves also the acquisition by the oocyte plasma membrane of the physiological conditions of the mature egg. This has been demonstrated by Dale et al.[53] in a study on the maturation of the starfish oocyte. It is known that in the starfish oocyte maturation can be obtained in vitro in about 30 min by the action of 1-methyl-adenine (1-MA) which has been identified as the natural maturation-inducing substance (MIS) produced by the follicle cells.[54] Within 15 to 30 min of adding 1-MA the membrane potential of the oocyte drops from -70 to -90 mV to -10 to -20 mV and this change is concomitant with the GVBD. The membrane also loses its K^+ selectivity, which means that the process of maturation involves the closure of the K^+ channels. No such switches occur in anucleated oocyte fragments; this strongly suggests that the presence of the GV, and hence of the admixture of its contents with the cytoplasm, is a necessary requirement for the oocyte plasma membrane to acquire the mature condition. Which nuclear component triggers the change is not yet known.

Let us now discuss briefly some of the cytoplasmic changes related to maturation.

Neither changes of the nuclear envelope nor chromatin dispersion occur in the sperm nuclei in the cytoplasm of sea urchin oocytes prior to the onset of meiosis (see Longo[51] also for a review of the literature). Full differentiation into male pronucleus is only observed in pronuclear oocytes (which by now may be called eggs). Hence, the condition of the oocyte

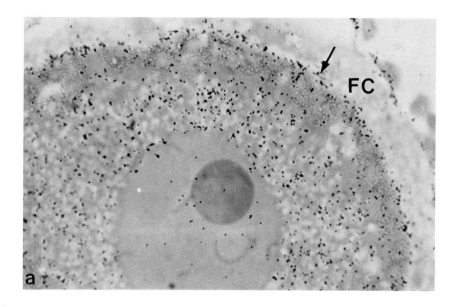

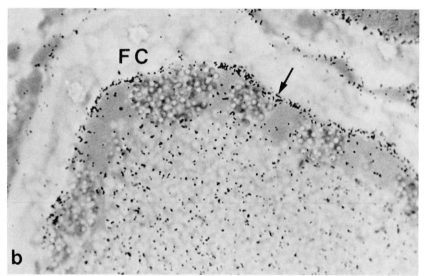

FIGURE 5. (a and b) Autoradiographs of sections of oocytes from gonads of *Ciona intestinalis* labeled with [³H]fucose in vivo for 18 hr. Grains present in the cytoplasm of the oocyte become greater in number in the area corresponding to the vitelline coat (arrow). No grains on the follicle cells (FC), while the concentration is highest at the surface of the oocyte where the vitelline coat is differentiating (arrow).

cytoplasm which permits the differentiation of the sperm nucleus into a pronucleus appears to arise following the admixture with the contents of the GV. Also, there is no cortical reaction to fertilization in the premeiotic oocytes; specifically, there is no cortical granule exocytosis and no fertilization potential[55] (see Section III.B). In the amphibians, at least in *Rana*,[56] removal of the GV does not interfere with progesterone-induced maturation of the oocyte. Indeed, the anucleated oocytes respond to activation by pricking and they also exhibit the same increase of the rate of protein synthesis in response to progesterone stimulation as the normal oocytes. It is interesting that cleavage of such anucleated pricking-activated oocytes is slow and irregular. Yet, if they are injected with the contents of a GV a certain percentage of them cleaves normally.[57]

The breakdown of the GV which results in the interaction of its contents with the oocyte cytoplasm thus appears to be a crucial event in the life history of the oocyte. The relevance of this interaction may be different in different animals; at the molecular level, its nature is still not understood. It is certainly a wide open and extremely interesting field for investigation. Let it suffice here to mention a few points.

Axolotl eggs, homozygous for the recessive gene *o*, if fertilized with sperm of normal animals, develop to the blastula stage and then die.[58] The injection of cytoplasm from wild type +/*o* into *o*/*o* eggs allows development to proceed to the late gastrula or early neurula stage. Even more dramatic results are obtained by injecting small amounts of the wild type GV contents; in this case, normal larvae develop.[59,60] Hence, at least in this case it looks as though a gene product controlling events of the postgastrula development is stored in the GV. Also, large amounts of RNA polymerase are stored in the GV of *Rana*; in fact, the quantity of the enzyme present in a single GV is nearly the same as that present in a whole embryo consisting of 4.10^5 cells.[61,62] The same is true of the *Xenopus* oocyte where almost all the DNA polymerase is contained in the GV.[63] Finally, an almost forgotten observation of Naora et al.[64] should be mentioned. These authors discovered that the GV of *Rana* oocyte accumulates enormous quantities of Na^+ in contrast with the very low Na^+ content of the cytoplasm. Whether Na^+ accumulation is a general property of the GV is not known. It is clear that the flooding of the oocyte cytoplasm with a large quantity of Na^+ at the time of the breakdown of the GV must involve dramatic changes, not only of the cytoplasmic organization and function, but also of the oocyte plasma membrane and, in particular, of its electrical properties.

These few data thus support the notion that the GV is a kind of "storehouse" of gene products, proteins, enzymes, ions, etc. whose compartmentilization in the GV is essential for oogenesis. Maturation, the condition that makes the oocyte responsive to activation, is triggered by the opening of the nuclear compartment and by the mixing of its contents with that of the cytoplasmic compartment. The "opening" does not necessarily coincide with the morphologically visible breakdown of the GV envelope; it may be the result of a selective alteration of its properties. Work in progress in our laboratory shows that in the starfish oocyte induced to mature by 1-MA, the actual breakdown of the GV occurs about 40 min after adding 1-MA. Yet, changes in the plasma membrane of the oocyte can be detected electrophysiologically about 15 min after 1-MA is added.[65] Fine structural studies have shown that at this time distinct alterations of the organization of the GV envelope, such as "swelling" of the nuclear pores, are already clearly visible. These observations may help to reconcile the apparently conflicting observations on the role of the GV in the process of maturation in different animals.[10] Indeed, it is not unreasonable to assume that some of the material contained in the GV may "leak" into the cytoplasm long before the onset of the visible breakdown of the GV.

2. Differentiation of Surface Antigens during Spermatogenesis

Let us now briefly discuss some aspects of the differentiation of the spermatozoon. To begin with it should be noted that the major difficulty here is the isolation of the various classes of spermatogenic cells, a difficulty that has been overcome only recently.[66,67] It now seems well established that gene expression occurs during the haploid phase of spermiogenesis and this involves both ribosomal RNA and putative mRNA. This is important among other things in relation to the expression in the mouse spermatozoa of the *T* mutation and of its normal allele and supports the notion that the *T* locus is expressed after meiosis.[66] The great interest in the immunological approach to the study of spermatogenesis is largely due to the interest aroused by the genetic studies on the *T* locus which have shown that it is expressed at the surface of the spermatozoa and that it may in fact play an important role in spermatozoon differentiation. Since it is also expressed at the surface of the egg after fertilization and mutations at this locus affect cell interactions in the blastocyst, it is suggested that genes in the *T* locus may play an important role in the early stage of embryonic development (in particular in the cell-cell contacts in the blastocyst).[68]

The study of the appearance of antigens specific of spermatogenic stages[67,69] has shown that membrane antigens specific to spermatogenetic cells first appear at the late pachytene stage of the first meiotic prophase. There is a marked cell growth during the first meiotic prophase and this is compatible with the insertion of new membrane components. The majority of antigens that are then retained on the spermatozoon are inserted into the plasma membrane after the spermatid stage. Most of the antigens that appear on pachytene spermatocytes are lost or masked in the spermatozoon;[69] the remainder are distributed over the entire plasma membrane of the spermatozoon, albeit somewhat more concentrated on the anterior head region.[69] The decrease of antigens that occurs at the time of the first meiotic prophase is accounted for by their concentration in the residual body.[67] This shows that the segregation of the spermatid components, and in particular of the membrane components, in the residual body is nonrandom. The late subclass of antigens which appears on differentiating spermatids is predominantly localized over the postacrosomal and middle piece regions of the spermatozoon.[69] These findings are also in accord with previous observations showing a differential localization of lectin-binding sites[70] and of serologically identified antigenic determinants[71] on the spermatozoon.

It is particularly interesting that the H-Y antigen is localized on the membrane overlaying the acrosomal cap.[72] This appears to suggest that this antigen plays a role in sperm-egg interaction. The use of monoclonal antibodies has also led to the recognition of two differentiation antigens[73] named XT-1 and XT-2. The spleen lymphocytes were obtained from mice immunized first with testis cells from 4-month-old animals and then from testis cells of 26-day-old animals (no spermatozoa were present in the testes of these animals). XT-2 is detectable in the 8-day testis (the majority of germ cells are pachytene spermatocytes).

XT-1 is not detectable in this stage, while it has been found in the 13-day testis (a large proportion of cells in early meiosis) and continues to increase at least through day 22 (spermatocytes and spermatids present). In the epididymal spermatozoa XT-1 is present in low concentrations, while XT-2 is not detectable at all.

Taken together these results show that in the course of spermatogenesis there is a sequential appearance and disappearance of surface-located antigens; these may thus be called *differentiation antigens*.

The spermatozoon is the most extreme case of a polar differentiated cell; this work shows that such a differentiation is also expressed in the molecular organization of its plasma membrane.

3. The Polarity of the Oocyte in Relation to the Site of Sperm-Egg Interaction

The acquisition of a differential organization of the egg components which results in the expression of different properties at the two opposite ends of the polarity axis is one of the most fundamental characters of the oocyte.[4] The aspect of polarity we discuss in this chapter is limited to the question as to whether or not this differential organization results in a restriction of the egg surface area where the interaction with the spermatozoon can take place.

The origin of polarity is a most untractable problem; while it is relatively easy to show the existence of a polar differentiation of the egg, it is quite a different matter to say how it has originated. The information available, though scanty, favors the notion of a prevailing epigenetic control. This notion rests largely on observations such as those of Wilson on *Cerebratulus*[74] and on *Dentalium*[75] and more recently of Raven[76] on *Lymnea* according to which the area of contact between the oocyte and the ovarian wall becomes the vegetal pole of the egg. Other observations show that the polar bodies are ejected from the animal pole; this is the case, e.g., of the sea urchin[77] and *Crepidula*.[79] The same holds true for ascidian and amphibian eggs. In the case of the sea urchin the animal-vegetal polarity does not appear to bear any relationship to the site of sperm penetration, which in fact may take place anywhere at the egg surface. On the other hand, in the amphibian egg, in which marked differences in the organization of the egg surface between the animal and the vegetal pole have been found,[80,81] sperm penetration is restricted to a limited area at the animal pole. In the ascidians, sperm penetration occurs at the vegetal pole (Conklin[82] and our own unpublished observations); however, thus far, no differences have been detected in the organization of the egg surface at the two poles.

We suggest that in these eggs the molecular organization of the plasma membrane or of the underlying cytoplasm is such that sperm-egg fusion is permitted only at certain sites. Two conditions (though not mutually exclusive) can be visualized. First, that the plasma membrane carries specific attachment sites for the reacted spermatozoa only in a limited area; or, there may be submembrane components which allow the engulfment of the spermatozoa only at certain sites. A system of microfilaments differentially attached to the inner aspect of the plasma membrane could modulate its ability to fuse with and/or allow the engulfment of spermatozoa. In any case, these observations show that in some animals there is, during oogenesis, a differentiation of the oocyte plasma membrane which specifies where it may fuse with the spermatozoon.

4. Synthesis of Gene Products during Oogenesis

As indicated in the introduction to this section, oogenesis is a period of intense transcriptional activity of the oocyte genome which results in the synthesis of a vast amount of gene products, most of which are stockpiled in the oocyte cytoplasm in a nontranslatable condition. Most of these ''maternal templates'' are meant to direct protein synthesis after fertilization and in fact for a certain period — which is different in different animal classes — of the early embryonic development. For example, it is well known (and in fact this was one of the earliest discoveries in this field) that the ribosomes that are used by the embryo until the gastrula stage (a notable exception is that of mammals in which synthesis of ribosomal RNA begins between the fourth and eighth cell stages) are maternal ribosomes; i.e., ribosomes that have been manufactured during oogenesis.

Recent studies on the rabbit oocyte using a high resolution two-dimensional gel electrophoresis have shown the synthesis of stage-specific proteins during maturation.[83] As will be discussed below, the synthesis of some of the oocyte proteins is arrested after fertilization[84] thus implying that these are proteins required either for maturation or for the early stages of development, and hence the transcription and translation of stage-specific templates. On the other hand, synthesis of some of the ''postfertilization'' proteins also occurs in nonfertilized aging oocytes[85] (see also Section III.F); this suggests that the *oocyte* program includes the transcription and translation of gene products to support early embryogenesis.

In some animals, and the best-known case is that of the insects, the nurse cells of the ovariole participate in the synthesis of the RNA which is stored during oogenesis. The significance of this ''cooperation'' between the oocyte and the cells of the ovarian follicle is not yet understood. It certainly strengthens the concept (Section II.A) of the oocyte and its follicle cells as forming a functional unity. In the specific case of the insects, it should be remembered that the oocyte and the nurse cells originate from a common stem cell. It would be interesting to know whether the transcription of the genome of the nurse cells involves the whole genome or only part of it. (For a more detailed discussion of this problem see Davidson[86] and Monroy and Moscona[10].)

The main question we want to address ourselves to is how are the oogenic templates made nontranslatable (the problem of their utilization after fertilization will be discussed in Section III.F).

Let us start by presenting some of the evidence suggesting that the protein synthesizing machinery of the oocyte, though potentially functional, is kept in a nonfunctional condition. The most elegant evidence comes from experiments in which rabbit globin messenger RNA (mRNA) injected into *Xenopus* oocytes[87,88] was able to direct the synthesis of rabbit-type globin in amounts proportional to that of the injected mRNA. The foreign template did not interfere with the utilization of the endogenous templates. Hence the conclusion that in the oocyte there is a block at some step of the protein synthesizing apparatus whereby the system is not used to the fullest. Indeed, it has been calculated[88] that out of a total of 50 to 100 mμg of RNA present in *Xenopus* oocyte,[86] only about 15 mμg is functional. Hence the block appears to be at the level of the template RNA.

It now appears sufficiently well established that the mRNA transcribed during oogenesis is made nontranslatable through its association with proteins into ribonucleoprotein particles (RNP). This was first suggested by Spirin[89] who named the particles "informosomes". The association of poly(A)-containing RNA with proteins in particles smaller than ribosomes was demonstrated in the *Xenopus* oocyte.[90]

Most of the studies have been carried out on the sea urchin egg. As will be discussed later, in this egg within a few minutes after fertilization the rate of incorporation of amino acids into protein starts to increase very rapidly.[91,92] This, however, is not dependent on a concomitant synthesis of mRNA; indeed, it even occurs in eggs in which RNA synthesis has been completely suppressed by actinomycin D.[93] This observation thus showed that in the early postfertilization stages of development, protein synthesis was directed by templates which were present in the egg at fertilization and hence were transcribed during oogenesis. That such templates do in fact exist was first indicated by experiments showing that RNA from unfertilized sea urchin eggs were able to stimulate amino acid incorporation into proteins in a cell-free system.[94,95] The question is then what are the mechanisms operating in the oocyte that prevent mRNA translation.

In the unfertilized sea urchin egg Gross et al.[96] identified a 20 S particle whose RNA directed in vitro synthesis of histones. The importance of this observation is that it proved that the RNA *from a well-defined particle* was able to direct the synthesis *of a specific protein* (and not just to stimulate the in vitro incorporation of amino acids into proteins). The work of Jenkins et al.[97] has shown that it is the association of the RNA with certain proteins that makes the RNA nontranslatable. Indeed, when prepared in the presence of a K^+ concentration equivalent to that of the cytoplasm of the egg (0.35 M) the RNP are unable to stimulate protein synthesis in an in vitro cell-free system. On the other hand, when the particles are destabilized by treatment with the same concentration of Na^+, they contain less protein than the K^+ particles and they are active in stimulating protein synthesis; in fact, they prove to be just as active as particles prepared from polyribosomes of embryos cultured in the presence of actinomycin D (which hence contained only oocyte templates). It may thus be concluded that (1) the association of the mRNA with certain proteins makes it nontranslatable and (2) in order for the mRNA to become translatable it does not need to be "naked".

Previtellogenic oocytes of *Xenopus* and of the fish, *Tinca*, also store large amounts of 5 S and tRNA in particles sedimenting at 42 and 7 S, respectively. The 7 S particle is made up of one molecule of 5 S RNA and one protein molecule[98] with the only difference being that the protein moiety is lower in the fish than in *Xenopus*. The 42 S particle has a more complex composition; it is made up of four subunits, each of which consists of *one* molecule of 5 S RNA and three molecules of tRNA, combined with two proteins of 50,000 M_r and one of 40,000 M_r.[99]

However, the nontranslatable mRNA may not be the only method of repressing protein synthesis in the oocyte. In the case of the sea urchin egg there are reports suggesting that a protein factor on the ribosomes may be involved in their translation inefficiency.[100-102] The most striking case is that of the previtellogenic oocytes of the lizard, *Lacerta*, in which

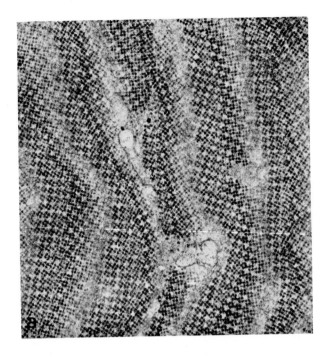

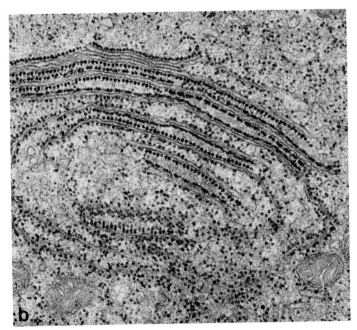

FIGURE 6. (a) Stacks of ribosomes crystallized in tetrameres and embedded in an electron dense matrix (ribosomal bodies) in a previtellogenic oocyte of a hibernating lizard. (b) Dispersal of the ribosomal bodies at the end of hibernation. (Magnification × 39,999.) (From **Taddei, C.,** *Exp. Cell Res.,* 70, 285, 1972. With permission.)

stacks of ribosomes crystallized in tetrameres have been observed during hibernation of the animals; they disperse at the end of hibernation and give rise to polyribosomes[103] (Figure 6). In vivo, these structures are inactive in protein synthesis. When, however, isolated and tested in vitro, they prove to be active in supporting incorporation of amino acids into proteins.[104] This may suggest that these "ribosomal bodies" are a storage form of programmed ribosomes, i.e., of ribosomes associated with mRNA, the complex being made inactive by a matrix that is dissolved during its isolation. Similar ribosomal stacks have been described in the oocytes of a number of mammals; these rapidly disperse after fertilization.[105]

That the inactivation of the ribosomes may be rather widespread in nature is also suggested by the case of the encysted embryos of the brine shrimp, *Artemia salina*. In this animal, protein synthesis is repressed and an inhibitor has been located on the 60 S ribosomal subunit.[106] Other factors may also be involved in the repression of protein synthesis in the oocyte before fertilization, such as, e.g., the low activity of the binding factor T_1 (the factor responsible for the formation of the ternary complex: mRNA-aminoacyl-tRNA-ribosome).[107] Hence, the situation appears to be quite complex and it is likely that although the major controlling factor of translation of the oocyte messages may be the complexing of the mRNA into nontranslatable RNP, blocks at the level of the ribosomes and other factors of the protein synthesizing machinery may play an important role.

III. THE EGG RESPONSE TO FERTILIZATION

Having described the condition of the egg at the time of fertilization and the method of sperm-egg recognition and fusion, we arrive at the heart of the problem: What is the nature of the reaction that triggers the activation of the egg?

It is generally assumed that the reactions that trigger the egg into development occur in its surface layers (often spoken of as the "cortex"), possibly within its plasma membrane. In fact, the sparking reaction of activation consists of a perturbation of the molecular organization of the plasma membrane or, at the least, of the surface layers of the egg. It is well to remember that Loeb[108] suggested that egg activation depends on "a transient surface cytolysis" of the egg. However, the connection between the initial "cortical" events and those of the initiation of development, such as, e.g., the acceleration of protein synthesis and the resumption of the egg's ability to replicate its DNA and to divide, are not yet understood.

A. The Changes of the Organization of the Egg Surface

Most of the relevant studies have been reviewed in several recent publications.[10,47,109,110] Hence, we will confine ourselves to a few points which are pertinent to the problem of egg activation in preparation for development.

The first direct evidence that fertilization brings about a reorganization of the egg surface came from the observations of Runnström[112] on the changes of the interference color of the surface of the sea urchin egg when studied under dark field illumination, and the work of Monroy and Montalenti[113] and of Monroy[114] who used the polarizing microscope. The advent of electron microscopy was a breakthrough in these studies.[47,111] One of the most interesting observations was that of the exocytosis of the cortical granules (CG), a typical component of the "cortex" of most eggs. It was found that the limiting membrane of the CG becomes incorporated into the plasma membrane of the zygote, being a mosaic of the original plasma membrane of the oocyte *plus* the newly inserted membrane of the CG. Application of the new technique of quick freezing and freeze fracture[32] has shown that the CG membrane fuses with the plasma membrane in a single "burst" (Figure 7); i.e., the multiple fusions originally described with the conventional fixation technique[115] may in fact be due to technical artifacts. Furthermore, the finding that the density of the intramembrane particles is lower in the membranes of the CG than in the plasma membrane led Chandler and Heuser[32] to establish that while the actual fusion of the two membranes is a very fast process, the mixing of their constituents is quite slow. CG membrane added to the plasma membrane causes an

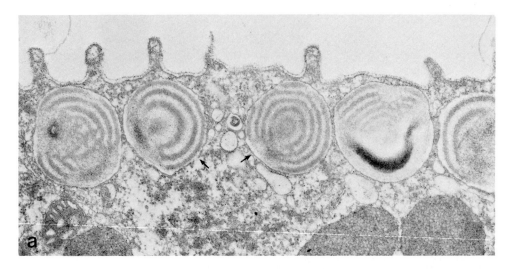

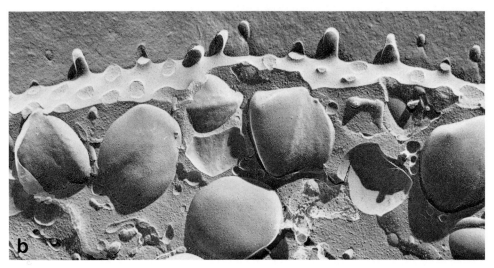

FIGURE 7.   (a) Electron micrograph of the cortex of an unfertilized sea urchin egg. Cortical granules lie in a single layer just beneath the plasma membrane (arrow). (b) Freeze fracture replica of the cortex of an unfertilized egg. Cortical granules lie just under the P face of the plasma membrane. The specimen was quick frozen. This freeze fracture plate has been photographically reversed and platinum deposits appear white. (Magnification × 25,000.) (Reproduced from *The Journal of Cell Biology*, 1980, Vol. 84, page 618. By copyright permission of The Rockefeller University Press, New York.)

approximate doubling of the surface area. This redundancy is resolved by the formation of endocytotic coated pits and of highly anastomotic microvilli (Figure 8).

Further information has come from studies of the egg plasma membrane using lectins and antibodies as probes. In the ascidian egg, which is fertilized while arrested at the metaphase of the first meiotic division, it has been shown that surface binding of Concanavalin A (Con A) undergoes a severalfold increase coincident with completion of meiosis following fertilization, i.e., at the time the egg reacquires the ability to replicate its DNA and to divide; in other words, when it becomes, for the first time, an embryonic cell.[116,117]

This observation is important because the ascidian egg lacks CG and no gross structural alterations of its surface have been observed at the time of fertilization.[118] And yet, in preparation for cleavage, the molecular organization of the egg undergoes a profound rearrangement resulting in the exposure of lectin binding sites which were not available before

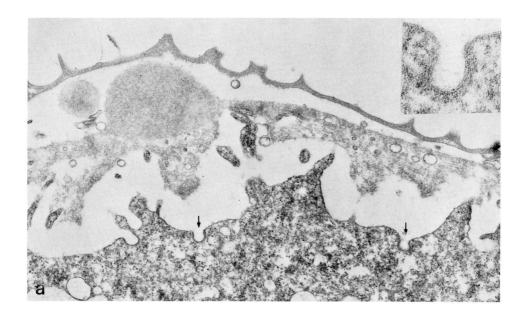

FIGURE 8. (a) Electron micrograph of the egg periphery at 2 min postinsemination showing highly arborized microvilli. Between the microvilli are numerous invaginations (arrows) which have dense coats typical of endocytosis (inset). [Magnification × 23,000; (inset) magnification × 120,000.] (b) Cross fracture of the egg periphery 2 min postinsemination. Numerous long- and highly branched microvilli have appeared. Specimen was fixed, glycerinated (30% glycerol), and frozen in Freon®. (μm × 22,000.) (Reproduced from *The Journal of Cell Biology,* 1980, Vol. 84, page 618. By copyright permission of The Rockefeller University Press, New York.)

fertilization. Even more striking is the changed binding of the lectin of *Dolichos* (whose receptor is the α-linked *N*-acetyl-D-galactosamine). In the unfertilized egg the lectin is diffusely bound to the egg surface with occasional patches. Following fertilization the binding is progressively restricted to the membrane of the vegetal-posterior compartment (the yellow crescent territory) from which the mesoderm of the embryo originates.[119] Hence, in this egg the segregation of morphogenetic compartments is associated with the selective segregation of plasma membrane components.

In the amphibian egg (in *Discoglossus*) dramatic changes have been observed at the surface of the animal dimple (the specific site at the animal pole where sperm penetration occurs[80]) in conjunction with maturation and fertilization. Binding sites for wheat germ agglutinin (WGA) and for soy bean agglutinin (SBA) are specifically localized on the walls of the dimple, whereas those for fucose binding protein (FBP) are only to be found at its very bottom. Shortly after fertilization and concurrently with the disappearance of the dimple, lectin binding becomes much fainter and in particular, the binding sites for FBP are no longer detectable.[120]

Similar studies have been conducted on mammalian oocytes. It has been observed that the Con A-induced agglutination of the mouse oocyte increases following fertilization. It is interesting that two inhibitors of protein synthesis, which cause parthenogenetic activation, puromycin and cycloheximide, while increasing the agglutinability of the unfertilized egg have no effect either on the zygote or on the blastocyst.[121]

Also, the lateral diffusion both of proteins and lipids of the mouse oocyte plasma membrane is strongly reduced after fertilization.[122] Of particular interest is an antigen, F 9, expressed at the surface of the cells of a line of mouse teratocarcinoma and at the surface of the spermatozoa (as well as of all stages of spermatogenesis).[123,124] The antigen is not expressed at the surface of the oocyte, but it makes its appearance some hours after fertilization during cleavage.[125] Accordingly, the anti-F 9 serum has no effect on cleavage but it prevents compaction (Babinet et al., quoted by Jacob[126]). A relationship between F 9 and the antigens expressed by the T/t locus has been considered.[126] In fact, some mutations at this locus affect compaction.[68] Of considerable interest is also the *disappearance* of surface proteins following fertilization;[127] in this connection, one of the most recent and puzzling observations is that acetylcholine receptors are present at the surface both of *Xenopus*[128] and mouse[129] oocytes. In mouse oocytes they disappear soon after fertilization. The interpretation of these observations is difficult. Do these receptors play a role in the sperm-egg interaction and/or in the process of egg activation? Or, are they just remnants of structures that were functional in the ovarian oocyte, for example, in connection with maturation and ovulation?

When taken together, these observations indicate that as a result of fertilization the egg surface may undergo a large-scale reorganization consisting of (1) a topological reorganization of already present molecules; (2) insertion of new molecules; (3) unmasking of molecular groups not exposed before fertilization; and (4) disappearance of preexisting molecules. Some of these changes are the immediate result of the interaction with the spermatozoon and are hence related to the process of activation per se. These include some of the fine structural changes in chemical composition and transport properties of the egg. The disappearance of surface components may also fall within this group. A second group of changes seems to be related to the ''events-to-come'' and in particular, to cleavage and blastula formation. These comprise the changed distribution of the lectin binding sites; the changed lectin agglutinability of the egg; the restriction of the lateral diffusion of membrane components; and the appearance of new membrane components such as antigens.

We suggest that these fertilization-initiated processes of the plasma membrane reorganization are related to the segregation of the morphogenetic territories and, in particular, to the encoding of information required for the differentiation of specific cell lines.[23,119,122,130] On the other hand, the alterations detected as changes in the lectin agglutinability and the appearance of new surface antigens may be relevant to the control of cell interactions and perhaps to the formation of junctions during cleavage and blastula formation. How these changes are controlled is not known.

An important role may be played by microfilaments attached to the inner side of the plasma membrane modulating the distribution and the insertion of specific molecules in the plasma membrane.[131,132] In fact, a dramatic increase of the subcortical actin filaments occurs in the sea urchin egg after fertilization[133,134] (Figure 9).

B. Electrophysiological Changes in the Egg Membrane as a Result of Fertilization

It has long been known that fertilization is accompanied by a depolarization of the egg plasma membrane which appears to be concurrent with the cortical reaction;[134a] this has been named fertilization potential (FP). Since most of these studies have been conducted

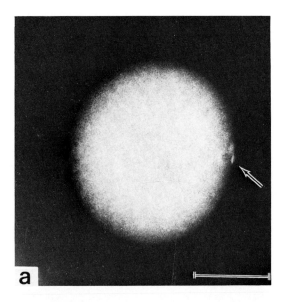

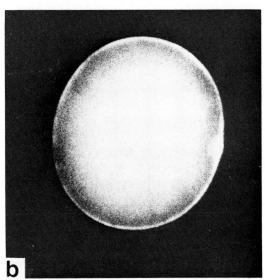

FIGURE 9. (a) Fluorescent image of an unfertilized egg which has been microinjected with 5-iodoacetanidofluorescein (IAF) actin showing the wound healing response at the site of needle insertion (arrow). The pronucleus is invisible. (b) After fertilization, eggs preloaded with IAF actin show a dramatic increase in fluorescence intensity near the plasma membrane, about 4 min after fertilization. Bar indicates 50 μm. (Reproduced from *The Journal of Cell Biology,* 1979, Vol. 82, page 672. By copyright permission of The Rockefeller University Press, New York.)

on the sea urchin egg, it is assumed that the FP is related to the exocytosis of the cortical granules. However, a FP has also been observed in eggs without CG, such as those of the ascidians[135] and in the fish egg it precedes the fusion of the cortical alveoli with the plasma membrane.[136] Hence the FP appears to be related to the molecular reorganization of the plasma membrane that follows fertilization rather than to the insertion of the membrane of the CG (or of the cortical alveoli) in the plasma membrane.

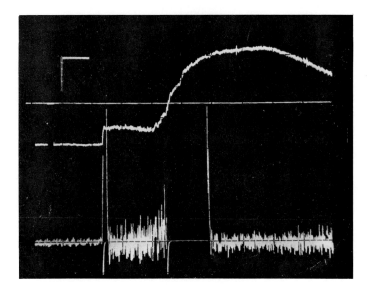

FIGURE 10.   Intracellular recording from an egg of *Paracentrotus lividus* during fertilization. The fertilization potential is the slow change in membrane potential shown in the latter half of the upper trace. The fertilization potential is preceded by a small step depolarization. The upper trace was d.c.-coupled, while the lower trace, a.c.-coupled, and demonstrates the increased voltage noise during the step event. Vertical bar = 5 sec. (Courtesy of Drs. B. Dale and A. De Santis.)

Detailed electrophysiological studies of fertilization of the sea urchin egg as well as of other marine invertebrates have been conducted in our laboratory. We were first able to show that both in the sea urchin[65] and the starfish[53] the membrane depolarizes during maturation while, at the same time, its permeability to K[+] is decreased.[53] It has been suggested[65] that the loss of K[+] permeability and the low potential-high resistance state of the membrane is a feature of the mature egg, possibly linked to its condition of metabolic repression. In fact, the high density of the ionic channels might be an expression of low cell metabolism.

Upon fertilization the first detectable electrophysiological change is a step-like depolarization of 1 to 2 mV accompanied by a marked transient increase in conductance of the ionic channels. Between 10 and 15 sec later the FP is initiated[137] (Figure 10). By analogy with similar events in other biological membranes, the step event accompanied by the transient increase in membrane noise has been interpreted as the introduction of ionic channels (the single-channel conductance has been calculated to be 33 pS) into the egg plasma membrane. It has been suggested that the step may be caused by a factor released from the acrosome when in close proximity to the egg plasma membrane.[137] It is an interesting possibility that the reaction causing the step event may be the release of "bindin" which results in the "gluing" of the spermatozoon to the egg plasma membrane[28] (Section II.B). Several minutes after fertilization the egg membrane attains the characteristics of a K[+] selective system.[65,138]

No step event has been observed in the starfish[73] or in the ascidians.[135] In the eggs of the former it is known that the acrosome reaction occurs at the surface of the jelly coat; in the eggs of the latter it occurs upon contact of the spermatozoon with the vitelline coat which is clearly separated from the egg plasma membrane[53] (Section II.B).

A comparison with the events of fertilization in the fish egg is also of interest. The fish spermatozoa lack an acrosome and in fact, the method of sperm-egg fusion in fishes is still poorly understood. In the egg of the teleosten *Oryzias*, sperm-egg fusion generates a small

FP which is not preceded by any step event.[136] It is also interesting to note that changes in the external Na^+ and Cl^- have no effect on the amplitude of the FP; hence, this appears to be due to a nonspecific leak resulting from sperm-egg fusion.[65,136]

It is interesting that mitogenic stimulation of spleen lymphocytes is accompanied by a depolarization of the plasma membrane followed by a slow repolarization and a final hyperpolarization phase. It is during the two latter phases that DNA synthesis is activated. On the other hand, the initial depolarization is not sufficient to induce blast formation; the intervention of a "growth factor" is required.[139]

C. The So-Called "Polyspermy-Preventing Reaction"

One of the earliest changes occurring in the egg following fertilization is that preventing the fusion of the egg with more than one spermatozoon. This has been considered as an active change occurring at the egg surface and has been associated with a fast depolarization of the egg plasma membrane.[140] The problem of polyspermy prevention has been discussed recently by Dale and Monroy;[141] the present discussion will be largely based on their review. According to these authors, the fundamental event in fertilization is the fusion of one *male* haploid genome with the haploid oocyte genome. The probability of more than one male nucleus interacting with the female nucleus is reduced to a minimum essentially by two mechanisms: (1) by the organization of the egg investments which limits the number of spermatozoa that effectively reach the egg and (2) by the changes that occur in the egg in response to fertilization and which are, in fact, part of the process of egg activation. Some of the questions related to the first point have been discussed in Section II.A. Among these, we shall remember the role of the jelly coat of the sea urchin egg in triggering the acrosome reaction — the molecular organization of the sperm receptors on the vitelline coat and on the plasma membrane of the spermatozoon also in relation to the acrosome reaction and to the sperm-egg fusion. As to the role of the follicle cells we should like to add a few words about our observations on the egg of *Ciona*.[35] The follicle cells cover the whole surface of the vitelline coat and the spermatozoa can reach the vitelline coat only through narrow passageways between cells whose opening and closing is probably controlled by an intricate system of microtubules and microfilaments at the base of the follicle cells.

Concerning the second point, one of the most important mechanisms is the inactivation of the sperm receptors on the vitelline coat which results in the loss of their sperm-binding activity. This is suggested as being part of the extensive reorganization of both the egg plasma membrane and the vitelline coat, and is usually spoken of as the "cortical reaction". Dale and Monroy[141] have proposed that the plasma membrane + the vitelline coat give rise to a "functional unit". In the sea urchin the morphological basis of the unit is the vitelline posts which connect the two structures to one another.[31,32] They have further suggested that the basic mechanisms for the sperm-egg interaction and also for the repelling of the supernumerary spermatozoa reside in this functional unit. Upon fertilization the posts are[32] probably broken by a protease which is released by the cortical granules;[142,143] this results in the breakdown of the functional unit and in an extensive reorganization of the plasma membrane. The interaction of the *zona pellucida* with the contents of the cortical granules results in its loss of the ability to bind spermatozoa. In the sea urchin egg the interaction of the vitelline coat with the protease appears to be the immediate cause of its loss of sperm-binding ability.[142,143]

The question is whether, at least in the sea urchin and in mammals, these changes can be considered as the *primary mechanism* which prevents multiple sperm-egg fusions. In most species of sea urchins the events occurring between the attachment of the spermatozoon and those of the cortical and cytoplasmic changes follow one another at such a rapid rate that it is very difficult to be sure of their temporal sequence. In this connection, the observations of Ginzburg[144] are of particular relevance. In *Strongylocentrotus droebachiensis* she found that the "polyspermy block" coincided temporally with the time of propagation of the breakdown of the cortical granules. Now the previously mentioned protease activity was detected *in the sea water* in which the eggs had been fertilized 30 sec after insemination. However, at the egg surface the enzyme concentration may reach very high levels much earlier. An important contribution to the understanding of the mechanisms preventing polyspermic fertilization in the sea urchin comes from the recent observation[145] that hydrogen peroxide — a powerful inhibitor of sperm activity — is released upon fertilization. Fertilization in the presence of catalase results in 100% polyspermy. Also, the soybean trypsin inhibitor — which is known to cause polyspermy[146] — prevents the release of hydrogen peroxide (exactly as from the white blood cells). Hence, one may suggest that the outburst of hydrogen peroxide release concurrently with the successful interaction of the first spermatozoon with the egg surface results in the inactivation of the supernumerary spermatozoa which had not yet interacted with the egg.

An important contribution to the understanding of the mechanism of the inactivation of the sperm receptors has come from the recent work of Bleil and Wassarman.[20] As mentioned above, these authors[19] have identified a glycoprotein component of the mouse *zona pellucida* endowed with sperm receptor activity. Now they have shown that a protein with the same chemical characteristics can be extracted from the *zona pellucida* of 2-cell embryos albeit without sperm receptor activity. Hence, as a result of fertilization this protein has undergone some subtle change that has resulted in its loss of sperm receptor activity. The nature of the change is not known. Nor is it known how soon after fertilization it occurs. The question of timing is important. In the hamster, for example, the *zona* reaction is effective after 15 min, whereas the vitelline reaction takes from 2 to 3.5 hr to develop;[147] hence, if more than one spermatozoon managed to pass the *zona pellucida* within 15 min one would expect the egg to become polyspermic.

Close examination of the time course of the early events that follow fertilization — that of the electrical changes, the release of the proteololytic enzyme, the onset of the lifting of the vitelline coat from the egg surface, and the Ca^{2+} release — shows that, in fact, they occur within the range required to reduce to a minimum the probability of multiple sperm-egg fusions. Dale and Monroy[141] have in fact suggested that these changes, though serving the function of preventing polyspermic fertilization, are part of the mechanism of the activation of the egg. An additional factor limiting sperm-egg fusion is the differential organization of the egg plasma membrane that has been discussed in Section II.C.3.

An interesting situation is that of the naturally polyspermic eggs (such as those of some mollusks, sharks, urodeles, reptiles, and birds). Although several spermatozoa enter these eggs, after fusion of the first spermatozoon with the female pronucleus, the movements and the changes of the supernumerary sperm nuclei are arrested and they eventually either degenerate or are utilized for different purposes. Very little is known of the mechanisms controlling these events; there are, however, some observations on the egg of the urodele, *Triturus*, which are worth mentioning. In *Triturus* the "blockade" effect takes place soon after the first sperm nucleus has merged with the female pronucleus to give rise to the zygote nucleus.[148] This suggests the release of some factor from the zygote nucleus that either directly or indirectly acts on the supernumerary sperm nuclei. In fact, if the part of the egg containing the zygote nucleus is isolated from the rest, e.g., by a constriction, the supernumerary sperm nuclei do not degenerate.[149] Hence, here we have a case in which the cortical reaction is in no way associated with the prevention of polyspermy. Indeed, polyspermy is the natural condition; and yet the egg has developed cytoplasmic mechanisms which prevent multiple conjugations of the female pronucleus with male nuclei.

In conclusion, it looks as though different animal groups have developed different mechanisms to protect the egg from the deleterious effect of polyspermy. Part of the mechanisms are intrinsic to the oocyte and although in some cases they may be more prominently

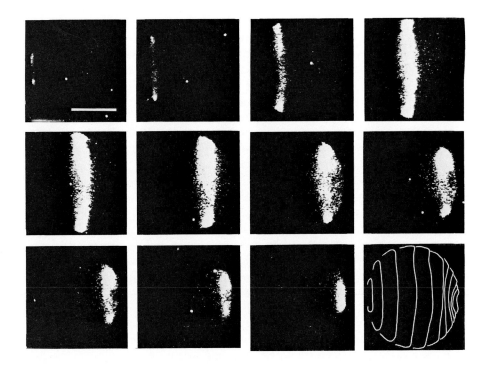

FIGURE 11. A free calcium wave propagating across a sperm-activated Medaka egg. Successive photographs taken every 10 sec. Egg axis horizontal with micropyle to the left. Last frame is a tracing showing the leading edges of the 11 illustrated wavefronts. Bar indicates 500 μm. (Reproduced from *The Journal of Cell Biology,* 1978, Vol. 76, page 448. By copyright permission of The Rockefeller University Press, New York.)

associated with the fertilization-dependent surface changes, in others they are cytoplasmic; in any case, they are part of the mechanisms of the activation of the egg. The intrinsic mechanisms are, however, assisted by a number of devices, morphological and functional, extrinsic to the oocyte proper which act on the spermatozoa and limit the number of effective sperm-egg collisions.

D. The Role of Calcium

The original observations that indicated a possibly important role of calcium in egg activation were by Mazia[150] who showed that in the eggs of *Arbacia* the concentration of free calcium changed from about 0.5 mM in the unfertilized egg to about 1 mM after fertilization. It was later shown by Monroy-Oddo[151] that both Ca^{2+} and Mg^{2+} are in fact lost by the egg: within the first 30 min of fertilization 48% of Ca^{2+} and 49% of Mg^{2+} are lost by the egg. During the next 30 min, i.e., when the egg is preparing for the first cleavage, the concentration of both ions again increases, although it does not reach the values found in the unfertilized egg.

Interesting results have recently been obtained by exploiting the property of the protein aequorin in emitting a flash of light in the presence of Ca^{2+}. Ridgway et al.[152] and Gilkey et al.[153] have injected aequorin into the unfertilized egg of the fresh water teleostean *Oryzias latipes*. Apart from its size (more than 1 mm in diameter) the advantage of this egg is that the site of sperm-egg fusion can be clearly seen as the spermatozoon reaches the egg through a micropyle in the chorion. These authors were thus able to show a wave of luminescence "exploding" from the site of sperm-egg fusion (between 0.8 and 2 min after insemination) and traveling to the opposite pole of the egg in about 128 sec (Figure 11).

The wave was observed also in eggs in which the cortical vacuoles had been displaced by centrifugation. Hence the source of free Ca^{2+} cannot be its release from the cortical vacuoles and it is likely to occur in the surface cytoplasmic layers of the egg. The wave of propagation is interpreted as being due to a calcium-stimulated calcium release.[153]

Similar results were obtained by Steinhardt et al.[154] in the sea urchin egg; the shortest latency time was 45 sec after sperm was added. By using a Vacquier-type[155] preparation of cortices from unfertilized eggs, the authors were able to show that the Ca^{2+} concentration required to induce the breakdown of the cortical granules is about five times greater than that estimated assuming that the calcium released at the time of fertilization is uniformly distributed in the egg cytoplasm. Hence these experiments as well as those of Ridgway et al.[152] and Gilkey et al.[153] suggest that the transient release of Ca^{2+} is confined to the egg surface. In eggs without cortical granules, such as those of the ascidians, a Ca-protein complex is released at the time of fertilization.[118] However, what occurs in this case may be an event *subsequent* to the release of Ca^{2+}, namely that Ca^{2+} is first released and then rapidly bound to a calcium-binding protein which is, in part, lost by the egg. Interestingly, large quantities of Ca^{2+} are released by the zoospores of *Blastocladiella* at the time of germination.[156] The study of the role of Ca^{2+} in cellular processes has entered a new exciting phase as a result of the discovery that in the cell the effect of Ca^{2+} is mediated through calmodulin-regulated enzymes.[157,158] The mode of action of this protein is schematically summarized in Figure 12. The transient increase in Ca^{2+} that occurs at the time of fertilization may result in the formation of an active complex of Ca^{2+} with calmodulin which will, in turn, combine with certain apoenzymes, thus triggering specific reactions.

An immediate effect of free-Ca^{2+} release at fertilization could be, for example, the inactivation of a "cytostatic factor" present in the mature oocyte. This factor appears in the cytoplasm of maturing frog oocyte shortly after the GVBD. The injection of a cytoplasmic extract of maturing oocytes (but not of immature oocytes or fertilized eggs) into blastomeres blocks cleavage and chromosome replication. The interesting point is that the factor requires Mg^{2+} for activity but is inactivated by Ca^{2+}.[159] Whether or not a similar factor is present in animals other than the frog is not known.

In Section II.B we have already discussed the role of calcium in the acrosome reaction. In particular, we have mentioned that the acrosome reaction is accompanied by a very large influx of Ca^{2+} in the sperm head. This implies that at the site of sperm-egg fusion the Ca^{2+} concentration is very high and in fact much higher than that required to trigger the cortical reaction and in particular the dehiscence of the cortical granules.[38]

By the way of analogy, it is worth mentioning that agents that depolarize synaptosomes stimulate Ca uptake by synaptosome *and* neurotransmitter release. The depolarization-dependent uptake of calcium into the nerve terminal is adequate to simultaneously stimulate both release of neurotransmitters and phosphorylation of synaptic vesicle proteins. Furthermore, Ca stimulation of norepinephrine release and of protein phosphorylation in synaptic vesicles requires a vesicle-bound, heat-stable calmodulin-like protein.

The high concentration of calcium at the site of sperm attachment to the plasma membrane of the egg may thus initiate a calmodulin-mediated reaction of the type observed in the synaptosomes.

It should be remembered that in the sea urchin egg, calmodulin makes up about 1% of the total proteins of the egg.[161] We suggest that the transient destabilization of the molecular organization of the plasma membrane is brought about by some factor released by the spermatozoon at the time of the acrosome reaction.[135] From the site of sperm attachment the reaction should propagate to the whole surface of the egg as a calcium-stimulated calcium release. The propagated opening of CG in isolated cortices of sea urchin eggs initiated by the artificial rupture of a few of them[155] may be a model of such a propagated response.

As to the nature of the processes underlying the destabilization of the plasma membrane, one may only offer some speculation on the basis of model experiments. In an in vitro

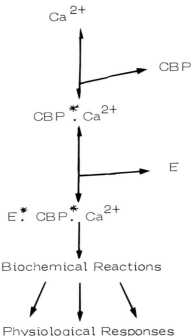

FIGURE 12. Schematic representation of the mechanism of Ca^{2+} action. CBP indicates a calcium binding protein; E indicates an apoenzyme or a receptor protein. The asterisk indicates a new conformation (activated states). Stimulation of the cell causes an increase of the internal Ca^{2+} concentration from 10^{-8}-10^{-7} to 10^{-6} M; this causes the formation of the active complex, CBP^*. Ca^{2+} regulates the activity of the appropriate enzyme which in turn controls a physiological response. (From **Cheung, W. Y.,** *Science,* 207, 19, 1980. Copyright © 1980. By permission of the American Association for the Advancement of Science, Washington, D.C.)

system in which the acrosome reaction was induced by a solution of jelly coat in the presence of phospholipids (to mimic the plasma membrane) Conway and Metz,[162] extending the earlier results of Maggio and Monroy,[163] observed a breakdown of phospholipids to lysophospholipids (due to the action of phospholipase A contained in the acrosome) and then to glycerolphosphorylcholine; the timing of the reaction coincides with that elapsing between the onset of the acrosome reaction and the sperm-egg fusion. The transient formation of lysophospholipids may be responsible for the destabilization of the plasma membrane which results in the fusion of the acrosome membrane of the spermatozoon with the egg plasma membrane. Lysophospholipids are known to cause membrane fusion of somatic cells[164] due to their destabilizing action of the phospholipid moiety of the plasma membrane. After fusion, the membrane should be restabilized as a result of the degradation of lysophospholipids to glycerylphosphorylcholine (by the phospholipase).[162] Besides its role in sperm-egg fusion this reaction may be thought of as the first trigger of the activation of the egg. Related to the reorganization of the plasma membrane is the activation of the Na^+-dependent amino acid transport.[165]

In the sea urchin egg fertilization or artificial activation (by ammonia) causes an increase of the intracellular pH from 6.84 to 7.26[166] (Figure 13) essentially due to a massive H^+

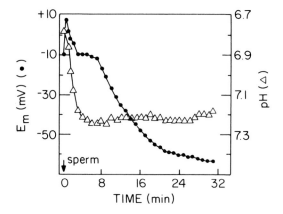

FIGURE 13.   Continuous recording of membrane potential
(E) (●) and intracellular pH (Δ) during fertilization. (Re-
printed by permission from *Nature (London)*, Vol. 272, page
253. Copyright © 1978, Macmillan Journals Limited.)

release (this is the old story of *acid release* at the time of fertilization)[110,165] associated to
the Na^+/H^+ exchange. The "alkalinization" of the egg may be responsible for inducing
the acceleration of protein synthesis[167] (see Section III.F). The increase in pH appears to
start 1 to 2 min after sperm-egg fusion, although it is difficult to establish exactly when the
change actually starts due to the slow response time (about 30 sec) of the pH-sensitive
intracellular microelectrodes.[166] It is interesting that in the spores of *Bacillus cereus* and of
B. megatherium the internal pH of the spores rises from 6.3 to 7.5 upon germination.[168]

E. Activation of DNA Synthesis

The most dramatic event that follows fertilization is the resumption of the egg's ability
to synthesize DNA and to divide. Indeed when the oogonia stop dividing and become oocytes
their DNA synthesizing machinery is turned off, to be then turned on as a result of fertilization. The molecular mechanism of the turning off and on are not known; there are,
however, observations which suggest that the oocyte reacquires the ability to synthesize
DNA in the course of maturation. Experiments of transplantation of somatic nuclei[169,170]
and of DNA[171] into amphibian immature and mature oocytes have shown that the former
do not allow DNA replication. On the other hand, when DNA is injected into mature oocytes
it is actively replicated. In the amphibians the acquisition of the ability of the oocyte cytoplasm
to permit DNA replication is concomitant with the appearance of a new molecular form of
DNA polymerase.[172] Also in vitro studies show that the cytoplasm of eggs and of young
embryos, but not of oocytes, contains a protein factor which stimulates DNA synthesis in
adult liver nuclei.[173] The unfertilized sea urchin egg contains all the enzymes for DNA
synthesis[174] and the deoxynucleotide pool, albeit small, is not limiting.[175] And yet, DNA
synthesis in both pronuclei can only be detected between 10 and 20 min after fertilization.
It has been shown[176,177] that the unfertilized egg is unable to phosphorylate thymidine either
to thymidine monophosphate or to triphosphate; in vivo, the phosphorylating activity appears
about 10 min after fertilization. However, homogenates of unfertilized eggs are able to
phosphorylate thymidine. These results may be interpreted in terms of "compartmentalization" of the enzymes in the unfertilized egg whereby the interaction with their substrates
is prevented. If this is true then the problem is the nature of the compartmentalization and
how decompartmentalization is brought about as a result of fertilization.

One should also consider the possibility that the nucleus of the oocyte or of the unfertilized
egg is unable to interact with the enzymes in the cytoplasm. For example, in the sea urchin
eggs and in other eggs which have completed meiosis at the time of fertilization, the

Table 1
INCORPORATION OF [³H]THYMIDINE (IN CPM/10³ EGGS OF EGG HALVES) IN THE mitDNA

	Incorporation	
	Nuclear DNA	**mitDNA**
Activated whole eggs	920,000	100
Activated anucleate	—	200,000
Fertilized anucleate halves	600,000	50

Note: After butyric acid activation the eggs were exposed to the radioactive precursor for 6 hr.

organization of the nuclear membrane may be such as to prevent the traffic of the enzymes between the cytoplasm and the nucleus.

Also, the organization of the chromatin of the unfertilized egg may be such as not to allow its interaction with the enzymes. In the case of spermatozoon chromatin several lines of evidence do in fact suggest that its organization is different from that of the differentiated cell nuclei. For example, sea urchin sperm chromatin is much more resistant to nuclease digestion than the chromatin from, e.g., blastula nuclei.[178,179] Sperm chromatin has a typical nucleosome arrangement; however, its repeat unit is of about 240 base pairs in contrast to the 210 base pairs of the embryonic chromatin.[179a] As for the chromatin of the nucleus of the unfertilized egg very little is known, mostly due to the technical difficulty of obtaining satisfactory preparations of the nuclei. Yet, some recent observations suggest that its organization may in fact be different from that of the embryonic nuclei. Indeed, in contrast with chromatin from embryos and *of oocytes,* digestion with micrococcal nuclease does not produce the usual pattern of chromatin with nucleosomes.[180]

In conclusion, also as far as DNA replication is concerned, the oocyte is a ready-to-go system. However, it is fertilization (or artificial activation) that triggers the interaction between the chromatin of the pronuclei and the enzyme system controlling DNA replication.

An interesting question is that of the role of mitochondrial DNA in development. Let us first say that in most of the eggs that have been investigated (with the possible exception of the mollusk, *Mytilus*),[181] the sperm mitochondrial DNA (mitDNA) of the embryo is entirely derived from the egg mitochondria.[182] Recent experiments have disclosed some new interesting aspects of the relationships between nucleus and mitDNA in the sense that the replication of mitDNA is under nuclear control. In the sea urchin embryo the number of mitochondria does not increase until quite late during development; in fact, no evidence of mitDNA synthesis has been detected until the prism stage.[183,184] It now appears that the nucleus exerts a negative control over the replication of the mitDNA.

Unfertilized sea urchin eggs can be broken by centrifugation into two halves;[185] the mitochondria are almost entirely concentrated in the centrifugal, heavy half, while the light half contains the nucleus and very few mitochondria. Both the nucleate and the anucleate halves can be activated parthenogenetically (e.g., by butyric acid or by hypertonic sea water). In these experiments, Rinaldi et al.[184] compared the incorporation of [³H]thymidine into mitDNA of parthenogenetically activated anucleate halves with that in the mitDNA of parthenogenetically activated whole eggs and of fertilized anucleate halves.

The results (Table 1) show a very intense synthesis of mitDNA in the anucleate halves and that the mere presence of a replicating haploid nucleus, as in fertilized anucleate halves, is sufficient to completely halt mitDNA replication.

These observations were confirmed by electron microscope analysis of mitDNA prepared

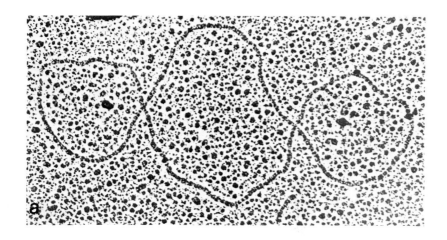

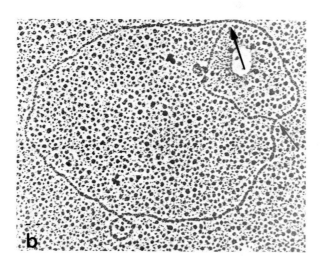

FIGURE 14.   Replicative forms of mitochondrial DNA of sea urchin eggs. (a) Nonreplicating molecule from a whole egg activated parthenogenetically. (b) A replicative molecule from the mitochondria of a parthenogenetically activated anucleate fragment. Arrows, D loops. (From **Rinaldi, A. M., De Leo, G., Arzone, A., Salcher, I., Storace, A., and Mutolo, V.,** *Proc. Natl. Acad. Sci. U.S.A.,* 76, 1916, 1979. With permission.)

from whole eggs and from the anucleate halves (Figure 14); numerous replication forms appeared in the latter, while none was found in the mitochondria from activated whole eggs.[184]

The negative control does not apply to transcription of mitDNA. In fact, it appears that, at least in the sea urchin egg, most of the RNA synthesized during the early postfertilization stages is mitochondrial. This again is shown by the comparison of the patterns of synthesis by parthenogenetically activated whole eggs and anucleate egg fragments.[186-188] Furthermore, as much as 32% of the RNA synthesized by parthenogenetically activate anucleate egg fragments anneals with mitDNA from eggs.[186]

F. Protein Synthesis after Fertilization

Most of these studies have been conducted on the sea urchin egg in which the rate of protein synthesis markedly accelerates a few minutes after fertilization. This has often led to the generalization that one of the major results of fertilization is the removal of a "block" to protein synthesis operating in the oocyte. Subsequent comparative studies have, however, shown that while this is true in the case of the sea urchin egg it is not a general event. For example, in *Rana*[189,190] and in the starfish[191] the rate of protein synthesis increases concomitantly with the onset of maturation and is not affected by fertilization; in the mouse[192] no change occurs until just prior to implantation, while in the surf clam,[193,194] and in the worm, *Urechis*,[195] only a modest increase is observed after fertilization.

Recent work which has taken advantage of the high resolution that can be achieved by gel electrophoresis has provided interesting information as to the *qualitative* changes in the pattern of protein synthesis during oogenesis and early development.

In this section we will consider two aspects of the problem. First, we will discuss the mechanisms underlying the acceleration of protein synthesis that follows fertilization in the sea urchin egg. Although largely based on the sea urchin, these studies have provided data relevant not only to the problem of fertilization, but also to the general problem of the control of protein synthesis. We will then discuss the qualitative aspects. These studies have largely utilized the mammalian embryo. In recent times there has been remarkable progress in the techniques for handling the mammalian embryo and very promising results are on the way.

The first suggestion that in the sea urchin egg the main block to protein synthesis may be at the level of the ribosomes came from the experiments of Hultin[196] which showed that the ribosomes of the unfertilized egg were sluggish in supporting amino acid incorporation into proteins in comparison with those of newly fertilized eggs. This, combined with the observation that RNA from unfertilized eggs was able to stimulate amino acid incorporation into proteins in a cell-free system,[94,95] suggested some inability of the ribosomes to interact with the mRNA. The presence of a trypsin-sensitive inhibitor on the ribosomes of the unfertilized egg has been a matter of controversy. In any case, what can be said is that in the sea urchin egg the efficiency of the translational machinery (defined by the number of the completed polypeptides released per unit time per message),[197] is increased after fertilization. This results from the increased availability of mRNA (so-called "unmasking"), while at the same time the transit time of the ribosomes is reduced by a factor of 2.5[198,199] and hence the time required to complete a polypeptide chain is also reduced.[200] Thus, the "recruitment" of ribosomes increases and this explains the observed increase in polyribosomes.[194,201,202] In this context, it is interesting to note that the unmasking of mRNA and the translational efficiency appear to be under separate control. Indeed, activation of the eggs by ammonia causes mRNA unmasking, whereas it has no influence on translational efficiency.[199]

Now the question arises as to what "unmasking" of the mRNA means. As we have seen before, in order for the mRNA in the RNP particles to become translatable it need not be naked;[97] in fact, what is required is the removal of some of the proteins of the particles (only three proteins appear to be removed).[203] How this is achieved and, more importantly, how the selection of the individual particles to be made translatable is controlled, is not known. A selective and specific proteolytic cleavage may be visualized.

Among the various mechanisms involved in the activation of the mRNA, polyadenylation and capping have been considered. As to the former, although in the unfertilized egg only about one half of the potential mRNA is polyadenylated, this amount is more than sufficient to support a rate of protein synthesis greater than that of the unfertilized egg.[204] Furthermore, inhibition of RNA polyadenylation does not interfere either with the fertilization-dependent increased rate of protein synthesis or with polyribosome formation.[206] Also, the lack of the mRNA cap of the unfertilized egg as a cause of its inactivity is unlikely since mRNA capping has not been observed until about 3 hr after fertilization.[206]

The use of a high-resolution gel electrophoresis is a powerful tool in answering the question regarding the kind of proteins synthesized at different times after fertilization as compared to those synthesized by the oocyte and the unfertilized egg. Let us start with the sea urchin egg. The question is whether the acceleration of protein synthesis that occurs after fertilization is caused by an acceleration of the translation of the same templates that were translated before fertilization, albeit at a lower rate, or whether new mRNAs begin to be translated. Should the latter prove to be the case, how large is their contribution? The answer[207,208] is that "with a few possible exceptions" the pattern of proteins synthesized by the fertilized egg (30 to 60 min after fertilization) is identical to that of the proteins synthesized by the unfertilized egg. Hence, one should conclude that the increased rate of protein synthesis appears to be primarily due to the acceleration of the translation of the same mRNA species utilized by the unfertilized egg. However, the situation is likely to be far more complex. The recent observations of Brandhorst[209] show that the haploid nucleus of the mature unfertilized sea urchin egg synthesizes mRNA at a rate comparable to that of the embryos and that this RNA is immediately translated on the egg polysomes. Following fertilization, most of the RNA on polysomes is oogenic (i.e., the stable RNA transcribed during oogenesis) while very little, if any, egg RNA enters the polysomes. Since, as mentioned above, the classes of (abundant) proteins synthesized before and after fertilization are the same, two populations of mRNA with the same coding sequences (and hence the product of the same structural genes) appear to coexist in the egg. One of them is synthesized during oogenesis, is more stable, and its translation begins after fertilization; the other is transcribed by the haploid genome of the unfertilized egg and is translated by the polysomes of the unfertilized egg.

The study of the single copy DNA sequences (structural genes) that are expressed during oogenesis and throughout development[210] has shown that many of the mRNA species that are synthesized by the oocyte continue to be synthesized, translated, and degraded throughout embryogenesis. As late as the pluteus stage about one half of the polysomal mRNA is composed of maternal-type single copy sequences. On the other hand, since from the blastula stage on most of the polysomal mRNA is newly transcribed (i.e., it is not maternal mRNA),[211] it follows that the maternal-type polysomal mRNA is being transcribed by the same structural genes that were transcribing during oogenesis. As the authors put it, "The whole period, from early oogenesis until well into embryological development, can thus be considered as one in which a set of structural genes presumably required for early morphogenesis is being transcribed, while their protein products slowly accumulate . . . . Following fertilization, the progressive utilization of the hypothetical morphogenesis proteins could be a sequential process, requiring a relatively minor number of new stage-specific proteins that appear only during embryogenesis."

The situation in the egg of the surf clam, *Spisula*, is totally different.[212] In this egg maturation is initiated by fertilization. The patterns of proteins synthesized by the oocyte and by the fertilized egg are quite different. Some of the proteins synthesized by the oocyte continue to be synthesized at the same rate after fertilization; some, which are scarcely synthesized by the oocyte, become very prominent after fertilization; others, whose synthesis was very active in the oocyte, become less representative in the fertilized egg. Yet in vitro translation of the total mRNA from the oocyte results in identical patterns. This means that some oocyte mRNAs which were nontranslatable became translatable after fertilization. As a matter of fact in vitro translation of mRNA from polysomes of unfertilized and fertilized eggs shows that different subsets of maternal mRNA are recruited by the polysomes and translated before and after fertilization. This raises the major problem of how the stage-specific selection of the various mRNAs is operated.

Interesting information has recently come from the study of the proteins synthesized during the early development of the mouse and of the rabbit.

In the mouse[192] as in the sea urchin, during *the early postfertilization* stages the pattern of proteins synthesized is very similar to that of unfertilized egg (Figure 15). However, late in day 1 or early in day 2 (one- to two-cell stage) the pattern changes drastically; most of

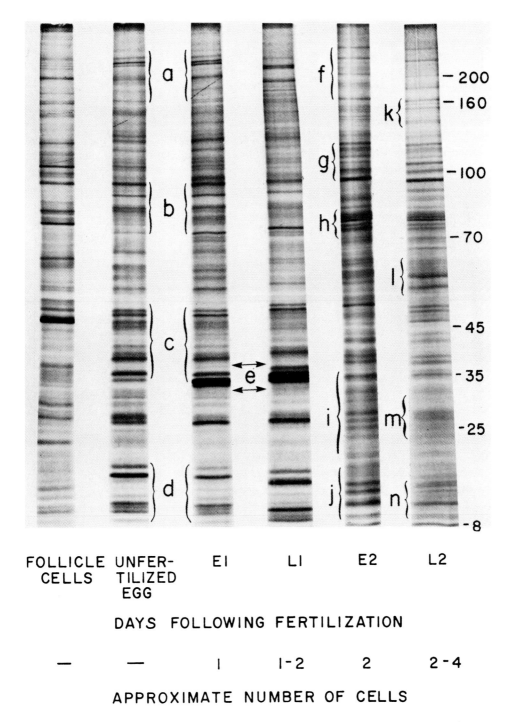

FIGURE 15. Autoradiographs of SDS gel electrophoresis showing protein synthesis in follicle cells, ovulated, unfertilized eggs, and early (E_1,E_2) and late (L_1,L_2) 1- and 2-day mouse embryos labeled with [^{35}S]methionine. The approximate number of cells per embryo was determined after labeling. Approximate molecular weight values × 10^{-3} are given on the right. (From **Van Blerkom, J. and Brockway, G. O.,** *Dev. Biol.,* 44, 148, 1975. With permission.)

the oocyte type proteins are no longer synthesized and new proteins appear. Hence, in the mouse embryo the translation of new templates (i.e., of templates not being translated in the oocyte) occurs at a much earlier stage than in the sea urchin. It is interesting to remember also that the pattern of macromolecules expressed at the surface of the mouse egg before and after fertilization undergoes sequential changes involving the appearance and disappearance of macromolecules (Section III.A). Hence, at least part of the changes observed in the whole egg may reflect changes in the cell membrane. In this respect, the analysis of some mutants in the T/t locus may prove to be very interesting. Also, in the rabbit egg a similar type of change, i.e., the disappearance of numerous oocyte-type proteins and the appearance of new ones, occurs during the early postfertilization stages and during cleavage.[82,213] Most interestingly, *some* of the polypeptides synthesized during the early postfertilization stages are also synthesized by unfertilized aged oocytes of the same chronological age as that of the embryos.[85] This observation is strengthened by the finding that in a cell-free system, mRNA from mouse oocytes is able to direct the synthesis of polypeptides specific for the one- or two-cell embryo.[214]

This implies a program of sequential translation of oocyte templates which are probably required to initiate development; a situation comparable to that of the sea urchin embryo.[210]

IV. CONCLUSIONS: SOME OF THE UNSOLVED PROBLEMS

Fertilization is one of the keystones of evolution; indeed, its phylogenetic history is linked to the evolution of all organisms. This makes the study of fertilization both fascinating and difficult. In the course of evolution, innumerable adaptations have taken place whereby generalizations from the study of one or a very few organisms can be rather misleading. There are, however, some characteristics of fertilization which are common to all organisms. The major one is the fusion of the genetic material of the two parental nuclei into the zygote nucleus. For this to occur, three requirements must be met. First, the two parental nuclei must be haploid. This implies that sexual reproduction in eukaryotes is linked to the invention of meiosis. To our knowledge, there are no clues, not even on which to speculate, as to the origin of meiosis. Second, mechanisms had to be devised to prevent the fusion of the oocyte with more than one nucleus of the male germ cells. There are a variety of mechanisms whereby this fundamental goal is achieved; however, their nature still escapes us. Third, fusion of germ cells belonging to the same sex (gametic type) has to be prevented. This latter mechanism, *surface exclusion*, operates even in bacteria; a protein coded for by a gene in the sexuality factor is inserted in the bacterial membrane. One of the differences between somatic and germ cells is that while the former are able to aggregate in a histotypical manner independently of their genetic sex, germ cells belonging to the same genetic sex fail to interact with one another and in fact, they can interact only with cells of the opposite sex.[4] How these processes, which must involve specific structures on the cell surface are controlled is not known.

Sex differentiation has resulted in an unequal share of responsibilities in the binding up of the new organism, the largest share resting with the oocyte. This is why the study of oogenesis has proven to be so important for the understanding of the basic events of fertilization and of embryonic development. One of the most fascinating results to emerge recently from the study of the expression of the oocyte genome is that even after the formation of the zygote nucleus there are maternal genes that continue to express themselves for some time after fertilization; in fact, there is a whole part of the embryonic developmental program that does not need to be triggered by fertilization.

Thus, the role of the male gamete seems to be limited to supplying the paternal genetic information and to triggering the egg into embryogenesis. Is this condition of the spermatozoon as a second class citizen the adaptive result of natural selection (something comparable to the situation of the parasite vs. the host) or is it a primary condition, primary meaning that in the ancestors of the present day (probably in a premeiotic time) the segregation of the germ cell from the somatic cell line involved only one female-like gemetic type, the differentiation of the male gamete being a late event resulting from the insertion of an episome-like factor in the genome or from gene duplication and transposition? There are cases which may encourage the view that the basic sex is female. One of the most interesting examples is that of *Bonellia viridis*. In this worm the female is about 1 m long, while the male is only a few millimeters and all its organs except the testes are quite rudimentary. If the round larvae are allowed to develop without any contact with the body of the female, they differentiate into females (in fact, a small percentage differentiates into males or intersexes). On the other hand, the larvae that attach to the probiscis of the female differentiate into males.[215]

Depending on the length of contact with the proboscis, the larvae develop into males or into intersexes.[215] The active factor is secreted by the proboscis;[216] it has recently been purified and has turned out to be an unusual chlorin, and has been given the name of *Bonellin*.[217]

Turning now more specifically to the problem of fertilization, the main unanswered question is what sparking reaction triggers the egg into development? Although many important observations have been made in the latter years we do not seem yet to be able to answer the question. Therefore, we shy away from venturing hypotheses as they would be speculations rather than useful working hypotheses. Hence, in these concluding remarks it seems more profitable to point out the events that appear to immediately follow sperm-egg interaction and which may be thought of as being part of the chain of events that lead to the activation of the egg. However, even within this narrow framework a word of warning is necessary. Most of the data have emerged from the study of fertilization of the sea urchin egg and only recently have data become available on other eggs such as the mammalian, the ascidian, and the amphibian egg. Hence, general conclusions should be drawn with great caution.

The earliest event which may have an immediate bearing on the activation of the egg is the result of a *perturbation of the molecular organization of the egg plasma membrane*. This, it should be acknowledged, is an updated version of the "trasient surface cytolysis" of Loeb.[108] It involves:

1. The release of a protease which, besides playing a role in the inactivation of the sperm receptors, is likely to be important both for the activation of some enzymes, for the "unmasking" of mRNA, and for the process of "decompartmentalization"
2. The release of Ca^{2+} which most likely occurs in the surface layers of the egg and, which again, is likely to play a key role in the process of activating the metabolic machinery of the egg
3. The alteration of the ionic channels of the egg plasma membrane which reflects in the change of its electrical and transport properties

These events are followed by:

1. The resumption of the meiotic process in those eggs which are fertilized before completion of meiosis
2. The reactivation of the mechanisms of DNA replication and cell division
3. A rise in the internal pH of the egg (this probably plays a role in the activation of the protein synthesis)
4. The activation, or acceleration, of the translation of the oogenic messages

The dependence of all these events on one another is difficult to assess. One of the major questions is how the perturbation of the organization of the plasma membrane is transduced to the egg cytoplasm and, in particular, to the nucleus. The cytoskeleton may be thought of as playing the key role in this process.

REFERENCES

1. **Maynard Smith, J.,** *The Evolution of Sex,* Cambridge University Press, London, 1978.
2. **Jacob, F., Brenner, S., and Cuzin, F.,** On the regulation of DNA replication in bacteria, *Cold Spring Harb. Symp. Quant. Biol.,* 28, 329, 1963.
3. **Kennedy, N., Beutin, L., Achtman, M., Rahnsdorf, U., and Herrlich, P.,** Conjugation proteins encoded by the F sex factor, *Nature (London),* 270, 580, 1977.
4. **Monroy, A. and Rosati, F.,** The evolution of the cell-cell recognition system, *Nature (London),* 278, 165, 1979.
5. **Wiese, L. and Hayward, P. C.,** On sexual agglutination and mating-type substances in isogamous dioecious *Chlamydomonas.* III. The sensitivity of sex cell contact to various enzymes, *Am. J. Botany,* 59, 530, 1972.
6. **Crandall, M.,** Mating type interactions in microorganisms, in *Receptors and Recognition,* Vol. 3, Series A, Cuatrecasas, P. and Greaves, M. E., Eds., Chapman & Hall, London, 101, 1977.
7. **Taylor, N. W. and Orton, W. L.,** Sexual agglutination in yeast. VII. Significance of the 1.75 compnent from reduced 5-agglutinin, *Arch. Biochem. Biophys.,* 126, 912, 1968.
8. **Goudenough, N. W. and Jurivich, D.,** Tipping and mating-structure-activation induced in *Chlamydomonas* gametes by flagellar membrane antisera, *J. Cell Biol.,* 79, 680, 1978.
9. **Rommrell, L. J. and O'Rand, M. G.,** Capping and ultrastructural localization of sperm surface isoantigens during spermatogenesis, *Dev. Biol.,* 63, 76, 1978.
10. **Monroy, A. and Moscona, A. A.,** *Introductory Concepts in Developmental Biology,* University of Chicago Press, Illinois, 1979.
10a. **McCarrey, A. J. and Abbott, U. K.,** Chick gonad differentiation following excision of primordial germ cells, *Dev. Biol.,* 66, 256, 1978.
11. **Ohno, S.,** *Major Sex-Determining Genes,* Springer-Verlag, Berlin, 1979.
12. **Ohno, S., Nagay, Y., and Ciccarese, S.,** Testicular cells lysostripped of H-Y antigen organize ovarian follicle-like aggregates, *Cytogen. Cell Genet.,* 20, 351, 1978.
13. **Zenzes, M. T., Wolf, U., Gunther, E., and Engel, W.,** Studies on the function of the H-Y antigen: dissociation and reorganization experiments on rat gonadal tissue, *Cytogen. Cell Genet.,* 20, 365 1978.
13a. **McLaren, A.,** *Germ Cells and Soma: A New Look at an Old Problem,* Yale University Press, New Haven, 1981.
14. **Hylander, B. L. and Summers, R. G.,** An ultrastructural analysis of the gametes and early fertilization in two bivalve molluscs, *Chama macerophylla* and *Spisula solidissima* with special reference to gamete binding, *Cell Tissue Res.,* 182, 469, 1977.
15. **Summers, R. G. and Hylander, B. L.,** Species-specificity of acrosome reaction and primary gamete binding in echinoids, *Exp. Cell Res.,* 96, 63, 1975.
16. **Hartman, J. F., Gwatkin, R. B. L., and Hutchinson, C. F.,** Early contact interactions between mammalian gametes *in vitro:* evidence that the vitellus influences adherence between sperm and zona pellucids, *Proc. Natl. Acad. Sci. U.S.A.,* 69, 2767, 1972.
17. **Rosati, F. and De Santis, R.,** Studies on fertilization in the ascidians. I. Self-sterility and specific recognition between gametes of *Ciona intestinalis, Exp. Cell Res.,* 112, 111, 1978.
18. **Rosati, F. and De Santis, R.,** The role of the surface carbohydrates in sperm-egg interaction in *Ciona intestinalis, Nature (London),* 283, 762, 1980.
19. **Bleil, J. D. and Wassarman, P. M.,** Structure and function of the zona pellucida: identification and characterization of the proteins of the mouse oocyte's zona pellucida, *Dev. Biol.,* 76, 185, 1980.
20. **Bleil, J. D. and Wassarman, P. M.,** Mammalian sperm-egg interaction: identification of a glycoprotein in mouse zona pellucidae possessing receptor activity for sperm, *Cell,* 20, 873, 1890.
21. **Epel, D. and Vacquier, V. D.,** Membrane fusion events during invertebrate fertilization, "membrane fusion", *Cell Surf. Rev.,* 5, 1, 1978.
22. **Yanagimachi, R.,** Sperm-egg association in mammals, in *Current Topics in Developmental Biology,* Vol. 12, Moscona, A. A. and Monroy, A., Eds., Academic Press, New York, 1978, 83.
23. **Monroy, A. and Rosati, F.,** Cell surface differentiation during early embryonic development, in *Current Topics in Developmental Biology,* Vol. 13, Moscona, A. A. and Monroy, A., Eds., Academic Press, New York, 1979, 45.
24. **Dan, J. C.,** Acrosome reaction and lysins, in *Fertilization,* Metz, C. B. and Monroy, A., Eds., Academic Press, New York, 1967, 1 and 237.
25. **Metz, C. B.,** Gamete surface components and their role in fertilization, in *Fertilization,* Vol. 1, Metz, C. B. and Monroy, A., Eds., Academic Press, New York, 1967, 163.
26. **Glabe, C. G. and Vacquier, V. D.,** Species specific agglutination of eggs by bindin isolated from sea urchin sperm, *Nature (London),* 267, 836, 1977.
27. **Vacquier, V. D. and Moy, G. W.,** Isolation of bindin: the protein responsible for adhesion of sperm to sea urchin eggs, *Proc. Natl. Acad. Sci. U.S.A.,* 74, 2456, 1977.

28. **Moy, G. W. and Vacquier, V. D.,** Immunoperoxidase localization of bindin during the adhesion of sperm to sea urchin eggs, in *Current Topics in Developmental Biology,* Vol. 13, Moscona, A. A. and Monroy, A., Eds., Academic Press, New York, 1979, 31.

29. **Vacquier, V. D.,** The fertilizing capacity of sea urchin sperm rapidly decreases after induction of acrosome reaction, *Dev. Growth Diff.,* 21, 61, 1979.

30. **Aketa, K. and Ohta, T.,** When do sperm of the sea urchin, *Pseudocentrotus depressus,* undergo the acrosome reaction at fertilization?, *Dev. Biol.,* 61, 366, 1977.

31. **Kidd, P.,** The jelly and vitelline coats of the sea urchin egg: new ultrastructural features, *J. Ultrastruct. Res.,* 64, 204, 1978.

32. **Chandler, D. E. and Heuser, J.,** The vitelline layer of the sea urchin egg and its modification during fertilization, *J. Cell Biol.,* 84, 618, 1980.

33. **Decker, G. L., Joseph, D. B., and Lennarz, W. J.,** A study of factors involved in induction of acrosomal reaction in sperm of the sea urchin, *Arbacia punctulata, Dev. Biol.,* 53, 115, 1976.

34. **Saling, P. M. and Storey, B. T.,** Mouse gamete interactions during fertilization *in vitro, J. Cell Biol.,* 83, 544, 1979.

35. **De Santis, R., Jamunno, G., and Rosati, F.,** A study of the chorion and of the follicle cells in relation to sperm-egg interaction in the ascidian, *Ciona intestinalis, Dev. Biol.,* 74, 490, 1980.

36. **Pinto, M. R., De Santis, R., D'Alessio, G., and Rosati, F.,** Fucosyl sites on the vitelline coat of *Ciona intestinalis, Exp. Cell Res.,* 132, 289, 1981.

37. **Tilney, L. G., Clain, J. G., and Tilney, M. S.,** Membrane events in the acrosomal reaction of Limulus sperm. Membrane fusion, filament-membrane particle attachment, and the source and formation of new membrane surface, *J. Cell Biol.,* 81, 229, 1979.

37a. **Tilney, L. G., Kiehart, D. P., Sardet, C., and Tilney, M.,** Polymerization of actin. IV. Role of Ca and H^+ in the assembly of actin and in membrane fusion in the acrosomal reaction of echinoderm sperm, *J. Cell Biol.,* 77, 536, 1978.

38. **Schackmann, R. W., Eddy, E. M., and Shapiro, B. M.,** The acrosome reaction of *Strongylocentrotus purpuratus* sperm. Ion requirements and movements, *Dev. Biol.,* 65, 483, 1978.

39. **Dan, J. C.,** Studies on the acrosome. III. Effect of calcium deficiency, *Biol. Bull.,* 107, 335, 1954.

40. **Takahashi, T. M. and Sugiyama, M.,** Relation between the acrosome reaction and fertilization in the sea urchin. I. Fertilization in sea water with egg-water treated spermatozoa, *Growth Diff.,* 15, 261, 1977.

41. **Yanagimachi, Y. and Usui, N.,** Calcium dependence of the acrosome reaction and activation of guinea pig spermatozoa, *Exp. Cell Res.,* 89, 161, 1974.

42. **Yanagimachi, R.,** Calcium requirement for sperm-egg fusion in mammals, *Biol. Reprod.,* 19, 949, 1978.

43. **Jones, H. P., Lenz, R. W., Palevitz, R. W., and Cormier, M. J.,** Calmodulin localization in mammalian spermatozoa, *Proc. Natl. Acad. Sci. U.S.A.,* 77, 2722, 1980.

44. **Satir, B. H.,** Membrane-fusion, in *Transport of Macromolecules in Cellular Systems,* Silverstein, S. C., Ed., Berlin/Dahlem Konfer., Germany, 1978, 431.

45. **Plattner, H.,** Fusion of cellular membranes, in *Transport of Macromolecules in Cellular Systems,* Silverstein, S. C., Ed., Berlin/Dahlem Konfer., Germany, 1978, 465.

46. **Colwin, L. H. and Colwin, A. L.,** Membrane-fusion in relation to sperm-egg association, in *Fertilization,* Vol. 1, Metz, C. B. and Monroy, A., Eds., Academic Press, New York, 1967, 295.

47. **Monroy, A.,** Fertilization and its biochemical consequences, in *Module in Biology,* No. 7, Addison-Wesley, Reading, Mass., 1973.

48. **Rosati, F., De Santis, R., and Monroy, A.,** Studies on fertilization in the ascidians. II. Lectin binding to the gametes of *Ciona intestinalis, Exp. Cell Res.,* 116, 419, 1978.

49. **Delage, Y.,** Etudes Experimentales Chez Les Echinodermes, *Arch. Zool. Exp. Gen. Ser.,* 3, 9, 285, 1901.

50. **Bleil, J. D. and Wassarman, P. M.,** Synthesis of zona pellucida proteins by denuded and follicle-enclosed mouse oocytes during culture *in vitro, Proc. Natl. Acad. Sci. U.S.A.,* 77, 1029, 1980.

50a. **Rosati, F., Cotelli, F., De Santis, R., Monroy, A., and Pinto, M. R.,** Synthesis of the fucosyl-containing glycoproteins in the oocytes of *Ciona intestinalis* (Ascidia), *Proc. Natl. Acad. Sci. U.S.A.,* 79, 1982.

51. **Longo, F. J.,** Insemination of immature sea urchin (*Arbacia punctulata*) eggs, *Dev. Biol.,* 62, 271, 1978.

52. **Usui, N. and Yanagimachi, R.,** Behavior of hamster sperm nuclei incorporated into eggs at various stage of maturation, fertilization and early development, *J. Ultrastruct. Res.,* 57, 276, 1976.

53. **Dale, B., De Santis, A., and Hoshi, M.,** Membrane response to 1-methyladenine requires the presence on the nucleus, *Nature (London),* 282, 89, 1979.

54. **Kanatani, H.,** Maturation-inducing substance in starfishes, *Int. Rev. Cytol.,* 35, 253, 1973.

55. **De Felice, L. and Dale B.,** Voltage response to fertilization and polyspermy in sea urchin eggs and oocytes, *Dev. Biol.,* 72, 327, 1979.

56. **Smith, L. D. and Ecker, R. E.,** Role of the oocyte nucleus in physiological maturation in *Rana pipiens, Dev. Biol.,* 19, 281, 1969.

57. **Smith, L. D. and Ecker, R. E.,** Regulatory processes in the maturation and early cleavage of amphibian eggs, in *Current Topics in Developmental Biology,* Vol. 5, Moscona, A. and Monroy, A., Eds., Academic Press, New York, 1970, 1.

58. **Humphrey, R. R.,** A recessive factor (o, for ova deficient) determining a complex of abnormalities in the Mexican axolotl *(Ambystoma mexicanum), Dev. Biol.,* 13, 57, 1966.

59. **Briggs, R. and Cassens, G.,** Accumulation in the oocyte nucleus of a gene product essential for embryonic development beyond gastrulation, *Proc. Natl. Acad. Sci. U.S.A.,* 55, 1103, 1966.

60. **Malacinski, C. M.,** Identification of a presumptive morphogenetic determinant from the amphibian oocyte germinal vesicle nucleus, *Cell Diff.,* 1, 253, 1972.

61. **Wassarman, P. M., Hollinger, T. G., and Smith, L. D.,** RNA polymerase in the germinal vesicle contents of *Rana pipiens* oocytes, *Nature New Biol.,* 240, 208, 1972.

62. **Hollinger, T. G. and Smith, L. D.,** Conservation of RNA polymerase during maturation of the *Rana pipiens* oocyte, *Dev. Biol.,* 51, 86, 1976.

63. **Grippo, P., Taddei, C., Locorotondo, G., and Gambino-Giuffrida, A.,** Cellular localization of DNA polymerase activities in full-grown oocytes and embryos of *Xenopus laevis, Exp. Cell. Res.,* 109, 247, 1977.

64. **Naora, H., Naora, H., Izawa, M., Allfrey, V. G., and Mirsky, A. E.,** Some observation on differences in composition between the nucleus and cytoplasm of the frog oocyte, *Proc. Natl. Acad. Sci. U.S.A.,* 48, 853, 1962.

65. **Dale, B. and De Santis, A.,** Maturation and fertilization of the sea urchin oocyte; an electrophysiological study, *Dev. Biol.,* 85, 474, 1981.

66. **Monesi, V., Geremia, R., D'Agostino, A., and Boitani, C.,** Biochemistry of male germ cell differentiation in mammals: RNA synthesis in meiotic and postmeiotic cells, in *Current Topics in Developmental Biology,* Moscona, A. and Monroy, A., Eds., Academic Press, New York, 1978, 11.

67. **Millette, C. F.,** Cell surface antigens during mammalian spermatogenesis, in *Current Topics in Developmental Biology,* Vol. 13, Moscona, A. and Monroy, A., Eds., Academic Press, 1979, 1.

68. **Artzt, K. and Bennett, D.,** Analogies between embryonic (T/t) antigens and adult major histocompatibility (H-2) antigens, *Nature (London),* 256, 545, 1975.

69. **O'Rand, M. G. and Romrell, L. J.,** Appearance of regional surface autoantigens during spermatogenesis: comparison of anti-testis and anti-sperm autoantisera, *Dev. Biol.,* 75, 431, 1980.

70. **Millette, C. F.,** Distribution and mobility of lectin binding site on mammalian spermatozoa, in *Immunobiology of Gametes,* Edidin, M. and Johnson, M. H., Eds., Cambridge University Press, London, 1976, 51.

71. **Koehler, J. K.,** Studies on the distribution of antigenic sites on the surface of rabbit spermatozoa, *J. Cell Biol.,* 67, 647, 1975.

72. **Koo, G. C., Stackpole, C. W., Boyse, E. A., Hammerling, U., and Lardis, M.,** Topographical location of H-Y antigen on mouse spermatozoa by immunoelectron-microscopy, *Proc. Natl. Acad. Sci. U.S.A.,* 70, 1502, 1973.

73. **Bechtol, K. B., Brown, S. C., and Kennett, R. H.,** Recognition of differentiation antigens of spermatogenesis in the mouse using antibodies from spleen cell-myeloma hybrids after syngeneic immunization, *Proc. Natl. Acad. Sci. U.S.A.,* 76, 363, 1979.

74. **Wilson, E. B.,** Experiments on cleavage and localization in the nemertine egg, W., *Roux Arch. Entw. Mech.,* 16, 44, 1903.

75. **Wilson, E. B.,** Experimental studies on germinal localization. I. The germ-regions in the egg of *Dentalium, J. Exp. Zool.,* 1, 1, 1904.

76. **Raven, C. P.,** The cortical and subcortical cytoplasm of the *Lymnea* egg, *Int. Rev. Cytol.,* 28, 1, 1970.

77. **Boveri, T.,** Die Polarität von Ovocyte, Ei und Larve des *Strongylocentrotus lividus, Zool. Jahrbuch., Abt. Anat. Ontog.,* 14, 630, 1901.

78. **Lillie, F. R.,** The organization of the egg of *Unio,* based on a study of its maturation, fertilization and cleavage, *J. Morphol.,* 17, 227, 1901.

79. **Conklin, E. G.,** Karyokinesis, and cytokinesis in the maturation, fertilization and cleavage of *Crepidula* and other gasteropoda, *J. Acad. Nat. Sci. (Philadelphia),* 12, 1, 1902.

80. **Campanella, C.,** The site of spermatozoon entrance in the unfertilized egg of *Discoglossus pictus* (Anura): an electron microscopy study, *Biol. Reprod.,* 12, 439, 1975.

81. **Monroy, A. and Baccetti, B.,** Morphological changes of the surface of the egg of *Xenopus laevis* in the course of development. I. Fertilization and early cleavage, *J. Utrastruct. Res.,* 50, 131, 1975.

82. **Conklin, E. G.,** The organization and cell lineage of the ascidian egg, *J. Acad. Nat. Sci. (Philadelphia),* 13, 1, 1905.

83. **van Blerkom, J. and McGaughey, R. W.,** Molecular differentiation of the rabbit ovum. I. During oocyte maturation *in vivo* and *in vitro, Dev. Biol.,* 63, 139, 1978.

84. **van Blerkom, J. and McGaughey, R. W.,** Molecular differentiation of the rabbit ovum. II. During the preimplantation development of *in vivo* and *in vitro* matured oocytes, *Dev. Biol.,* 63, 151, 1978.

85. **van Blerkom, J.,** Molecular differentiation of the rabbit ovum. III. Fertilization-autonomous polypeptide synthesis, *Dev. Biol., 72,* 188, 1979.

86. **Davidson, E. H.,** *Gene Activity in Early Development, 2nd ed.,* Academic Press, New York, 1976.

87. **Lane, C. D., Marbaix, G., and Gurdon, J. B.,** Rabbit haemoglobin synthesis in frog cells: the translation of reticulocyte 9S RNA in frog oocytes, *J. Mol. Biol., 61,* 73, 1971.

88. **Moar, V. A., Gurdon, J. B., Lane, C. D., and Marbaix, G.,** Translational capacity of living frog eggs and oocytes as judged by messenger RNA injection, *J. Mol. Biol., 61,* 93, 1971.

89. **Spirin, A. S.,** On masked forms of messenger RNA in early embryogenesis and in other differentiating systems, in *Current Topics in Developmental Biology,* Vol. 1, Moscona, A. A. and Monroy, A., Eds., Academic Press, New York, 1966, 1.

90. **Rosbash, M. and Ford, P. J.,** Polyadenylic acid containing RNA in *Xenopus laevis* oocytes, *J. Mol. Biol., 85,* 87, 1974.

91. **Nakano, E. and Monroy, A.,** Incorporation of ^{35}S methionine in the cell fractions of sea urchin eggs and embryos, *Exp. Cell Res., 14,* 236, 1958.

92. **Epel, D.,** Protein synthesis in sea urchin eggs: a ''late'' response to fertilization, *Proc. Natl. Acad. Sci. U.S.A., 57,* 899, 1967.

93. **Gross, P. R. and Cousineau, G. H.,** Macromolecule synthesis and the influence of actinomycin on early development, *Exp. Cell Res., 33,* 368, 1964.

94. **Maggio, R., Vittorelli, M. L., Rinaldi, A. M., and Monroy, A.,** *In vitro* incorporation of amino acids into proteins stimulated by RNA from unfertilized sea urchin eggs, *Biochem. Biophys. Res. Commun., 15,* 436, 1964.

95. **Slater, D. W. and Spiegelman, S.,** An estimation of genetic messages in the unfertilized echinoid egg, *Proc. Natl. Acad. Sci. U.S.A., 56,* 164, 1966.

96. **Gross, K. W., Jacobs-Lorena, M., Baglioni, C., and Gross, P. R.,** Cell-free translation of maternal RNA from sea urchin eggs, *Proc. Natl. Acad. Sci. U.S.A., 70,* 2614, 1973.

97. **Jenkins, N. A., Kaumayer, J. F., Young, E. M., and Raff, R. A.,** A test for masked message: the template activity of messenger ribonucleoprotein particles isolated from sea urchin eggs, *Dev. Biol., 63,* 279, 1978.

98. **Picard, B. and Wegnez, M.,** Isolation of a 7S particle from *Xenopus laevis* oocytes: A 5S-RNA protein complex, *Proc. Natl. Acad. Sci. U.S.A., 76,* 241, 1979.

99. **Denis, H., Picard, B., Le Maire, M., and Clerot, J. L.,** Biochemical research on oogenesis. The storage particles of the teleost fish *Tinca tinca, Dev. Biol., 77,* 218, 1980.

100. **Monroy, A., Maggio, R., and Rinaldi, A. M.,** Experimentally induced activation of the ribosomes of the unfertilized sea urchin egg, *Proc. Natl. Acad. Sci. U.S.A., 54,* 107, 1965.

101. **Metafora, S., Felicetti, L., and Gambino, R.,** The mechanism of protein synthesis activation after fertilization of sea urchin eggs, *Proc. Natl. Acad. Sci. U.S.A., 68,* 600, 1971.

102. **Hille, M. B.,** Inhibitor of protein synthesis isolated from ribosomes of unfertilized eggs and embryos of sea urchins, *Nature (London), 249,* 556, 1974.

103. **Taddei, C.,** Ribosome arrangement during oogenesis of *Lacerta sicula* Raf., *Exp. Cell Res., 70,* 285, 1972.

104. **Taddei, C., Gambino, R., Metafora, S., and Monroy, A.,** Possible role of ribosomal bodies in the control of protein synthesis in previtellogenic oocytes of the lizard *Lacerta sicula* Raf., *Exp. Cell Res., 78,* 159, 1973.

105. **Burkholder, G. D., Comings, D. E., and Okata, T. A.,** A storage form of ribosome in mouse oocytes, *Exp. Cell Res., 69,* 361, 1971.

106. **Huang, F. L. and Warner, A. H.,** Control of protein synthesis in brine shrimp embryos by repression of ribosome activity, *Arch. Biochem. Biophys., 163,* 716, 1974.

107. **Felicetti, L., Metafora, S., Gambino, R., and Di Matteo, G.,** Characterization and activity of the elongation factors T1 and T2 in the unfertilized egg and in the early development of sea urchin, *Cell Diff., 1,* 265, 1972.

108. **Loeb, J.,** in *Artificial Parthenogenesis and Fertilization,* Chicago University Press, Illinois, 1913.

109. **Austin C. R.,** in *Fertilization,* Prentice-Hall, New York, 1965.

110. **Monroy, A.,** in *Chemistry and Physiology of Fertilization,* Holt, Rinehart & Winston, New York, 1965.

111. **Schuel, H.,** Secretory function of egg cortical granules in fertilization and development: a critical review, *Gamete Res., 1,* 239, 1979.

112. **Runnström, J.,** Uber die Veranderung der Plasmakolloide bei der Entwicklung-serrgung des Seeigeleies, *I. Protoplasma, 4,* 388, 1928.

113. **Monroy, A. and Montalenti, G.,** Variation of the submicroscopic structure of the cortical layer of fertilized and parthenogenetic sea urchin eggs, *Biol. Bull., 92,* 151, 1947.

114. **Monroy, A.,** Further observations on the fine structure of the cortical layer of the unfertilized and fertilized sea urchin eggs, *J. Cell Comp. Physiol., 30,* 105, 1947.

115. **Millonig, G.,** Fine structure analysis of the cortical reaction in the sea urchin egg: after normal fertilization and after electrical induction, *J. Submicr. Cytol., 1,* 69, 1969.

116. **Monroy, A., Ortolani, G., O'Dell, D. S., and Millonig, G.,** Binding of concanavalin A to the surface of unfertilized and fertilized ascidian eggs, *Nature (London),* 242, 409, 1973.

117. **O'Dell, D. S., Ortolani, G., and Monroy, A.,** Increased binding of radioactive concanavalin A during maturation of *Ascidia* egg, *Exp. Cell Res.,* 83, 408, 1974.

118. **Rosati, F., Monroy, A., and De Prisco, P.,** Fine structural study of fertilization in the ascidian, *Ciona intestinalis, J. Ultrastruct. Res.,* 58, 261, 1977.

119. **Ortolani, G., O'Dell, D. S., and Monroy, A.,** Localized binding of *Dolichos* lectin to the early *Ascidia* embryo, *Exp. Cell Res.,* 106, 402, 1977.

120. **Denis-Donini, S. and Campanella, C.,** Ultrastructural and lectin-binding changes during the formation of the animal dimple in oocytes of *Discoglossus pictus* (Anura), *Dev. Biol.,* 61, 140, 1977.

121. **Siracusa, G., De Felici, M., Coletta, M., and Vivarelli, E.,** Inhibitors of protein synthesis increase the agglutinability mediated by concanavalin A of the unfertilized mouse oocyte, *Dev. Biol.,* 62, 530, 1978.

122. **Johnson, M. and Edidin, M.,** Lateral diffusion in plasma membrane of mouse egg is restricted after fertilization, *Nature (London),* 272, 448, 1978.

123. **Arzt, K., Dubois, P., Bennett, D., Condamine, H., Babinet, C., and Jacob, F.,** Surface antigens common to mouse cleavage embryos and primitive tetracarcinoma cells in culture, *Proc. Natl. Acad. Sci. U.S.A.,* 70, 2988, 1973.

124. **Gachelin, G., Fellous, M., Guenet, J. L., and Jacob, F.,** Developmental expression of an early embryonic antigen common to mouse spermatozoa and cleavage embryos and to human spermatozoa: its expression during spermatogenesis, *Dev. Biol.,* 50, 310, 1976.

125. **Buc-Caron, M. H., Gachelin, G., Hofmung, M. and Jacob, F.,** Presence of a mouse embryonic antigen on human spermatozoa, *Proc. Natl. Acad. Sci. U.S.A.,* 71, 1730, 1974.

126. **Jacob, F.,** Cell surface and early stages of mouse embryogenesis, in *Current Topics in Developmental Biology,* Vol. 13, Moscona, A. A. and Monroy, A., Eds., Academic Press, New York, 1979, 117.

127. **Johnson, L. V. and Calarco, P. G.,** Electrophoretic analysis of cell surface proteins of preimplantation mouse embryos, *Dev. Biol.,* 77, 224, 1980.

128. **Kusano, K., Miledi, R., and Stinnakre, J.,** Acetylcholine receptors in the oocyte membrane, *Nature (London),* 270, 739, 1977.

129. **Eusebi, F., Mangia, F., and Alfei, L.,** Acetylcholine elicited responses in primary and secondary mammalian oocytes disappear after fertilization, *Nature (London),* 277, 651, 1979.

130. **Catalano, G., Eilbeck, C., Monroy, A., and Parisi, E.,** A model for early segregation of territories in the ascidian egg, in *Cell Lineage, Stem Cells and Cell Determination,* Le Douarin, N., Ed., Elsevier/North-Holland, Amsterdam, 1979, 15.

131. **Edelman, G. M.,** Surface modulation in cell recognition and cell growth, *Science,* 192, 218, 1976.

132. **Nicolson, G. L.,** Transmembrane control of the receptions on normal and tumor cells. I. Cytoplasmic influence over cell surface components, *Biochim. Biophys. Acta Rev. Biomembr.,* 457, 57, 1976.

133. **Wang, Y. L. and Taylor, D. L.,** Distribution of fluorescently labeled actin in living sea urchin eggs during early development, *J. Cell Biol.,* 82, 672, 1979.

134. **Spudich, A. and Spudich, J. A.,** Actin in triton-treated cortical preparations of unfertilized and fertilized sea urchin eggs, *J. Cell Biol.,* 82, 212, 1979.

134a. **Dale, B. and De Santis, A.,** Maturation and fertilization of the sea urchin oocyte: a electrophysiological study, *Dev. Biol.,* 85, 474, 1981.

135. **Dale, B., Denis-Donini, S., De Santis, R., Monroy, A., Rosati, F., and Taglietti, V.,** Sperm-egg interaction in the ascidians, *Biol. Cellulaire,* 32, 129, 1978.

136. **Nuccitelli, R.,** The electrical changes accompanying fertilization and cortical vesicle secretion in the medaka eggs, *Dev. Biol.,* 76, 483, 1980.

137. **Dale, B., De Felice, L., and Taglietti, V.,** Membrane noise and conductance increase during single spermatozoon-egg interactions, *Nature (London),* 275, 217, 1978.

138. **Shen, S. and Steinhardt, R. A.,** Intracellular pH controls the development of new potassium conductance after fertilization of the sea urchin egg, *Exp. Cell Res.,* 125, 55, 1980.

139. **Kiefer, H., Blume, A. J., and Kaback, H. R.,** Membrane potential changes during mitogenic stimulation of mouse spleen lymphocytes, *Proc. Natl. Acad. Sci. U.S.A.,* 77, 2200, 1980.

140. **Jaffe, L. A.,** Fast block to polyspermy is sea urchin eggs in electrically mediated, *Nature (London),* 261, 68, 1976.

141. **Dale, B. and Monroy, A.,** How is polyspermy prevented? *Gamete Res.,* 4, 151, 1981.

142. **Vacquier, V. D., Tegner, M. J., and Epel, D.,** Protease release from sea urchin eggs at fertilization alters the vitelline membrane layer and aids in preventing polyspermy, *Exp. Cell Res.,* 80, 111, 1973.

143. **Carroll, E. J. and Epel, D.,** Isolation and biological activity of the proteases released by sea urchin eggs following fertilization, *Dev. Biol.,* 44, 22, 1975.

144. **Ginzburg, A. S.,** On the mechanisms of egg protection against polyspermy in echinoderms, *Doklady Akad. Nauk U.S.S.R.,* 152, 501, 1963.

145. **Coburn, M., Schuel, H., and Troll W.,** Hydrogen peroxide release from sea urchin eggs during fertilization: importance in the block to polyspermy, *Biol. Bull.,* 157, 362, 1979.

146. **Vacquier, V. D., Tegner, M. J., and Epel, D.,** Protease activity establishes the block against polyspermy in sea urchin eggs, *Nature (London),* 240, 352, 1972.

147. **Barros, C. and Yanagimachi, R.,** Polyspermy preventing mechanisms in the golden hamster egg, *J. Exp. Zool.,* 180, 251, 1972.

148. **Fankhauser, G.,** Analyse der Physiologischen Polyspermie des Triton Eies auf Grund von Schnürunpexperimenten, *Arch. Entw. Mech.,* 105, 501, 1925.

149. **Spemann, H.,** Ueber Versögerte Kernversorgung von Keimteilen, *Verh. Deutsch. Zool. Ges.,* 24, 216, 1914.

150. **Mazia, D.,** The release of calcium in arbacia eggs on fertilization, *J. Cell Comp. Physiol.,* 10, 291, 1937.

151. **Monroy-Oddo, A.,** Variations in Ca and Mg contents in arbacia eggs as a result of fertilization, *Experientia,* 2, 371, 1946.

152. **Ridgway, E. B., Gilkey, J. C., and Jaffe, L. F.,** Free calcium increases explosively in activating medaka eggs, *Proc. Natl. Acad. Sci. U.S.A.,* 74, 623, 1977.

153. **Gilkey, J. C., Jaffe, L. F., Ridgway, E. B., and Reynolds, G. T.,** A free calcium wave traverses the activating egg of the medaka, *Oryzias latipes, J. Cell Biol.,* 76, 448, 1978.

154. **Steinhardt, R., Zucker, R., and Schatten, G.,** Intracellular calcium release at fertilization in the sea urchin egg, *Dev. Biol.,* 58, 185, 1977.

155. **Vacquier, V. D.,** The isolation of intact cortical granules from sea urchin eggs: calcium ions trigger granule discharge, *Dev. Biol.,* 43, 62, 1975.

156. **Gomes, S. L., Mennucci, L., and da Costa Maia, J. C.,** Calcium efflux during germination of *Blastocladiella emersonii, Dev. Biol.,* 77, 157, 1980.

157. **Cheung, W. Y.,** Calmodulin plays a pivotal role in cellular regulation, *Science,* 207, 19, 1980.

158. **Means, A. R. and Dedman, J. R.,** Calmodulin — an intracellular calcium receptor, *Nature (London),* 285, 73, 1980.

159. **Mayerhof, P. G. and Masui, Y.,** Ca and Mg control of cytoplasmic factors from *Rana pipiens* oocytes which cause metaphase and cleavage arrest, *Dev. Biol.,* 61, 214, 1978.

160. **De Lorenzo, R. J., Freedman, S. D., Yohe, W. B., and Maurer, S. C.,** Stimulation of Ca^{2+}-dependent neurotransmitter release and presynaptic nerve terminal protein phosphorylation by calmodulin and calmodulin-like protein isolated from synaptic vesicles, *Proc. Natl. Acad. Sci. U.S.A.,* 76, 1838, 1979.

161. **Head, J. F., Mader, S., and Kaminer, B.,** Calcium-binding modulator protein from the unfertilized egg of the sea urchin *Arbacia punctulata, J. Cell Biol.,* 80, 211, 1979.

162. **Conway, A. F. and Metz, C. B.,** Phospholipase activity of sea urchin sperm: its possible involvement in membrane fusion, *J. Exp. Zool.,* 198, 39, 1976.

163. **Maggio, R. and Monroy, A.,** Some experiments pertaining to the chemical mechanisms of the cortical reaction in fertilization of sea urchin eggs, *Exp. Cell. Res.,* 8, 840, 1955.

164. **Lucy, J. A.,** The fusion of biological membranes, *Nature (London),* 227, 815, 1970.

165. **Epel, D.,** Mechanisms of activation of sperm and egg during fertilization of sea urchin gametes, *Curr. Top. Dev. Biol.,* 12, 185, 1978.

166. **Shen, S. S. and Steinhardt, R. A.,** Direct measurement of intracellular pH during metabolic depression of the sea urchin egg, *Nature (London),* 272, 253, 1978.

167. **Grainger, J. L., Winkler, M. M., Shen, S. S., and Steinhardt, R. A.,** Intracellular pH controls protein synthesis rate in the sea urchin egg and early embryo, *Dev. Biol.,* 68, 396, 1979.

168. **Setlow, B. and Setlow, P.,** Measurements of the pH within dormant and germinated bacterial spores, *Proc. Natl. Acad. Sci. U.S.A.,* 77, 2474, 1980.

169. **Gurdon, J. B. and Woodland, H. R.,** On the long-term control of nuclear activity during cell differentiation, in *Current Topics in Developmental Biology,* Vol. 5, Moscona, A. A. and Monroy, A., Eds., Academic Press, New York, 1970, 39.

170. **Di Berardino, M. E.,** Genetic stability and modulation of metazoan nuclei transplanted into eggs and oocytes, *Differentiation,* 17, 17, 1980.

171. **Laskey, R. A. and Gurdon, J. B.,** Induction of polyoma DNA synthesis by injection into frog-egg cytoplasm, *Eur. J. Biochem.,* 37, 467, 1973.

172. **Grippo, B., Locorotondo, G., and Caruso, A.,** Characterization of the two major DNA polymerase activities in oocytes and eggs of *Xenopus laevis, FEBS Lett.,* 51, 137, 1975.

173. **Benbow, R. M. and Ford, C. C.,** Cytoplasmic control of nuclear DNA synthesis during early development of *Xenopus laevis:* a cell-free assay, *Proc. Natl. Acad. Sci. U.S.A.,* 72, 2437, 1975.

174. **De Petrocellis, B. and Monroy, A.,** Regulatory processes of DNA synthesis in the embryo, *Endeavour,* 33, 92, 1974.

175. **De Petrocellis, B. and Rossi, M.,** Enzymes of DNA biosynthesis in developing sea urchins. Changes in ribonucleotide reductase, thymidine and thymidilate kinase activities, *Dev. Biol.,* 48, 250, 1976.

176. **Longo, F. J. and Plunkett, W.,** The onset of DNA synthesis and its relation to morphogenetic events of the pronuclei in activated eggs of the sea urchin, *Arbacia punctulata, Dev. Biol.,* 30, 56, 1973.

177. **Ord, M. G. and Stocken, L. A.,** Thymidine uptake by *Paracentrotus* eggs during the first cell cycle after fertilization, *Exp. Cell Res.,* 83, 411, 1973.

178. **Spadafora, C. and Geraci, G.,** The submit structure of sea urchin sperm chromatin: a kinetic approach, *FEBS Lett.,* 57, 79, 1975.

179. **Spadafora, C., Noviello, L., and Geraci, G.,** Chromatin organization in nuclei of sea urchin embryos. Comparison with the chromatin organization of sperm, *Cell Diff.,* 5, 225, 1976.

179a. **Spadafora, C., Ballard, M., Compton, L. J., and Chambon, P.,** The DNA lengths in chromatins from sea urchin sperm and gastrula cells are markedly different, *FEBS Lett.,* 69, 281, 1976.

180. **Albanese, I., Di Liegro, I., and Cognetti, G.,** Unusual properties of sea urchin unfertilized egg chromatin, *Cell Biol. Int. Rep.,* 4, 201, 1980.

181. **Longo, F. J. and Anderson, E.,** Cytological aspects of fertilization in the lamellibranch, *Mytilus edulis.* II. Development of the male pronucleus and the association of the maternally and paternally derived chromosomes, *J. Exp. Zool.,* 172, 97, 1969.

182. **Dawid, I. B. and Blackler, A. W.,** Maternal and cytoplasmic inheritance of mitochondrial DNA in *Xenopus, Dev. Biol.,* 29, 152, 1972.

183. **Bresch, H.,** Mitochondrial profile densities and areas in different developmental stages of the sea urchin *Sphaerechinus granularis, Exp. Cell Res.,* 111, 205, 1978.

184. **Rinaldi, A. M., De Leo, G., Arzone, A., Salcher, I., Storace, A., and Mutolo, V.,** Biochemical and electron microscopic evidence that cell nucleus negatively controls mitochondrial genomic activity in early sea urchin development, *Proc. Natl. Acad. Sci. U.S.A.,* 76, 1916, 1979.

185. **Harvey, E. B.,** *The American Arbacia and Other Sea Urchins,* Princeton University Press, New Jersey, 1956.

186. **Rinaldi, A. M., De Leo, G., Arzone, A., Salcher, I., Storace, A., and Mutolo, V.,** Biochemical and electron microscopic evidence that cell nucleus negatively controls mitochondrial genomic activity in early sea urchin development, *Proc. Natl. Acad. Sci. U.S.A.,* 76, 1916, 1979.

187. **Chamberlain, J. P.,** RNA synthesis in anucleate egg fragments and normal embryos of the sea urchin, *Arbacia punctulata, Biochim. Biophys. Acta,* 213, 183, 1970.

188. **Chamberlain, J. P. and Metz, C. B.,** Mitochondrial RNA synthesis in sea urchin embryos, *J. Mol. Biol.,* 64, 593, 1972.

189. **Smith, L. D., Ecker, R. E., and Subtelny, S.,** The initiation of protein synthesis in eggs of *Rana pipiens, Proc. Natl. Acad. Sci. U.S.A.,* 56, 1724, 1966.

190. **Ecker, R. E. and Smith, L. D.,** Protein synthesis in amphibian oocytes and early embryos, *Dev. Biol.,* 18, 232, 1968.

191. **Houk, M. S. and Epel, D.,** Protein synthesis during hormonally induced meiotic maturation and fertilization in starfish oocytes, *Dev. Biol.,* 40, 298, 1974.

192. **Van Blerkom, J. and Brockway, G. O.,** Qualitative patterns of protein synthesis in the preimplantation mouse embryo. I. Normal pregnancy, *Dev. Biol.,* 44, 148, 1975.

193. **Bell, E. and Reeder, R.,** The effect of fertilization on protein synthesis in the egg of the surf clam, *Spisula solidissima, Biochim. Biophys. Acta,* 142, 500, 1967.

194. **Firtel, R. A. and Monroy, A.,** Polysomes and RNA synthesis during early development of the surf clam, *Spisula solidissima, Dev. Biol.,* 21, 87, 1970.

195. **Gould, M. C.,** A comparison of RNA and protein synthesis in fertilized and unfertilized eggs of *Urechis caupo, Dev. Biol.,* 19, 482, 1969.

196. **Hultin, T.,** Activation of ribosomes in sea urchin eggs in response to fertilization, *Exp. Cell Res.,* 25, 405, 1961.

197. **Palmiter, R. D.,** Quantitation of parameters that determine the rate of ovalbumin synthesis, *Cell,* 4, 189, 1975.

198. **Hille, M. B. and Albers, A. A.,** Efficiency of protein synthesis after fertilization of sea urchin eggs, *Nature (London),* 278, 469, 1979.

199. **Brandis, J. W. and Raff, R. A.,** Elevation of protein synthesis is a complex response to fertilization, *Nature (London),* 278, 467, 1979.

200. **Brandis, J. W. and Raff, R. A.,** Translation of oogenetic mRNA in sea urchin eggs and early embryos. Demonstration of a change in translational efficiency following fertilization, *Dev. Biol.,* 67, 99, 1978.

201. **Rinaldi, A. M. and Monroy, A.,** Polyribosome formation and RNA synthesis in the early post-fertilization stages of the sea urchin egg, *Dev. Biol.,* 19, 73, 1969.

202. **Kedes, L. H. and Gross, P. R.,** Synthesis and function of messenger RNA during early embryonic development, *J. Mol. Biol.,* 42, 559, 1969.

203. **Moon, R. T., Moe, K. S., and Hille, H. B.,** Polypeptides of nonpolyribosomal messenger ribonucleoprotein complexes of sea urchin eggs, *Biochemistry,* 19, 2723, 1980.

204. **Slater, I., Gillespie, D., and Slater, D. W.,** Cytoplasmic adenylylation and processing of maternal RNA, *Proc. Natl. Acad. Sci. U.S.A.,* 70, 406, 1973.

205. **Mescher, A. and Humphreys, T.,** Activation of maternal mRNA in the absence of poly(A) formation in fertilized sea urchin eggs, *Nature (London),* 249, 138, 1974.

206. **Sconzo, G., Roccheri, M. C., di Liberto, M., and Giudice, G.,** Studies on the structure and possible function of the RNA "cap" in developing sea urchins, *Cell Diff.,* 5, 323, 1977.

207. **Brandhorst, B.,** Two-dimensional gel patterns of protein synthesis before and after fertilization of sea urchin eggs, *Dev. Biol.,* 52, 310, 1976.

208. **Tufato, F. and Brandhorst, B.,** Similarity of proteins synthesized in isolated blastomeres of early sea urchin embryos, *Dev. Biol.,* 72, 390, 1979.

209. **Brandhorst, B. P.,** Simultaneous synthesis translation and storage of mRNA including histone mRNA in sea urchin eggs, *Dev. Biol.,* 79, 139, 1980.

210. **Hough-Evans, B. R., Wold, B. J., Ernst, S. G., Britten, R. J., and Davidson, E. H.,** Appearance and persistence of maternal RNA sequence in sea urchin development, *Dev. Biol.,* 60, 258, 1977.

211. **Galau, G. A., Lipson, E. D., Britten, R. J., and Davidson, E. H.,** Synthesis and turnover of polysomal mRNAs in sea urchin embryos, *Cell,* 10, 415, 1977.

212. **Rosenthal, E. T., Hunt, T., and Ruderman, J. V.,** Selective translation of mRNA controls the pattern of protein synthesis during early development of the surf clam, *Spisula solidissima, Cell,* 20, 487 1980.

213. **Van Blerkom, J. and Manes, C.,** Development of preimplantation rabbit embryos *in vivo* and *in vitro.* II. Comparison of qualitative aspects of protein synthesis, *Dev. Biol.,* 40, 40, 1974.

214. **Braude, P. R. and Pelham, H. R. B.,** A microsystem for the extraction and *in vitro* translation of mouse embryo mRNA, *J. Reprod. Fert.,* 56, 153, 1980.

215. **Baltzer, F.,** Die Bestimmung des Geschlects nebst einer Analyse des Geschlechtsdimorphismus bei *Bonellia, Mitt. Zool. St. Neapel,* 22, 1914.

216. **Baltzer, F.,** Ueber die Vermünulichung Indifferenter *Bonellia* — Larnen durch *Bonellia* — Extrakte, *Rev. Suisse Zool.,* 33, 359, 1926.

217. **Pelter, A., Ballantine, J. A., Ferrito, V., Jaccarini, V., Psaila, A. F., and Schembri, P. J.,** Bonellin, a most unusual chlorin, *J. Chem. Soc. D,* 999, 1976.

Chapter 3

MATERNAL RNA AND THE EMBRYONIC LOCALIZATION PROBLEM

William R. Jeffery

TABLE OF CONTENTS

I. INTRODUCTION

The specialized cell types of adult metazoans arise as the consequence of a series of embryonic events collectively known as determination. These events fix the morphogenetic fate of the cell lineages at the appropriate times during embryogenesis. Although very little is known of the nature of the events involved in determination, particularly at the molecular level, it has been understood for some time that factors localized in the cytoplasm of the egg are of great significance. These cytoplasmic factors have been termed morphogenetic determinants and the evidence for their existence in a wide variety of animals has been critically reviewed first by Wilson[1] in 1925 and more recently by Davidson.[2] According to current opinion the morphogenetic determinants are synthesized or accumulated during oogenesis and become distributed in a specific pattern in the mature egg. The distributional pattern is often finalized by the extensive cytoplasmic movements, known as ooplasmic segregation, which occur in many eggs during early development. Once the positional pattern of the determinants is fixed they are thought to be segregated into the proper embryonic cell lineage during cleavages.

Although the existence of morphogenetic determinants in the egg cytoplasm has been known since the turn of the century almost everything else about these factors including their biochemical identity, the mechanisms by which they are localized, and their mode of action, is still a mystery. In order to solve the localization problem it will be important to understand the composition of the determinants. This question has attracted much speculation in recent years and several substances have emerged as possible candidates. The most popular have been low molecular weight or ionic substances, such as the animalizing and vegetalizing factors of sea urchin embryos,[3,4] proteins like the o^+ corrective substance of axolotl eggs,[5,6] and informational RNA. The possible role of maternal RNA molecules localized in the cytoplasm of eggs in the localization phenomena has recently received considerable attention. This is due to the wealth of information accumulated over the past few decades concerning the behavior of maternal messenger RNA during early development[2] and the inherent potential of the RNA molecule as a general purpose regulatory substance. RNA is an especially attractive candidate for a regulatory molecule because of its ability to code for the synthesis of proteins, its potential to interact either with complementary DNA sequences or proteins, and its relatively high turnover rate in the cell. In this paper we wish to consider the possible role of informational RNA molecules of maternal origin as morphogenetic determinants.

Since the involvement of maternal RNA in cell determination implies that these molecules must be subject to differential distribution in the egg cytoplasm, we begin by examining the known cases of positional RNA deposition in eggs. We shall see that maternal RNA is spatially localized in certain putative germ cell determinants and other regionalized dense granular organelles of the egg cytoplasm, in distinctive ooplasmic regions, and in specific embryonic cell lineages. We then consider the possible ways through which these RNA localizations might arise. We next review the experimental evidence for the involvement of maternal RNA in cell determination. Finally the information which we have considered is integrated into working hypotheses for the possible action of localized maternal RNA in the regulation of embryonic gene expression.

II. MATERNAL RNA LOCALIZATIONS

A. Putative Germ Cell Determinants

The primordial germ cells are precociously segregated from the somatic cell lineages during the early embryonic life of a number of animals.[7,8] Germ line determination is often correlated with the presence of a distinctly staining cytoplasm rich in RNA which is passed primarily to the primordial germ cells during early development. The elegant cytoplasmic

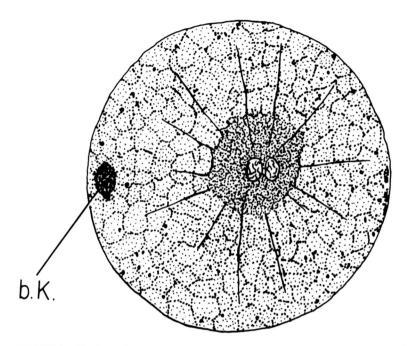

b.K.

FIGURE 1. The *besondere Körper* (bK) germ cell determinant in the uncleaved egg of the chaetognath, *Sagitta* Redrawn from Buchner.[13]

transplantation experiments engineered by Okada et al.[9] and Illmensee and Mahowald[10] with *Drosophila melanogaster* embryos conclusively show that some of these regions contain the germ cell determinants. Germ cell determinants were also demonstrated to exist in the vegetal cytoplasm of *Rana pipiens* eggs by similar techniques.[11] The unique staining properties of these specialized cytoplasmic regions are undoubtably caused by the presence of polarized aggregations of organelles. Exactly which of the visable inclusions, if any, represent the germ cell determinants is still uncertain. A class of dense granular organelles, however, are the best candidates. Strong correlations exist between the presence of organelles with this morphology and germ line segregation during early development. In the chaetognaths *Sagitta* and *Spadella,* for instance, a relatively large, distinctly staining organelle of this type is localized in the uncleaved egg (Figure 1). It was called the *besondere Körper* by its discoverer, Elpatiewsky,[12] to emphasize its unique life history. Although the oogenetic origin of this organelle is controversial its embryonic history is quite well known.[12-18] It passes exclusively to a single blastomere during the first five cleavages of the egg until the sixth cleavage when it is partitioned between two cells which form only the germ line. In the chaetognaths, which are balanced hermaphrodites, one of the two primordial germ cells of the 64-cell embryo gives rise to all the spermatogonia while all of the oogonia are derived from the other. Some investigators have observed that the *besondere Korper* is unequally partitioned to the primordial germ cells. The cell that receives the greater portion becomes the female stem cell.[12,14] Thus the *besondere Körper* not only shows a high correlation with germ cell development, but with sex determination as well.

   Another strong correlation which exists between the dense granular organelles of the germinal cytoplasm and germ cell determination occurs in certain polyembryonic hymenopterans. In the embryo of one of these, *Litomatrix truncatellus,* Sylvestri[19] observed a distinctly staining granule localized in the oosome, the region of germinal cytoplasm at the posterior pole of the egg. He called this structure a *nucleolo* presumably because of its resemblance in morphology and staining properties to the nucleolus. Unlike other higher insects the egg of *Litomatrix* becomes segmented during the early cleavages and each

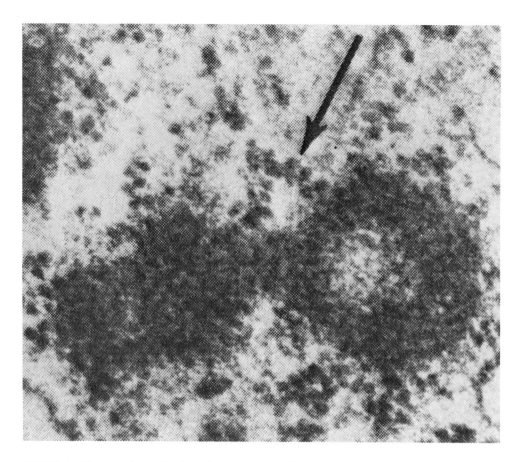

FIGURE 2. Electron micrograph of a polar granule in the *Drosophila* egg showing a helical polysome-like structure (arrow) at its periphery. (From Mahowald, A. P., *J. Exp. Zool.*, 151, 201, 1962. With permission.)

segmented portion becomes one of many embryos derived from the same zygote. During this period of segmentation a portion of the *nucleolo* is passed to each part of the embryonic mass which will develop into a fertile larva.[19] Embryos of the asexual larva, which lack the reproductive tract as well as other vital organ systems, do not obtain any part of the *nucleolo*.[19] Despite these and other suggestive correlations between the dense granular organelles and germ cell development their designation as determinants can only be tentative until they are isolated and assays are developed to test for their morphogenetic significance. In this article they are therefore referred to simply as putative germ cell determinants.

The best-known putative germ cell determinants are probably the polar granules. These organelles are localized in the oosome of many kinds of holometabolous insects. Mahowald and associates[20-22] have been responsible for the extensive structural characterization of polar granules from various species of *Drosophila*. Little is known about their origin except that they first appear in vitellogenic oocytes. They could conceivably be derived from the oocyte itself, from the nurse cells, or from both. They occur in mature oocytes as a compact, disc-shaped mass located immediately beneath the plasma membrane at the posterior pole. At first they show intimate associations with the mitochondria which are also present in this region, but later these contacts disappear and after fertilization are replaced by peripheral interactions with ribosomes. At this time the polar granules also become more dispersed in the oosome and helical polyribosome-like structures are often seen at their periphery (Figure 2). As discussed below this observation along with cytochemical evidence has led some investigators to hypothesize that the polar granules are storage sites for informational RNA.

The surrounding ribosomes are eventually lost or greatly reduced in number when the polar granules fragment into smaller units during later development. This occurs about the time that they are enclosed within the forming pole cells at the posterior tip of the embryo.

At the ultrastructural level the polar granules appear as extremely electron-dense bodies, polymorphic in size and shape, and free of a limiting membrane.[23-26] They consist of an interwoven meshwork of fibrillar and granular elements. The polar granules of *Drosophila* and other insects bear a truly remarkable resemblance in morphology to the *besondere Körper* of chaetognath eggs (Figures 3 and 4) and to the putative germ cell determinants of anuran eggs (Figure 5) which we shall consider in more detail later.

Polar granules or other putative germ cell determinants have never been isolated from eggs in a pure form so their biochemical composition remains essentially unknown. Cytochemical tests, however, have repeatedly supported the contention that they contain RNA and protein but not DNA. The cytochemical evidence for the presence of RNA in the putative germ cell determinants is summarized in Table 1. Most but not all of the staining reactions were carried out with RNase controls. A possible criticism of most of the cases summarized in Table 1 is that since most of the cytochemical analyses were carried out with material prepared for light microscopy, the surrounding ribosomes rather than the polar granules themselves may be the source of the positive signal for RNA. This objection might even hold for the polar granules of the oocyte which, although not encircled by ribosomes, are in close contact with mitochondria, another potential source of RNA. Mahowald[41] has met this objection by staining the polar granules with indium trichloride, a substance which binds to phosphate groups in nucleic acids.[42] Electron micrographs show that indium trichloride very intensely stains the polar granule matrix and the surrounding ribosomes and that the staining is sensitive to perchloric acid hydrolysis, indicative of the involvement of nucleic acids. Several lines of evidence suggest that the polar granules contain informational RNA. First, the electron density exhibited by these structures after indium trichloride staining is homogeneous rather than granular in appearance as might be expected if polar granules consisted only of tightly packaged ribosomes. The ribosomes surrounding the polar granules, in contrast, show a granular type of staining. Second, some of the fibillar elements of the polar granules, which might themselves be mRNA or mRNA associated with proteins as an mRNP, appear to be connected to polyribosomes within the cytoplasm of the oosome. Finally, ribosomal proteins are not major constituents of the crude polar granules recently isolated from *Drosophila* embryos.[43,44] Although none of these data are proof for the existence of informational RNA in the polar granules they do not support the presence of large amounts of ribosomal RNA.

The presence of RNA in the polar granules of *Drosophila* appears to be a transient phenomenon.[41] Cytochemical tests, both at the light and electron microscopic levels, indicate that basophilic staining is greatest in the mature oocyte, declines gradually during intravitelline cleavage and blastoderm formation, and finally disappears completely after pole cell formation. Since no RNA component can be detected by biochemical means in crude polar granule preparations derived from isolated pole cells,[43] it is likely that the RNA has actually been lost by this time rather than merely masked by changes in the polar granule organization. The polar granules appear to consist primarily of protein after their enclosure into the pole cells and recently they have been shown to contain a 93,000 molecular weight polypeptide which has a basic isoelectric point.[44] It is possible that this substance interacts with the polar granule RNA and is responsible for its localization in this organelle during early development. It will be interesting to determine if the 93,000 molecular weight polypeptide is capable of recognizing the polar granule RNA in vitro.

Mahowald[23,41] has postulated that the polar granules may be storage sites for localized informational RNA molecules of maternal origin which code for proteins involved in germ cell determination. According to his hypothesis these molecules are synthesized in the nurse

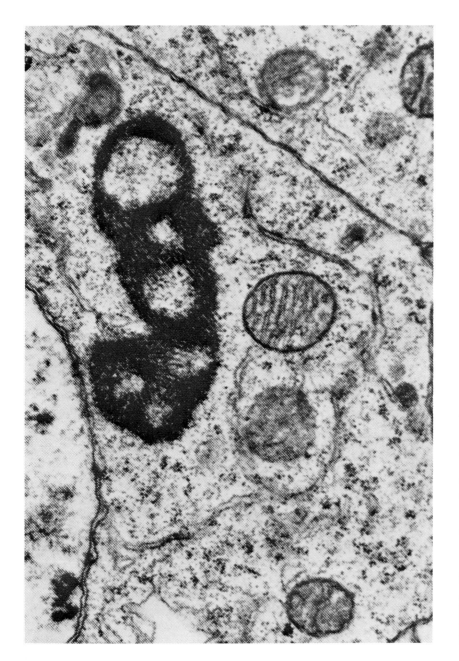

FIGURE 3.   Electron micrograph of a polar granule aggregate in the pole cells of a *Drosophila* embryo. (From Mahowald, A. P., *J. Exp. Zool.*, 167, 237, 1968. With permission.)

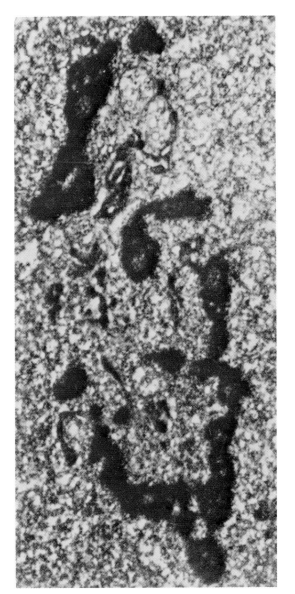

FIGURE 4.   Electron micrograph of the *besondere Körper* germ cell determinant in the egg of the chaetognath, *Spadella cephaloptera*. (From Ghirardelli, E., *Arch. Zool. Ital.*, 51, 841, 1966. With permission.)

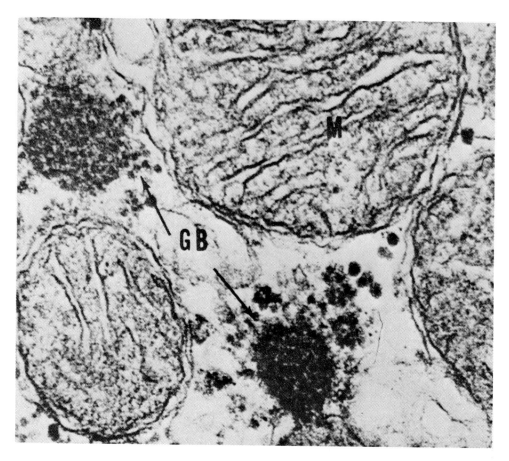

FIGURE 5. Electron micrograph of germinal granules (GB) surrounded by mitochondria (M) in the germ plasm of *Rana pipiens* eggs. (From Kessel, R. G., *Z. Zellforsch.*, 112, 313, 1971. With permission.)

cells of *Drosophila* ovaries during oogenesis and transported to the growing oocyte where they become localized in the developing oosome by their association with the polar granules. They are sequestered in an inactive form until fertilization when the polar granules disperse and develop polyribosomes at their surface, presumably due to the translational activation of their stored mRNA. At present it cannot be decided whether the polar granule RNA is degraded after pole cell formation or if it continues to be translated in the cytoplasm of the pole cells after its release from the polar granules. The fact that pole cells are known to be very active in protein synthesis almost immediately after their formation, however, is consistent with the latter possibility.[45] If RNA from the polar granules is conserved in the pole cells they can be conceived of as a kind of localization matrix functioning in the deposition of specific RNA molecules in particular embryonic cell lineages.

RNA has also been detected by cytochemical means within the dense granular organelles in the germinal cytoplasm of anuran eggs (Table 1). Bounoure first showed that the primordial germ cells of *Rana temporaria* embryos were descended from cells originally located in the vegetal hemisphere of the egg.[46] These cells contain specially staining cytoplasms rich in RNA.[38] The specialized cytoplasmic regions, which are initially positioned at the vegetal pole of the egg, migrate up the sides of the first cleavage furrow during early development and are eventually incorporated into a few cells near the floor of the blastocoel. During later embryogenesis they migrate to the genital ridges where they can be identified as primordial germ cells. As mentioned before, strong evidence for the existence of germ cell determinants in the vegetal cytoplasm of *Rana pipiens* eggs was obtained by Smith who was able to rescue

Table 1
A SUMMARY OF THE CYTOCHEMICAL EVIDENCE FOR RNA IN PUTATIVE GERM CELL DETERMINANTS

Animal	Germ cell determinant	Cytochemical assay	Ref.
Chaetognaths			
Sagitta and *Spadella*	"besondere Körper"	Pyronin stain ± RNase and HCl	16,17
Nematode			
Parascaris equorum	Unnamed	Pyronin stain ± RNase	27
Crustacean			
Cyclops strenuus	Unnamed	Toluidine blue stain	28
Insects			
Acanthoscelis obtectus	Polar granules	Pyronin stain	29
Miastor sp.	Polar granules	Pyronin B stain ± RNase	30
Mayetiola destructor	Polar granules	Pyronin stain ± RNase	31
Musca vicina	Oosome	Pyronin stain ± RNase	32
Heteropeza pygmaea	Oosome	Pyronin stain	33
Drosophila melanogaster, *D. willistoni, D. hydei,* *D. immigrans*	Polar granules	Azure B, toluidine blue, or hematoxylin stain, ± RNase; indium trichloride stain ± perchoric acid	23,34, 35,41
Pimpla turionellae	Oosome	Gallocyanin or toluidine blue stain ± RNase	36
Apanteles glomeratus	Oosome	Methyl green stain	37
Amphibians			
Rana temporaria, R. esculenta, Bufo bufo, Xenopus laevis	Germinal cytoplasm	Pyronin stain ± RNase	38,39
Rana pipiens		Indium trichloride stain ± perchloric acid	40

UV-sterilized embryos by the microinjection of vegetal cytoplasm from untreated controls.[11] Many investigators believe that the densely staining organelles — often called germinal granules — which lie in the germinal cytoplasm of anuran eggs (Figure 5) represent the determinants. This contention is supported by the ability of egg fractions which are enriched in these organelles to promote some germ cell development when injected into UV-irradiated eggs.[47]

The germinal granules of anuran eggs resemble other putative germ cell determinants, particularly the insect polar granules and chaetognath *besondere Körper* (Figures 3 to 5), in their morphological and behavioral characteristics. Like the polar granules, for instance, they show some mitochondrial affinities in oocytes, are surrounded by ribosomes during early development and appear to lose their RNA staining properties during later development.[38,40] At the ultrastructural level they consist of very electron dense fibillar and granular elements which stain intensely with indium trichloride[40] and are not bounded by a limiting membrane (Figure 5).[48-51] As mentioned above, ribosomes surround the granules and occasionally fine fibrils, possibly consisting of mRNA or mRNP, appear to connect them to the granule surface.[49]

Our examination of the putative germ cell determinants indicates that they share an impressive set of structural and behavioral similarities in a phylogenetically diverse group of animal embryos. Based on cytochemical evidence they all appear to be sites for the localization of maternal RNA molecules. Although it is possible that these molecules include informational RNA, further studies (particularly on the isolated and purified organelles when they become available) will be necessary to adequately demonstrate this point.

B. Other Unique Cytoplasmic Organelles
There is considerable evidence that the somatic cell lineages, like the stem cell lineages

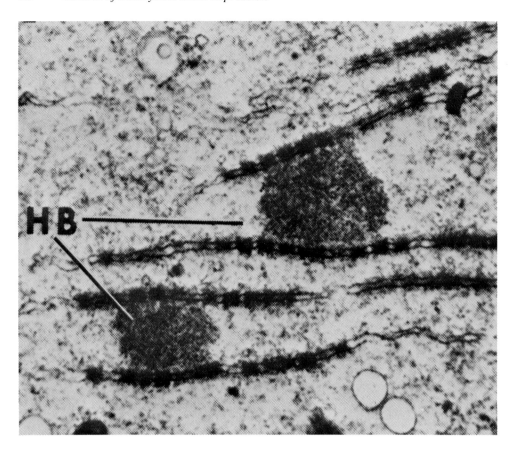

FIGURE 6. Electron micrograph of heavy bodies (HB) associated with annulate lamellae in the egg of *Arbacia punctulata*. (From Conway, C. M., *J. Cell Biol.*, 51, 889, 1971. With permission.)

of the embryo, are dependent upon localized cytoplasmic determinants for their particular developmental expressions.[1,2] Thus morphogenetic determinants may also exist in a visable form in nongerm cell-forming regions of the egg cytoplasm. A number of specialized organelles which contain RNA and, at least at present, have not been correlated with germ cell formation, occur in the cytoplasm of animal eggs. Some of these are densely staining organelles which we will refer to collectively in this article as *nuage*. The *nuage* closely resemble putative germ cell determinants in their morphology. They are often found in both sperm and egg and some of them could be germ cell determinants. At this time, however, they are discussed separately since they have not been traced to the germ cell lineages. Only the *nuage* bodies present in the female gamete will be considered here. We will also discuss RNA localization in unique organelles which are structurally different from *nuage* and the putative germ cell determinants, the vesicular structures of molluscan polar lobes and the crystalline polyribosomal assembledges of rodent oocytes.

1. Nuage

   Most of the unique organelles in the cytoplasm of oocytes and eggs classified as *nuage* contain RNA. Probably the best example of an RNA containing *nuage* is the so-called heavy bodies commonly observed in echinoid eggs. They were given this name by Afzelius because aside from the yolk platelets, they are the most rapidly sedimenting organelles of the egg cytoplasm.[52] The heavy bodies are very basophilic, Feulgen negative structures usually surrounded by and at many positions in intimate contact with stacks of annulate lamellae (Figure 6). Like other *nuage* they probably originate within the nucleus or at the outer nuclear membrane.[52,53] Since they first appear in the cytoplasm of the egg following the second maturation division they could either be derived from the membrane of the germinal vesicle (GV), from the nucleoplasm after GV breakdown, or from the female pronucleus which forms immediately after maturation.[54-57] The heavy bodies consist of aggregates of electron dense granules which are somewhat smaller in diameter than mature ribosomes.[52-54,58-60] The granular elements stain deeply with indium trichloride and this reaction is prevented by pretreatment with RNase but not DNase, suggesting they contain RNA (also see Table 2).[61] Significant electron density remains in the structure following exhaustive RNase exposure indicating that other constituents, possibly proteins, are represented in their composition. Harris has proposed that the heavy bodies, like the putative germ cell determinants, are storage receptacles for translationally inactive, maternal mRNA which exist in the form of mRNP particles and is translationally activated after fertilization.[62] Although there is little evidence for this viewpoint the developmental fate of the heavy bodies is consistent with it. They can be observed in fertilized eggs until the two-cell stage when they appear to fragment with the pieces being dispersed within the cytoplasm.[53] Some investigators claim that the heavy bodies can be detected in a dispersed form by special staining techniques until as late as the morula stage;[63] however, it is not known if they are segregated into particular embryonic cell lineages.

Another example of an RNA-rich *nuage* is the dense bodies discovered by Kessel and Beams in dragonfly oocytes.[64] These peculiar organelles arise from the nucleus of previtellogenic oocytes, migrate to the oocyte cortical regions during vitellogenesis, and persist in the cytoplasm at least until the time of maturation. Their developmental fate and function are unknown, although it has been postulated that, like the heavy bodies, they are sites of maternal mRNA storage for use during later development.[64,65] When the dense bodies reach the oocyte cortex they appear to elaborate numerous stacks of annulate lamellae, perhaps in order to package the RNA which they carry,[64] and obtain an even more electron dense core (Figure 7). The dragonfly dense bodies are often associated with microtubules.[65] The microtubules along with other cytoskeletal elements, as discussed in subsequent sections, could be involved in the positioning of these organelles in the cytoplasm of the oocyte.

Nuage bodies similar to those described above have been reported in female gametes of diverse phylogenetic origin. Like heavy bodies and dense bodies, they appear to be associated with stacks of annulate lamellae at certain developmental stages. They are known by a variety of names including nuclear and nucleolar extrusion bodies, nuclear emission bodies, accessory nuclei, yolk nuclei, and Balbiani bodies. In Table 2 we cite the examples of egg *nuage* which contain RNA and indicate the methods, mostly cytochemical, through which their composition has been studied. The large number of *nuage* bodies which have been reported makes it impossible to discuss each one of them in an article of this scope. They do seem to have many features in common and these are summarized below.

The *nuage* usually arise prior to vitellogenesis although exceptions to this generalization, namely in the echinoid heavy bodies, exist. They are all probably derived from the oocyte nucleus, or perhaps from the nucleus of nurse cells in the case of insects with meroistic ovaries, and are often altered in morphology and cytoplasmic localization during oogenesis. In some instances the *nuage* can be labeled following the addition of radioactive RNA precursors to growing oocytes,[75,90] further evidence that they contain RNA constituents. The identity of the *nuage* transcripts is unknown but in at least one case, that of ascidian oocytes, they probably contain a class of informational RNA. Using an *in situ* hybridization technique developed for sectioned material, Jeffery and Capco have shown that the yolk nuclei of *Styela* oocytes form complexes with radioactive poly(U) probes (Figure 8).[84] The poly(U) binding sites were sensitive to alkali and to RNase administered in buffers of low ionic strength which suggests that they represent RNA molecules containing poly(A) sequences. Thus this type of *nuage* may be the first which has been shown to contain maternal mRNA.

Table 2

**A SUMMARY OF THE *NUAGE* WHICH HAS BEEN SHOWN TO CONTAIN
RNA IN THE OOCYTES AND EGGS OF VARIOUS ANIMALS**

Animal oocyte or egg	Organelle name	Techniques for RNA localization and controls	Ref.
Trematode			
Gorgoderina attenuata	Nucleolus-like cytoplasmic body	Azure B stain ± RNase	66
Annelid			
Nereis pelagica	Nuclear extrusion body	EDTA-uranyl acetate-lead stain	67
Molluscs			
Molpalia mucosa, Chaeto-pleura apiculata	Oocyte basophilia areas	Korson's trichrome stain ± RNase	68
Spisula solidissima	Yolk nucleus	Azure B stain ± RNase	69,70
Otalia lactea	Yolk nucleus	Azure B stain ± RNase	69
Arthropods			
Arachnids			
Tegenaria domestica	Yolk nucleus	UV absorbance ± RNase, pyronin stain ± RNase	71
Tegenaria tridentina	Yolk nucleus	UV absorbance ± RNase, pyronin stain ± RNase	72
Phocus phalangioides	Balbiani body	Pyronin stain ± RNase	73
Crustacean			
Labinia marginata	Unnamed	Toluidine blue, pyronin, Azure A, and acridine orange stains ± RNase	74
Myriapods			
Polydesmus angustus	*Nuage*	EDTA-uranyl acetate-lead stain ± RNase	75
Schizophylum rutilans	Yolk nucleus	UV absorbance ± RNase, pyronin stain ± RNase	72
Insects			
Folsomia candida	Accessory nucleus	EDTA-uranyl acetate-lead stain	76
Blatta orientalis	Nucleolar extrusion body	Pyronin stain ± RNase	77
Periplaneta americana	Nuclear emission body	Korson's trichrom stain ± RNase	78
Acheta domesticus	Facicles	Azure B stain ± RNase	79
Libellula puchella	Yolk nucleus	Korsons trichrom, toluidine blue, and methylene blue stains ± RNase	64
Cordulia aenea	Dense masses	EDTA-uranyl acetate-lead stain	65
Bombus terrestris	Accessory nucleus	Pyronin and gallocyanin stains ± RNase and 1N HCl ±	80
Habrobracon juglandis	Accessory nucleus	Azure B stain	81
Ophion leteus	Accessory nucleus	Pyronin stain ± RNase	82
Apanteles glomeratus	Accessory nucleus	Pyronin stain ± RNase	82
Echinoderms			
Echinus esculentus	Heavy body	Pyronin stain ± RNase	52
Arbacia punctulata	Heavy body	Azure A and toluidine blue stains ± RNase	61
Lyechinus variegatus	Heavy body	Indium stain ± RNase and DNase	61
Paracentrotus lividus	Heavy body	Pyronin stain ± RNase	58
Ascidians			
Ciona intestinalis	Nuclear extrusion body	Pyronin stain ± RNase	83
Styela plicata	Nuclear extrusion body, yolk nucleus	Poly(U) *in situ* hybridization ± RNase, ± base hydrolysis	84
Vertebrates			
Teleost			
Xiphophorus helleri	Yolk nucleus	Pyronin and toluidine blue stains ± RNase	85

Table 2 (continued)
A SUMMARY OF THE *NUAGE* WHICH HAS BEEN SHOWN TO CONTAIN RNA IN THE OOCYTES AND EGGS OF VARIOUS ANIMALS

Animal oocyte or egg	Organelle name	Techniques for RNA localization and controls	Ref.
Amphibians			
Rana clamitans	Unnamed	UV absorbance	86
Xenopus laevis	Yolk nucleus	Pyronin stain ± RNase	87
Triturus alpestris	Unnamed	Toluidine blue stain ± RNase	88
Mammals			
Guinea pig	Yolk nucleus	UV absorbance ± RNase, pyronin stain ± RNase	72
Golden hamster	Granular fibrillar body	Pyronin stain ± RNase	89

The studies of Jeffery and Capco, however, were done at the light microscopic level and the possibility that polyribosomal or mRNA aggregations which surround the yolk nuclei are the actual sources of the poly(A) signal cannot be excluded.[84] In any case these studies do show that a form of informational RNA is localized at or near the position of the yolk nuclei in *Styela* oocytes.

If maternal informational RNA is deposited in the *nuage* during oogenesis, as seems possible from the *in situ* hybridization experiments, their developmental fate is of great significance. Unfortunately, as mentioned earlier, the fate of only a few of these organelles has been carefully studied. According to the available information they appear to have several developmental fates. They fragment in some oocytes after the completion of oogenesis, migrate to the periphery, and appear to release their contents into the cortical cytoplasm.[91,92] This is true in *Rana pipiens,* for example, in which the yolk nuclei disintegrate releasing a very basophilic substance, presumably RNA, into the region immediately beneath the plasma membrane.[93] In some species the *nuage* do not fragment until the early cleavages. We have already seen an example of this in the heavy bodies of echinoid eggs. In the mud snail *Nassarius reticulatus,* they remain intact for longer periods of time during early embryogenesis and their fate can be followed in particular cell lineages.[94] One yolk nucleus occurs in each blastomere of the 16-cell embryo of *Nassarius,* but at the 32-cell stage yolk nuclei are easily observed in the macromeres, but only very infrequently in the micromeres.[94] Although much more work needs to be done on this subject, we conclude that in some cases the *nuage* bodies are passed to particular embryonic cell lineages.

2. Vesicular Organelles

One of the most spectacular examples of RNA localization in an embryonic organelle of possible developmental significance is found in the polar lobe of the freshwater snail, *Bithynia tentaculata.*[95,96] Polar lobes are transitory cytoplasmic protrusions which form in the vegetal hemisphere of some molluscan and annelid eggs during the maturation divisions and the first two cleavages. They have been shown to contain morphogenetic agents responsible for the development of certain mesodermal and mesodermal-induced organs in the larva.[97] In the larger polar lobes, such as those which form in the eggs of the molluscan genera *Nassarius (Ilyanassa)* and *Dentalium,* there are no obvious visable structures which could be construed as the determinants. In *Bithynia,* however, the very small polar lobe is almost filled with an organelle shaped like an inverted cup, the vegetal body (Figure 9). There is reasonable indication that this organelle contains the morphogenetic substances characteristic of the polar lobe cytoplasm. If it is centrifuged out of the vegetal hemisphere prior to the first cleavage the polar lobe often forms without it. No significant developmental defects can be detected in the larva derived from centrifuged eggs when polar lobes lacking the vegetal body are amputated.[98]

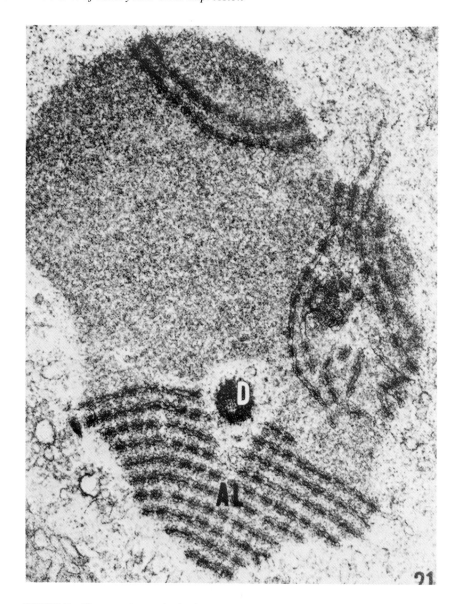

FIGURE 7.   Electron micrograph of a dense RNA body in the dragonfly oocyte. Note the stacks of annulate lamellae (AL) embedded in the dense body and its very dense core (D). (From Kessel, R. G. and Beams, H. W., *J. Cell Biol.*, 42, 185, 1969. With permission.)

The vegetal body of *Bithynia tentaculata* first appears as a disc-shaped structure localized immediately under the plasma membrane at the vegetal pole of the oocyte.[95] The disc remains in this position until just prior to the first cleavage when it is elevated slightly to form the typical inverted cup-shaped organelle and is eventually enclosed in the polar lobe. At this stage the vegetal body stains very intensely with pyronin, acridine orange, and Hoechst 33258 under conditions which suggest the presence of large amounts of RNA.[98] After the conclusion of the first cleavage the polar lobe and its contents are resorbed by the CD blastomere where the vegetal body then disintegrates releasing its contents into the cytoplasm. The major elements of the vegetal body appear to be quite distinct from *nuage* at the ultrastructural level.[95] They consist of groups of membrane-bound vesicles filled with an amorphous substance, possibly RNA. Between the vesicles, however, lie electron dense

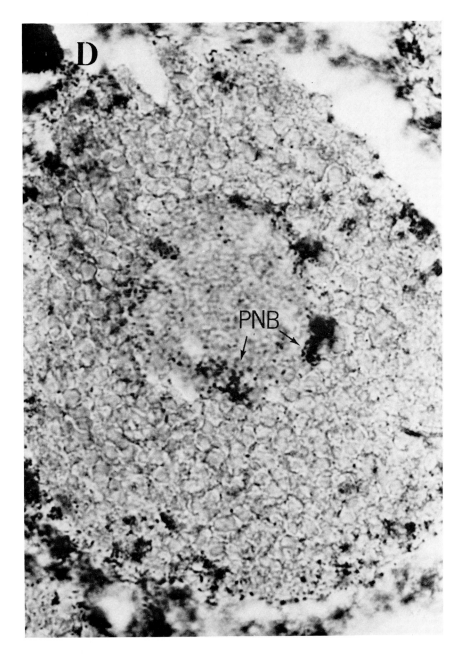

FIGURE 8.   A vitellogenic oocyte of *Styela partita* in which the perinuclear extrusion bodies (PNB) or yolk nuclei are heavily labeled following *in situ* hybridization with [³H]-poly(U). (From Jeffery, W. R. and Capco, D. G., *Dev. Biol.*, 67, 157, 1978. With permission.)

granules, mitochondria, and ribosomes. At present it is not exactly certain which of the vegetal body elements is responsible for its basophilic character since the appropriate cytochemical tests have only been done with material prepared for light microscopy. The mitochondria and ribosomes, however, are probably not abundant enough to entirely account for the intensity of basophilic staining registered by the vegetal body.

The aggregation of localized RNA in the polar lobe of *Bithynia tentaculata* is not an exceptional case since a vegetal body also exists in the eggs of *Amnicola*,[99] another freshwater snail, and the polar lobe of the oyster, *Gryphaea angulata*, is reported to stain intensely for RNA with basophilic dyes.[100]

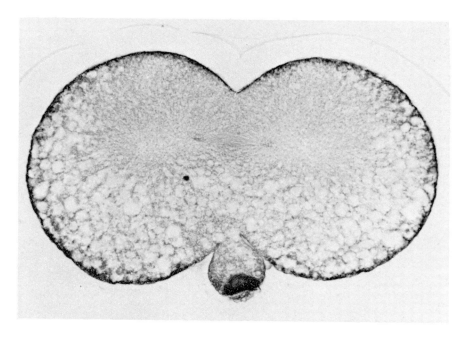

FIGURE 9.   The RNA-rich vegetal body in the small polar lobe of *Bithynia tentaculata*. (From Dohmen, M. R. and Verdonk, N. H., *J. Embryol. Exp. Morphol.*, 131, 423, 1974. With permission from Cambridge University Press.)

The vegetal bodies of the polar lobe-bearing molluscs may be homologous to the ectosomes, RNA-containing granules found in the embryos of the pulmonate snails, *Lymnaea, Arion,* and *Physa*.[101-105] According to Raven[101] and Minganti[104] at the eight-cell stage of the *Lymnaea* embryo precursors of the ectosomes appear as small granules in the most vegetal parts of the macromeres. At the 24-cell stage the granules migrate through the cortical cytoplasm to a position at the inner side of the macromeres where they aggregate into larger bodies. The ectosomes, as these bodies are called, appear to consist of membrane-bound vesicles, ribosomes, and endoplasmic reticulum.[106] After their initial aggregation they gradually disperse again and finally disappear in most of the macromeres. They persist, however, in three of the macromeres and are eventually segregated into the micromeres which form from these during the next cleavage. The function of the ectosomes is unknown although they are often assumed to be associated with bilateral symmetry which originates at this time in pulmonate embryos.[101]

3. Ribosome and mRNA Assemblages

Lattice-like whorls or linear arrays of organelles with a bead-like substructure are often seen in the cytoplasm of mouse, rat, and hamster oocytes (Figure 10).[107-110] These unique assemblages, which appear to be aggregates of ribosomes and fibrillar material, persist in the perinuclear regions of the ooplasm until the early cleavage stages when according to one group of investigators they begin to diminish in number.[195] They persist, however, at least until the blastocyst stage.[195,196] Unfixed preparations of these structures have been made into whole mounts and examined by electron microscopy (Figure 11).[111] The lattice structures in these whole mounts are sensitive to RNase and trypsin but not to DNase suggesting that they consist of ribonucleoprotein. Treatment with dilute acid converts them into free particles with about the same dimensions as monoribosomes. Breakdown of the lattices also occurs after their prolonged incubation in buffered media. The lattices disrupted in this way disperse into ribosome-like spheres connected by thin fibers (Figure 12). Despite the resemblance of

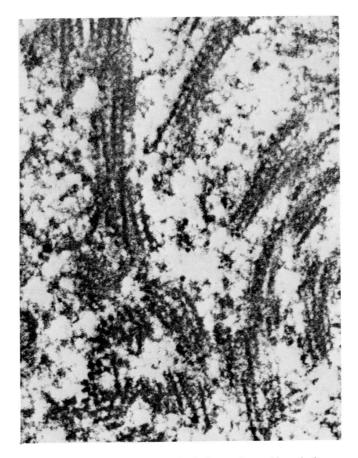

FIGURE 10. Electron micrograph of polyribosomal assemblages in the oocyte of the mouse. Section through the oocyte cytoplasm showing lattice structures. (From Burkholder, G. D., Comings, D. E., and Okada, T. A., *Exp. Cell Res.*, 69, 361, 1971. With permission.)

the components of the lattice to polyribosomes, they are insensitive to EDTA and puromycin.[197,198] Recent biochemical analysis of cellular fractions containing the lattices indicate they contain 18S, 28S, 5S, and 4S RNA species as well as poly(A)-containing RNA.[198] It has been postulated that these structures represent a storage assemblage for inactive oocyte ribosomes.[199] The presence of poly(A)-containing RNA in the lattices also suggests that they sequester mRNA. Clusters of ribosome-sized particles, which could be elaborated for a similar purpose, have also been observed in the cortex of *Ilyanassa* eggs,[112] in insect oocytes,[113] and in overwintering lizard oocytes.[114] The organization of these assemblages may represent a paradigm for the structure of the more complex cytoplasmic ribonucleoprotein aggregates such as the *nuage*.

C. Unique Ooplasmic Regions

Mature eggs commonly contain localized cytoplasmic regions (ooplasms) which are particularly enriched in RNA.[115] These regions appear to originate, at least in part, from the uniformly basophilic cytoplasm present in the previtellogenic oocyte. As oogenesis advances, however, the basophilia is reduced by dilution or is progressively restricted to certain oocyte regions by yolk deposition. In oocytes with large GV, the remaining cytoplasmic basophilia is sometimes enhanced by the addition of RNA-rich nucleoplasm following the maturation divisions. As we have already seen, the cytoplasmic RNA content may also be increased

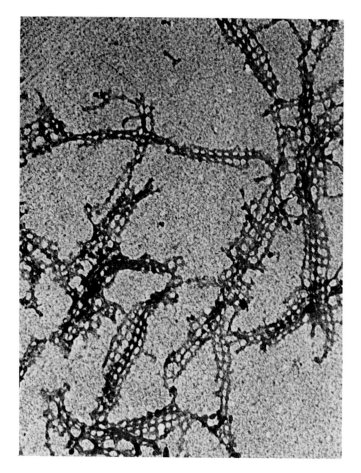

FIGURE 11.   Electron micrograph of a whole mount of isolated lattice-like structures from mouse oocytes. (From Burkholder, G. D., Comings, D. E., and Okada, T. A., *Exp. Cell Res.*, 69, 361, 1971. With permission.)

by the addition of material contributed by dissolving *nuage* bodies. The progressive localization of RNA-rich ooplasms during oogenesis is nicely illustrated in the oocytes of the beetle, *Acanthoscelis obtectus* (Figure 13).[116] Cytochemical analysis indicated that the cytoplasm of previtellogenic oocytes in this animal is very rich in RNA. After yolk deposition begins, however, the RNA-rich cytoplasm is laterally compressed between the anterior and posterior poles of the oocyte and this situation is attenuated during vitellogenesis. In postvitellogenic oocytes RNA-rich cytoplasm is obvious only in a distinct ooplasmic region at the anterior pole. At this time the GV plasm also becomes noticeably rich in RNA, and when it ruptures its constituents are added to the cortical regions of the egg. In the mature egg RNA-rich ooplasms are restricted to a narrow film of cortex, a wider region near the position of the maturation spindle and the posterior polar plasm. Specialized ooplasmic regions are formed in this fashion, with some variation depending on the species, in the eggs of a number of animals.[115] The outcome is a strikingly differential distribution of the maternal RNA complement. The mature eggs of sea urchins,[117] the teleost fish *Libistes*,[118] mouse,[119] and rabbit,[119] and the oocytes of *Rana*[93] and *Ciona intestinalis*[120] also contain RNA-rich cortical cytoplasms. In the case of *Rana* the cortical basophilia disappears after vitellogenesis,[93] but in *Ciona* it is conserved in the specific ooplasmic region which becomes the mesodermal crescent of the fertilized egg.[120]

Examples of RNA-rich ooplasms are not entirely restricted to those formed in the cortex.

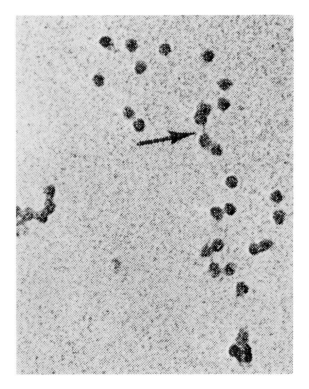

FIGURE 12. Electron micrograph of part of a mouse oocyte lattice structure which has broken down in vitro. The arrow indicates the position of a fiber linking the putative ribosomes. (From Burkholder, G. D., Comings, D. E., and Okada, T. A., *Exp. Cell Res.*, 69, 361, 1971. With permission.)

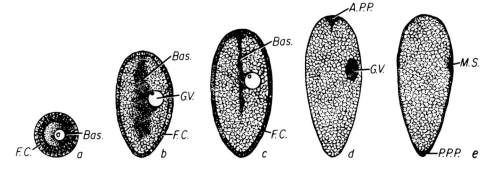

FIGURE 13. The distribution of RNA in the oocytes and egg of the beetle *Acanthoscelis obtectus* as determined by basophilic staining reactions. (a) In the early spherical oocyte intense basophila fills the cytoplasm. (b) At the beginning of vitellogenesis the oocyte has elongated and the basophilic area has become laterally compressed between the poles. (c) As vitellogenesis proceeds the compression and diminution of the basophilic cytoplasm continues. (d) In the postvitellogenic oocyte the remainder of the basophilic cytoplasm is in the anterior pole plasm and the germinal vesicle becomes very basophilic. (e) In the ripe egg the intense basophilic areas are restricted to the cortex, the region surrounding the maturation spindle, and the posterior polar plasm. F.C., follicle cells; Bas., basophilic area; G.V., germinal vesicle; A.P.P., anterior pole plasm; P.P.P., posterior pole plasm; and M.S., maturation spindle. Redrawn from Mulnard.[15]

A central yolk-free cytoplasm has been described in the mature oocytes and unfertilized eggs of the anuran *Discoglossus pictus* which is rich in RNA and is subject to redistribution during ooplasmic segregation (Figure 14).[121] In *Styela partita* a basophilic cytoplasm pri-

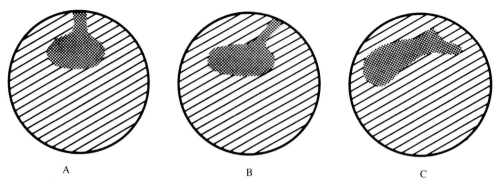

A   B   C

FIGURE 14.   Diagrammatic representation of the movements of a cytoplasm rich in RNA (dark areas) in the eggs of the anuran *Discoglossus pictus*. (A) 15-min postovulation; (B) 75-min post ovulation; and (C) 135-min postovulation. Redrawn from Klag and Ubbels.[121]

marily derived from the GV is deposited in the animal hemisphere of the egg during early development and most of it enters the ectodermal cell lineages during embryogenesis.[122] In the oligochaete worm *Tubifex* basophilic pole plasms originally localized near the animal and vegetal poles of the egg are segregated to the D lineage during early development.[123] The pole plasms of *Tubifex* and other oligochaetes are eventually separated into different blastomeres.

It is clear that maternal RNA is subject to localization in particular ooplasmic regions and that these regions are often distributed to specific cell lineages during embryogenesis. The important question is whether this is also true for informational RNA of maternal origin. The cytochemical methods utilized in the examples cited above detect only the abundant classes of maternal RNA, probably for the most part ribosomal RNA, transfer RNA, and perhaps RNA of mitochondrial origin. Until very recently there have been no methods available for the cytological detection of informational RNA. Recognition of this type of molecule would require detection methods for specific sequences such as the 3′ terminal poly(A) tract. As mentioned earlier, methods have recently been developed for analysis of the distribution of poly(A)-containing RNA in sectioned cells by *in situ* hybridization with radioactive poly(U).[124,200] A number of control studies conducted in conjunction with the new methods indicate that the distribution of poly(A) sequences in the egg can be assayed in a highly specific fashion by *in situ* hybridization. Since free poly(A) sequences are not found in eggs,[125,126] the data collected by *in situ* hybridization with radioactive poly(U) are relevant to the spatial distribution of poly(A)-containing RNA molecules.

The first developmental system which was examined by *in situ* hybridization with radioactive poly(U) was oogenesis and embryogenesis in the milkweed bug, *Oncopeltus fasciatus*.[124,127] In previtellogenic oocytes the poly(A)-containing RNA molecules are evenly distributed throughout the cytoplasm. The distribution remains uniform until late vitellogenesis when distinct accumulations of poly(A)-containing RNA are formed in the anterior and posterior polar plasms. It is proposed that these accumulations represent localized mRNA molecules coding for protein products which are themselves localized at the poles of the oocyte.[127] This interpretation is favored instead of one in which the localized mRNA molecules are stored in the egg and used to direct translational events during embryogenesis because the polar localizations of poly(A)-containing RNA cannot be detected in mature oocytes, freshly ovulated eggs, or developing embryos.[124,127] Thus the maternal poly(A)-containing RNA molecules which are localized in the polar plasms of *Oncopeltus* oocytes must be degraded or dispersed or their poly(A) sequences could be detached or masked following the completion of vitellogenesis so that they cannot subsequently react with poly(U).

The distribution of maternal poly(A)-containing RNA has also been analyzed during oogenesis and early development of the ascidian *Styela* by *in situ* hybridization with poly(U).[84]

The fertilized eggs of this organism contain pigmented ooplasmic regions which are rearranged during ooplasmic segregation and eventually segregated into embryonic cells of specific developmental fate.[122] The pigmented ooplasmic regions can be traced back to precursor regions in the mature oocyte. The yellow (mesodermal) crescent cytoplasm, which marks the position of substances involved in the determination of larval muscle cells, is initially located in the cortex of the oocyte. As mentioned earlier, the clear cytoplasm — which is distributed to the ectodermal cells of the larva — originates mainly from the GV and is released into the oocyte cytoplasm at the time of maturation. The interior yolky cytoplasm of the mature oocyte is the precursor of the gray plasm of the egg, most of which enters the endodermal lineage cells. Jeffery and Capco have discovered that poly(U) binding activity is concentrated about eightfold in the GV plasm of *Styela* oocytes (Figure 15A) relative to the yellow and yolky cytoplasms. This finding suggests the poly(A)-containing RNA accumulates in the nucleoplasm during oogenesis.

In order to determine the developmental fate the poly(A)-containing RNA localized in the GV plasm and other areas of *Styela* oocytes poly(U) hybridization was extended to fertilized eggs and embryos.[84] Sperm penetration triggers ooplasmic segregation in which the cytoplasmic regions of the egg are extensively rearranged. The cortical cytoplasm rapidly streams to the site of sperm entry in the vegetal hemisphere forming the yellow crescent. The clear cytoplasm, now liberated from the ruptured GV, follows in its wake, temporarily forming a clear cap around the yellow crescent and then flowing back to its final site of deposition in the animal hemisphere. These movements displace the yolky cytoplasm first into the animal hemisphere and then into the vegetal hemisphere. The poly(A)-containing RNA released from the GV plasm streams through the egg with the clear cytoplasm and eventually comes to rest in the animal hemisphere where it is preferentially partitioned to the ectodermal cell lineages (Table 3). The other plasms also retain their characteristic titers of poly(A)-containing RNA during early development (Table 3). The results of the *in situ* hybridization studies with the *Styela* egg not only suggest that informational RNA can be localized in particular ooplasmic regions, but also imply that these molecules can migrate with their associated plasms during ooplasmic segregation. Thus the localized poly(A)-containing RNA must possess a strong affinity for unknown binding sites in the cytoplasm.

The distribution of poly(A)-containing RNA has also been investigated during oogenesis and early embryogenesis in *Xenopus laevis* by *in situ* hybridization with poly(U).[201] Poly(A)-containing RNA was uniformly distributed in the cytoplasm of previtellogenic oocytes. Distinct cytoplasmic regions exhibiting high concentrations of poly(A)-containing RNA first appeared in the region located between the germinal vesicle and the cortex of early vitellogenic oocytes. By mid-vitellogenesis, the region of intense localization of poly(A)-containing RNA was present exclusively in the cortex. In postvitellogenic oocytes it was reduced in size and attenuated into the cortex of the vegetal hemisphere. Presumably the poly(A)-containing RNA detected by *in situ* hybridization is synthesized in the germinal vesicle early during vitellogenesis and migrates into the cortex during the period of yolk deposition.[201,202] It is dispersed or destroyed after oocyte maturation since localized poly(A)-containing RNA could not be detected in mature eggs and early embryos.[201] The only localization of poly(A)-containing RNA present in the early embryo seems to be a slight elevation in the concentration of these molecules in the animal hemisphere.[201,203]

Another very striking case of quantitative localization of poly(A)-containing RNA appears to occur in *Chaetopterus* eggs. In this marine annelid extensive cytoplasmic rearrangements occur during the maturation divisions and the first cleavage when a small polar lobe is formed near the vegetal pole. *In situ* hybridization with poly(U) has shown that poly(A)-containing molecules are primarily located in the ectoplasmic layer (cortex) of the egg.[128] During cleavage they initially accumulate in the polar lobe and then in the region of the spindle apparatus between the telophase nuclei, perhaps due to an association with materials

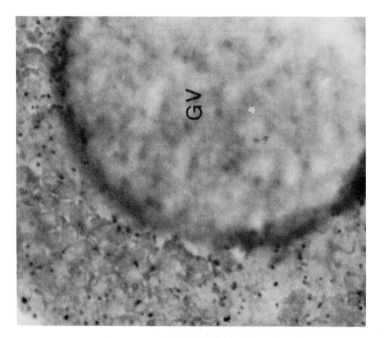

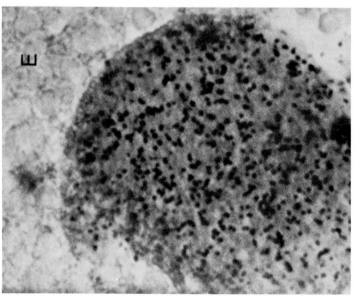

A

FIGURE 15.   (A)   Accumulation of poly(A)-containing RNA in the germinal vesicle of mature *Styela* oocytes as determined by *in situ* hybridization with (³H)-poly(U).   Autoradiographs with focus over the germinal vesicle (GV) and the surrounding cytoplasm (E).   (From Jeffery, W. R. and Capco, D. G., *Dev. Biol.*, 67, 157, 1978.   With permission.)   (B)   Autoradiographs of *Chaetopterus* embryos at first cleavage showing the accumulation of poly-(A)-containing RNA in the region between the telophase nuclei as determined by *in situ* hybridization with poly(U).   Unpublished results of Jeffery.[128] (C)   Detection of histone RNA sequences in the oocyte (left) and egg (right) of *Spisula solidissima* by *in situ* hybridization with a cloned histone DNA probe labeled with [125]I.   Left, autoradiograph of oocytes with heavily labeled GVs.   Right, autoradiograph of mature egg with label over the cytoplasm but not the female pronucleus (densely stained spot).   Unpublished results of Peterson and Jeffery.[129]

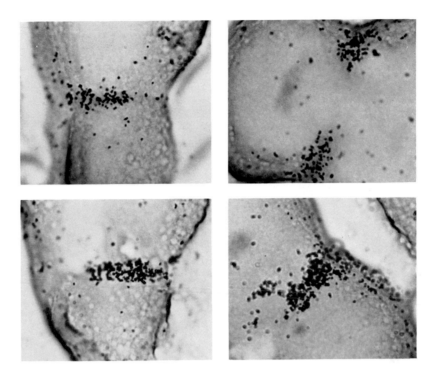

FIGURE 15B.

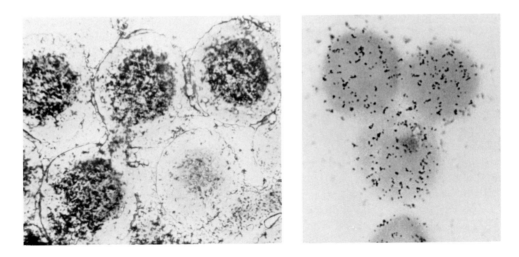

FIGURE 15C.

concentrated by the advancing cleavage furrow (Figure 15B). It is possible that much of this mass of poly(A)-containing RNA enters the larger CD blastomere during the first cleavage.

In contrast to the studies discussed above, there are two cases in which *in situ* hybridization with poly(U) has not revealed marked quantitative differences in the cytoplasmic distribution of poly(A)-containing RNA. Angerer and Angerer found no gross localization of poly(A)-containing RNA in the eggs or early embryos of the sea urchin, *Strongylocentrotus pur-puratus*, with the exception of a slight reduction in these molecules in the micromeres.[200] Likewise, no localizations of poly(A)-containing RNA have been detected in the cytoplasm

Table 3

**THE DISTRIBUTION OF POLY(A)-CONTAINING RNA IN THE
CYTOPLASMIC REGIONS IN *STYELA* EGGS AS ANALYZED BY *IN SITU*
HYBRIDIZATION WITH [³H]-POLY(U)**

Developmental stage	Grain density (grains ± SD)		Clear cytoplasm: yolky plasm ratio
	GV plasm or clear plasm	Yolky endoplasm	
Mature oocyte	22.7 ± 3.0	3.1 ± 0.6	7.3
GV breakdown and ooplasmic segregation	15.5 ± 4.2	3.6 ± 1.5	4.3
Two-cell embryo	13.6 ± 4.2	3.3 ± 1.6	4.1
Four-cell embryo	15.2 ± 1.7	4.1 ± 2.3	3.7

From Jeffery, W. R. and Capco, D. G., *Dev. Biol.,* 67, 157, 1978. With permission.

of mouse oocytes and embryos.[204,205] Thus the quantitative localization of poly(A)-containing RNA seen in the cytoplasm of some eggs may not be a general phenomenon during oogenesis or embryonic development.

Although poly(A)-containing RNA molecules appear to be localized in the GV of *Styela* and mouse oocytes,[84,205] this is also not a general situation in the oocytes of other animals. The large GV of *Xenopus* and *Chaetopterus,* for instance, are known to be relatively poor in poly(A)-containing RNA.[128,201] Using cloned histone DNA sequences as *in situ* hybridization probes, however, Peterson and Jeffery[129] have found that the GV of *Spisula solidissima* oocytes is markedly enriched in maternal histone mRNA (Figure 15C). After maturation and GV breakdown these sequences are released into the cytoplasm. In a similar study recently conducted with fertilized sea urchin eggs, Venezsky et al. have observed relatively high concentrations of histone mRNA in the female pronucleus.[206] These histone mRNA sequences pervade the cytoplasm at the time of the first mitotic division and do not enter the reconstituted nuclei of the two-cell embryo. These studies suggest that nuclei can serve as storage receptacles for specific mRNA species during early development.

In situ hybridization techniques using cloned DNA probes have also been recently developed in order to follow the distribution of specific mRNA molecules in the cytoplasm during early development. The eggs and embryos of the ascidian *Styela* have been a useful model system for these studies because, as mentioned above, they possess colored cytoplasmic regions of specific morphogenetic fate. Using these techniques it has been shown that a differential distribution of maternal actin mRNA occurs in the egg and embryo. The major concentration of actin mRNA appears to reside in the ectoplasm and in the cytoplasm of the yellow crescent region.[207] The actin mRNA sequences remain concentrated in these regions during ooplasmic segregation and early development where they are partitioned primarily to the ectodermal and tail muscle cells of the larva. Neither total mRNA, detected using a cDNA probe to egg poly(A)-containing RNA, nor histone mRNA, detected using cloned histone DNA sequences, show a distribution like that seen for actin mRNA. These results suggest that specific mRNA sequences are preferentially localized in cytoplasms of specific morphogenetic fate in *Styela* eggs. It is of interest in this light that each of the cell lineages which receive a large part of the complement of actin mRNA precociously form actin filaments in their cytoplasms during early development. The ectodermal cells form microfilaments in the epidermis which are thought to be responsible for cell movements during metamorphosis[208] whereas the tail muscle cells, of course, will produce the actin filaments involved in contraction.[209] It is possible that the localization of actin mRNA in these regions serves to deliver the capacity to produce large amounts of actin in a short period of time to the ectodermal and tail muscle cells.

While the *in situ* hybridization techniques have contributed valuable information on the distribution of total mRNA and the more prevalent maternal mRNA species, their efficiencies have not yet attained the levels necessary for the detection of the rare species of mRNA. In another approach to this problem Carpenter and Klein have recently divided *Xenopus* eggs and gastrulae into thirds and used RNA prepared from the three sections, animal pole, middle third, and vegetal pole, to prepare labeled cDNA probes.[210] The kinetics of hybridization of a cDNA probe enriched for vegetal pole poly(A)-containing RNA with animal pole RNA suggested that a minor fraction of rare mRNA sequences is enriched in the vegetal pole region of the *Xenopus* egg.

D. Specific Cell Lineages

The localization of maternal RNA in unique organelles and ooplasmic regions which are subject to segregation during early embryogenesis brings up the possibility that these molecules might exhibit qualitative differences in the various cell lineages. One way in which this problem has been approached is through a comparison of the proteins synthesized by separated blastomeres. This was done by Donohoo and Kafatos with the embryo of *Ilyanassa*.[130] The AB and CD blastomeres, as the first two cleavage products are termed, were separated shortly after the maximal protrusion of the polar lobe. This ensured the resorption of the polar lobe cytoplasm into the CD cell. The isolated blastomeres, which continued to cleave after their separation, were cultured until the first quartet stage (equivalent to the 12-cell stage of the intact embryo) and then exposed to amino acids which were radioactively labeled with ^{3}H or ^{14}C. After a sufficient time the labeling was terminated, the labeled proteins were extracted from each culture, mixed, and electrophoresed together on polyacrylamide gels. Qualitative differences between the proteins synthesized by the progeny of the AB and CD blastomeres could be distinguished by this double-labeling technique. Although these results are consistent with the possibility that maternal mRNAs are qualitatively segregated at the first cleavage they could also be explained by the differential transcription of new mRNA molecules since the labeling period occurred during a relatively advanced embryonic stage and no attempt was made to inhibit RNA synthesis.

The problem of maternal mRNA segregation in *Ilyanassa* embryos was approached in a slightly different way by Newrock and Raff.[131] They removed the polar lobe at its maximal state of protrusion during the first cleavage and then compared the electrophoretic mobility of proteins synthesized by the lobeless and control embryos using the double-labeling method. Significant differences were observed in the pattern of protein synthesis beginning at about 24 hr of development. These differences could be ascribed to the activity of maternal mRNA since new RNA synthesis was blocked by the addition of actinomycin D. The results of these studies, like those of Donohoo and Kafatos,[130] support maternal mRNA segregation but could also be explained by an uneven distribution of specific translational factors. If unique translational factors are segregated into the AB and CD lineages, qualitative differences in the pattern of protein synthesis could be obtained in the presence of the same mRNA species.

More recently the question of maternal mRNA segregation in *Ilyanassa* embryos has been approached using the high resolution afforded by two-dimensional gel electrophoresis.[211-213] The results of all these studies indicate that few, if any, qualitative differences exist between the prevalent proteins synthesized by the isolated polar lobes and lobeless embryos, although there are a number of quantitative differences.

A novel approach to the localization and differential segregation of morphogenetic determinants in ascidian eggs has been pioneered by Whittaker.[133-135] He has shown, using histochemical detection methods, that three tissue-specific enzymes in the tadpole larva of *Ciona intestinalis* can be elaborated precociously in the appropriate embryonic lineage when cleavage is arrested at various developmental stages with cytochalasin B. The situation observed for one of these enzymes, alkaline phosphatase, is of importance to this discussion. Alkaline phosphatase was shown to be synthesized *de novo* exclusively by the cells of the endodermal lineage.[133] Alkaline phosphatase synthesis also occurred in the absence of new RNA synthesis.[133] Thus it is possible that maternal mRNA molecules coding for alkaline phosphatase are localized in the egg and partitioned to the endodermal cells although in this case it is also impossible to exclude the selective segregation of specific translational factors. It is clear from these studies, however, that a protein typical of a specific cell lineage can be manufactured in a specific spatial pattern under the direction of a stored maternal mRNA in *Ciona* embryos.

The echinoid embryo also serves as very suitable material for studying the segregation of maternal RNA. The fourth cleavage of these embryos is unequal resulting in the formation of a 16-cell embryo composed of 8 animal mesomeres, 4 subequatorial macromeres, and 4 vegetal micromeres. There is also a segregation of developmental potential associated with this cleavage since at this time the micromeres become determined to form the primary mesenchyme. The small size of the micromeres relative to the other blastomeres of the 16-cell embryo allows them to be separated from mesomere-macromere mixtures on a density gradient. Consequently, direct comparisons of the sequence complexity of RNA derived from the separated cell types are possible by molecular hybridization. Mizuno et al., the first investigators to take advantage of this system for the purpose of RNA comparison, were unable to detect differences in reiterated RNA sequences between blastomeres of the 16-cell sand dollar embryo.[136] This result, of course, does not preclude the existence of differences in the unique sequence RNA classes. More recently Rogers and Gross have compared the sequence complexity of unique RNA molecules derived from the micromeres and mesomere-macromere mixtures of sea urchins by hybridization with radioactive single copy DNA probes.[137] It was discovered that the micromere RNA sequences form 20 to 30% fewer hybrids with DNA tracers complementary to total egg RNA or mesomere-macromere RNA than RNA molecules derived from the other blastomeres (Figure 16). A difference of this magnitude would correspond to about 10^7 nucleotides or, assuming all these sequences represent mRNAs, a total coding capacity for about 10^4 average-sized polypeptides. It is, of course, not shown that these sequences actually represent mRNAs. In fact the overall pattern of protein synthesis in the separated blastomere populations is qualitatively identical in two-dimensional polyacrylamide gels where as many as 10^3 different polypeptides can be resolved.[138] Thus if the single copy RNA sequences which are preferentially segregated to the mesomere-macromere population are mRNAs, they are either of low abundance or are translationally activated at a later time during embryogenesis. It is also possible that these molecules are regulatory factors rather than mRNAs. This is supported by the recent finding that the RNA sequences in question reside in the postpolysomal fraction of the meso- and macromeres.[139]

An interesting phenomenon has recently been reported for sea urchin embryos which may be relevant to the differential segregation of single copy RNA sequences observed by Rogers and Gross. Cytological studies have shown that the peripheral yolk-filled cytoplasm of the uncleaved egg is distributed entirely to the meso- and macromeres at the fourth cleavage.[140] If the maternal RNA molecules considered above are localized in this region of the egg this phenomenon would explain their absence from the micromeres.

III. MECHANISMS OF RNA LOCALIZATION

The examples discussed in the previous section suggest that informational RNA molecules of maternal origin may be localized in the cytoplasm of eggs and that this distributional pattern may be responsible for their segregation into the various embryonic lineages during development. The mechanisms which guide this process are naturally of great interest. In

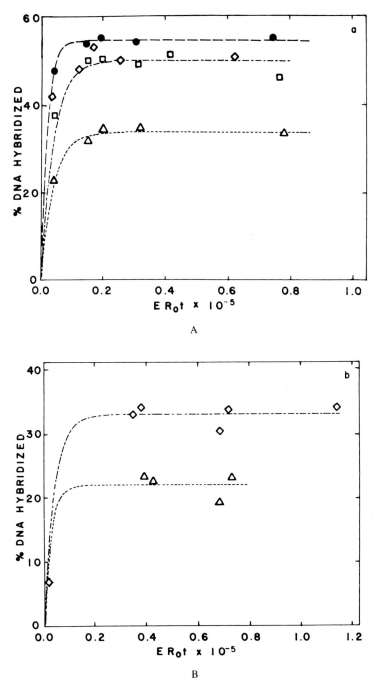

FIGURE 16. A demonstration by molecular hybridization methods of differential maternal RNA localization in the blastomeres of 16-cell sea urchin embryos. (A) Hybridization of a total egg, single copy DNA tracer with total egg RNA (●), total 16-cell embryo RNA (□), mesomere-macromere RNA (◇), and micromere RNA (△) drivers. (B) Hybridization of a total mesomere-macromere, single-copy DNA tracer with mesomere-macromere RNA (◇) and micromere RNA (△) drivers. (From Rogers, W. H. and Gross P. R., *Cell*, 14, 279, 1978. With permission from MIT Press, Cambridge, Mass.)

order to establish a preformed distributional pattern it would seem that the RNA molecules involved must contain specific binding sites capable of recognizing a regionalized localization matrix. The localization matrix must also have sites which are in some way complementary to those of the RNA. Furthermore the interaction of the RNA with the localization matrix should be strong enough to stabilize the complex during extensive cytoplasmic rearrangements, such as those which occur during ooplasmic segregation, and also be reversible or at least flexible enough to permit the RNA to relay its signal after reaching its final destination.

The most obvious way in which the maternal RNA molecules could specifically interact with a localization matrix would be by signal sequences present in their primary structure. Noncoding sequences exist in the 5′ and 3′ terminal regions of eukaryotic mRNA.[141-143] The 5′ regions apparently have a role in the binding of ribosomes or translational factors during the initiation of protein synthesis.[144] The 3′ terminal sequences, however, are excellent candidates for signal sequences. The poly(A) sequences present at the 3′ terminus of many mRNA would probably not contain the degree of specificity required of an RNA localization sequence. The variable sequence regions which lie adjacent to poly(A), however, might possess the specificity required to serve this purpose. In fact there is some indication that these regions may be involved in the binding of mRNA to cellular membranes.[145] It is also possible that the putative signal sequences do not associate with the localization matrix in a direct way, but since eukaryotic mRNA exists as an mRNP complex,[146] are linked instead via their associated protein components. Many different polypeptide species are represented in the free cytoplasmic mRNP of eukaryotic cells.[147] Thus the diversity required to recognize a number of different cellular binding sites potentially exists in mRNP.

Specific localization signals in maternal RNA could also be generated by posttranscriptional modifications. Modification of transcripts by 5′ capping, base and ribose methylations, or the positioning of spliced inserts would all be reasonable possibilities. During early oogenesis transcripts could be processed to form the binding sites specific for one kind of localization matrix while later during oogenesis the processing events could be modified in order to produce binding sites which recognize matrices in other parts of the oocyte. A differential processing mechanism like this could explain the accumulation of poly(A)-containing RNA molecules in the GV of mature *Styela* oocytes and their eventual localization in a specific ooplasm of the egg.[84] In animals which contain ovarian nurse cells, specialized RNA modifications could occur in the nurse cell nucleus or cytoplasm and result in the binding of the modified transcripts to a specific localization matrix after their delivery to the oocyte. A mechanism like this might be responsible for the localization of RNA in the polar granules of insect eggs.

Unfortunately there are few studies of specialized sequences in maternal RNA. Lifton and Kedes have compared the structure of sea urchin maternal and zygotic histone mRNAs using polyacrylamide gels capable of resolving RNA molecules which differ in only six nucleotides.[148] They could not find any significant differences in the electrophoretic mobility of the two types of histone mRNA. More recently Constantini et al. have shown that maternal mRNA in sea urchin eggs contains short repeat transcript elements linked to the coding regions of the molecules.[132] These repeats are not found in most of the RNA of the gastrula and hence are thought to be specific for maternal mRNA. Although these repeat sequences may serve a number of functions, it is possible that they represent specific signals for mRNA localization.

We next consider the possible nature of the localization matrices themselves. We have already mentioned one possible matrix of this kind, the 93,000 molecular weight basic protein which is the major constituent of isolated polar granules and may serve to bind polar granule RNA.[44] Classical centrifugation studies also suggest that the plasma membrane or the egg cortex are important factors in the localization phenomena,[149] but no studies have been conducted to test whether these structures contain specific RNA binding sites. Raff suggests that the subcellular fiber systems of eggs including the microtubules and the microfilaments may be important localization matrices.[150] These organelles along with the intermediate filaments and perhaps certain nonfilamentous cellular elements have been shown to constitute a cytoskeleton in many eukaryotic cells. The cytoskeleton is left as a residue when cells are extracted with nonionic detergents.[151] Recent studies have indicated that filamentous elements of the cytoskeleton contain binding sites for polysomal mRNA.[151] Heterogeneous nuclear RNA also appears to be associated with a skeletal system within the nucleus.[152-154] There is some evidence that the cytoskeleton-mRNA complex may contain a certain degree of specificity. In Hela cells infected with poliovirus, for instance, the viral-specific mRNA displaces host-specific mRNA from its normal cytoskeletal binding sites.[155] Thus, the binding of cellular messages to the cytoskeleton is subject to modulation. A number of studies have recently appeared on the cytoskeleton of eggs and early embryos. A cytoskeletal complex consisting of a submembrane network of actin filaments and associated components is specifically localized in the yellow crescent region of *Styela* eggs.[214] The yellow crescent cytoskeletal complex participates in ooplasmic segregation and is specifically partitioned to the tail muscle cells during early embryogenesis. A cytoskeleton is also present in the eggs and early embryos of sea urchins and Moon et al. have recently demonstrated that, like the situation previously reported for mammalian cells, the sea urchin egg cytoskeleton contains translatable mRNA.[215] More mRNA was found in the cytoskeletal fraction of bastulae than in that of the egg suggesting that mRNA-cytoskeletal interactions may increase during early development. In future studies it will be necessary to determine whether cytoskeletal interactions are responsible for the polarized distribution of maternal mRNA sequences observed in many eggs.

Gurdon has recently suggested a way to study how the distributional patterns of macromolecules are established in eggs.[156] He proposes that molecules from a specific egg region could be purified, labeled, and microinjected into other eggs where their eventual fate could be followed during development. An approach similar to this has been used to examine the distribution of globin mRNA after its microinjection into *Xenopus laevis* eggs.[157] The injected globin mRNA becomes distributed along a gradient from the animal to the vegetal pole and is preferentially segregated to the animal hemisphere blastomeres during early embryogenesis. The localization of globin mRNA in the animal hemisphere is not considered to be a passive phenomenon since poly(A) becomes evenly distributed throughout the egg after microinjection.[157] In experiments of a similar nature it has recently been shown that a large portion of the poly(A)-containing RNA extracted from the vegetal pole region of *Xenopus* eggs returns to the vegetal hemisphere after labeling in vitro and microinjection into fertilized eggs.[158] In contrast, the poly(A)-lacking RNA classes of the vegetal pole become evenly distributed in the egg cytoplasm following microinjection. This suggests that vegetal pole poly(A)-containing RNA molecules are capable of finding a specific localization matrix in the vegetal cytoplasm. Further experiments along these lines should prove helpful in understanding the basis for RNA localization in eggs.

IV. LOCALIZED RNA AS A MORPHOGENETIC DETERMINANT

If we accept the possibility that maternal RNA molecules are localized in a preformed pattern in the egg, the next logical question is whether these localizations play a significant role in embryonic determination. An alternate possibility would be that RNA localizations are the effect rather than the cause of the localization phenomena. This problem has been difficult to resolve or even approach by direct experimental means, mainly due to the lack of suitable assay systems for morphogenetic determinants. Despite these difficulties, several lines of evidence are now available which point to a causal role for localized RNA molecules in embryonic determination.

A. Analysis of Mutants

A reasonable way to search for the factors involved in determination would be to obtain

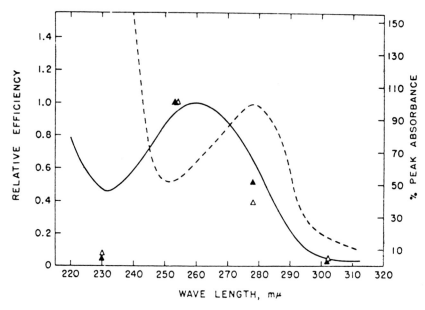

FIGURE 17.   Action spectrum for the effect of UV irradiation of the vegetal cytoplasm on primordial germ cell formation of *Rana pipiens* eggs. The filled and open triangles represent repeats of UV irradiation at the given wavelength. The solid line represents an idealized absorption spectrum for a nucleic acid. The broken line represents an idealized absorption spectrum for a protein. (From Smith, L. D., *Dev. Biol.*, 14, 330, 1966. With permission.)

developmental mutants containing defects in morphogenetic determinants and analyze the nature of the genetic lesion. A number of female sterile mutants have been described in *Drosophila* which affect germ cell development.[22] Among these are the well-known grand-childless (gs) mutants of *D. subobscura*[159,160] which at full penetrance contain a reduced number of apparently normal polar granules.[22] The agametic[161] and Nasrat (fs[1]N)[162] mutants of *D. melanogaster* are probably more interesting subjects for this discussion. They contain wild-type levels of polar granules, but many of them show morphological and perhaps biochemical abnormalities.[22] Cytochemical studies have been reported with eggs from the mutant (fs[1]N). Kern[163] has demonstrated that only about 60% of their polar granules contain cytochemically detectable RNA. The remainder either lack RNA or contain a defective or masked form of RNA. There are other defects manifested in (fs[1]N) mutants, however, thus it would be premature at this time to ascribe their phenotype to alterations in polar granule RNA. Presumably, as more mutants like these are examined cytochemically and others become available at full penetrance, further information concerning the role of polar granule RNA in germ cell determination will become available.

B. UV Irradiation and Photoreactivation

The first and still perhaps the strongest evidence for the active role of nucleic acids in germ cell determination is derived from UV irradiation experiments. As pointed out earlier, UV light causes complete or partial sterility if focused on the vegetal hemisphere of frog eggs at a wavelength of 254 mn.[164-167] Smith was the first to obtain what were then thought to be completely sterile tadpoles in this way.[11] He also constructed a crude action spectrum for primordial germ cell lability in *Rana pipiens* embryos. This was a significant achievement because it provided insight into the nature of the germ cell determinants. There is considerable evidence that the action spectrum of a UV-related effect is precisely correlated with the absorption spectrum of the molecular component involved in the phenomenon.[168,169] The action spectrum obtained by Smith resembles that of a nucleic acid (Figure 17), but since data points in the region between wavelengths of 254 and 278 nm were not obtained, a shoulder in the area of the spectrum in which proteins show their maximal UV absorption cannot be excluded. There is, however, some evidence against the possible role of proteins as the active substance in anuran germ cell determinants. By differential centrifugation Wakahara has prepared a subcellular fraction from the vegetal hemisphere of *Rana* eggs which is enriched in the dense granular elements of the germinal plasm and, if microinjected into living eggs, can induce the formation of supernumerary germ cells.[170] The germinal granule-rich fraction does not lose its activity when it is heated or when it is pretreated with protease, suggesting that proteins are not responsible for its action. Unfortunately, Wakahara did not test the effect of nucleases on the active fraction and consequently his experiments are not enlightening with respect to the possibility that nucleic acids are the active components of the germinal cytoplasm. In order to unequivocally prove the involvement of RNA in anuran germ cell determination it will be necessary to perform rescue experiments in which maternal RNA extracted from the vegetal hemisphere, or better yet from isolated germinal granules, is microinjected into UV-irradiated eggs.

Recently it was suggested by Züste and Dixon that UV irradiation of *Xenopus laevis* embryos delays the migration of the primordial germ cells to the genital ridges rather than causing their elimination.[171] It was found that the germ cells were not missing in UV-irradiated animals but appeared in the genital ridges at later times during development. Smith and Williams have confirmed this observation with *Rana pipiens* embryos, but they consider the effect of UV on germ cell migration to be separate from that which alters the determinants themselves.[172] They conclude that the anuran germ plasm theory and the possibility that it is manifested by localized maternal RNA are still valid.

It has been reported that UV irradiation prevents primordial germ cell formation in insect embryos[9,10,34,173-179] and also inactivates cytoplasmic elements (determinants?) which afford the germ line nuclei protection from chromosome elimination or diminution.[31,180] The latter effect of UV light has been reported in both insect and *Ascaris* eggs. It is concluded from some of these studies that, since UV light at a wavelength of 254 nm was utilized, the destruction of nucleic acids is responsible for these effects. In the absence of action spectra, however, these claims are unjustified.

Probably the most effective use of UV-irradiation technique to probe the molecular nature of morphogenetic determinants has been made by Kalthoff and his collaborators using the eggs of *Smittia*, a small chironomid midge.[181] UV irradiation and certain other manipulations at the anterior pole of this egg prevent the development of the normal pattern of embryonic head segmentation. Instead the anterior portion of the embryo develops a set of mirror image abdominal segments including all normal posterior constituents except the germ cells. The embryos which show this embryonic abnormality are known as double abdomens. Double abdomen formation is thought to involve an inactivation of the morphogenetic determinants which specifies the nature of anterior segmentation.[181] Kalthoff has conducted a detailed action spectrum for the induction of double abdomens by UV irradiation.[182] This action spectrum was corrected both for potential shielding effects mediated by the chorion and cytoplasmic components of the egg (Figure 18). The result is a biomodal curve with peaks at about 265 and 285 nm, near the absorption maxima of nucleic acids and proteins, respectively. Thus nucleic acids and proteins were potential candidates for the morphogenetic determinants involved in double abdomen formation.

Photoreactivation experiments provided further evidence for the molecular nature of the anterior determinants. When pyrimidine dimers are formed in DNA or RNA by UV irradiation they can often be enzymatically excised and repaired in the presence of visable light, a process known as photoreactivation.[169,183,184] Reversal of UV-induced double abdomens can be obtained by a subsequent exposure of the irradiated eggs to visable light at wavelengths between 310 and 460 nm.[185] This result strongly suggests that the factors responsible for

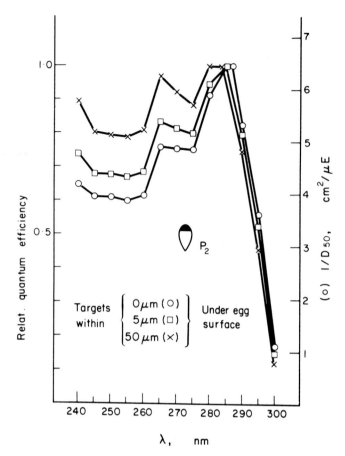

FIGURE 18.   Action spectrum for the formation of double abdomen embryos following UV irradiation of the anterior pole region of *Smittia eggs*. (From Kalthoff, K., *Photochem. Photobiol.*, 18, 355, 1973. With permission.)

normal head segmentation are nucleic acids. If so, then why does the action spectrum also include a peak in the region of maximal protein absorption? One explanation is that the morphogenetic determinant may be a nucleoprotein complex. This possibility has been subjected to an experimental test by Kalthoff.[181] He reasoned that since photoreactivation is unknown for UV-damaged proteins, visable light should be a more effective rescue agent following UV irradiation at 265 nm than after irradiation at 285 nm provided that two independently positioned UV targets exist. If the target is a nucleoprotein, however, photoreactivation should be as effective after UV irradiation at 285 nm as at 265 nm due to the transfer of excitation energy through the macromolecular complex. It turns out that visable light treatment is equally effective in rescuing double abdomen after UV irradiation at 265 or 285 nm. These data suggest that the target is a macromolecular complex, probably a nucleoprotein. Based on these and other experiments Kalthoff has suggested that the anterior morphogenetic determinant of *Smittia* eggs is an mRNP complex of maternal origin. The designation of maternal RNA as the active component of this complex is consistent with information discussed in the next section showing that the anterior determinants are specifically inactivated by RNase.

C. Local Injection of Enzymes

Another line of evidence for the identity of morphogenetic determinants in *Smittia* eggs involves the local injection of various enzymes.[186] These experiments are carried out by exposing the anterior cytoplasm of the egg to RNases, DNase, or proteases, either by microinjection or via gentle puncture of the anterior pole while the egg is immersed in aqueous solutions of enzymes. About half of the eggs treated with RNases, but none of those exposed to DNase or protease, became double abdomens. In order to prove that the RNase activity rather than the presence of the enzyme molecules per se was responsible for double abdomen formation, the eggs were treated with fragments of RNase S, either individually or after being mixed together. RNase S molecules can be proteolytically cleaved into two fragments, which are enzymatically inactive alone, but become active again after mixture. The enzyme fragment mixtures, but not the individual fragments, proved competent in double abdomen formation.[186] Thus RNase digestion of a substance localized in the cytoplasm at the anterior tip of the *Smittia* egg leads to a gross alteration in the normal pattern of head segmentation in the embryo. The active component of the morphogenetic determinant involved in the specification of a normal cephalic region in *Smittia* embryos is thus very likely to be maternal RNA.[181]

V. MODE OF ACTION OF LOCALIZED RNA

The UV irradiation, photoreactivation, and RNase injection experiments considered above together support the contention that some of the morphogenetic determinants are composed of maternal RNA molecules. If maternal RNA is causally involved in the spatial pattern of cell determination then at some stage of development the segregated RNA molecules must interact with the genetic apparatus of the cell. Although exactly how this is accomplished is still a mystery, three possibilities have been introduced as theories during the past few years. They are diagrammatically summarized in Figure 19.

In the first two hypotheses localized RNA molecules, either as regulatory or mRNA molecules, are proposed to modulate gene expression at the transcriptional, nuclear post-transcriptional, or translational levels. The concept that regulatory RNA molecules (Figure 19A) might play important roles in cell determination was introduced by Britten and Davidson[187] as a part of their general model for the control of gene activity in eukaryotic cells. They suggested that activator RNA (regulatory RNA) molecules produced by a regulatory gene could interact with genetic elements which in turn control the transcription of batteries of structural genes. In this way the transcriptional activity of the structural genes could be regulated coordinately. Subsequently this model was extended by Davidson and Britten to include a theory for the control of embryonic determination.[188] It was proposed that the regulatory genes transcribe activator RNAs during oogenesis which are subsequently localized in specific regions of the egg cytoplasm. After the activator RNAs are segregated to particular cell lineages during embryogenesis, they are thought to enter the nuclei and affect transcription by binding to the elements which control the structural gene batteries. The model of Davidson and Britten[188] provides the basic plan for the hypothesis of localized regulatory RNAs depicted in Figure 19A. Since it is now believed that genetic regulation can occur at points beyond the level of transcription, however, we have added control steps at the level of pre-mRNA processing and translation. These levels were chosen because there is some evidence that RNA may be involved in them as a control molecule. It has been suggested that small cytoplasmic RNAs, termed translational control RNAs (tcRNA), can selectively modulate the efficiency of mRNA translation.[189,190] Small nuclear RNA molecules (snRNA) have also been isolated and characterized and very recently it was suggested that they could be of importance in the splicing of nuclear pre-mRNA.[191-193] Thus it is possible that tcRNA and snRNA (derived from the GV?) could be included among the maternal regulatory RNAs which are subject to localization in eggs. Unfortunately there has been no systematic search for the presence of maternal tcRNA and snRNA in the cytoplasm of eggs.

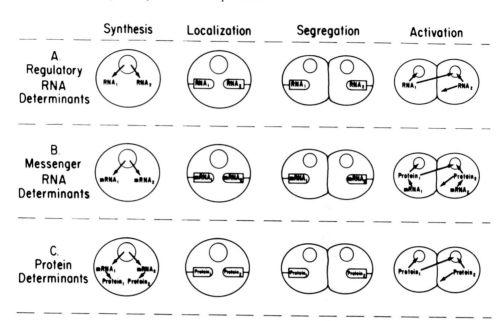

| | Synthesis | Localization | Segregation | Activation |

FIGURE 19.   Three possible modes of action of maternal RNA in localization phenomena and cell determination. (A) Regulatory RNA determinants. In this model regulatory RNA molecules are synthesized during oogenesis (arrow to cytoplasm), localized in the egg cytoplasm, segregated to particular cell lineages, and proposed to directly modulate gene expression by affecting transcription (arrow to nucleus), nuclear pre-mRNA processing (arrow to nucleus), or translation (arrow to cytoplasm). The regulatory RNAs could also pass from one cell to another (arrow from cell to cell). (B) mRNA determinants. In this model mRNAs synthesized during oogenesis (arrow to cytoplasm) are localized in the egg cytoplasm, segregated to particular cell lineages, and there indirectly affect gene expression by coding for regulatory proteins (arrow between mRNA and protein). The proteins affect gene expression by direct modulation of transcription (arrow to nucleus), nuclear pre-mRNA processing (arrow to nucleus), or translation (arrow in cytoplasm). Regulatory proteins may also be passed from one cell to another (arrow from cell to cell). (C) Protein determinants. In this model mRNAs synthesized during oogenesis (arrow to cytoplasm) are immediately translated (arrow from mRNA to protein) to form regulatory proteins, which are localized, segregated to particular cell lineages, and directly affect gene expression by modulation of transcription (arrow to nucleus), nuclear pre-mRNA processing (arrow to nucleus), and translation (arrow in cytoplasm). These regulatory proteins may also be passed from one cell to another (arrow from cell to cell).

The second hypothesis is that morphogenetic determinants are composed of localized maternal mRNA molecules. Proteins coded for by these messages are postulated to control gene expression at the levels of transcription, nuclear pre-mRNA processing, or translation (Figure 19B). As discussed previously, there is some evidence for the specific segregation of maternal mRNAs during early embryogenesis provided by *in situ* hybridization studies with actin DNA sequences in *Styela* eggs[207] and solution hybridization experiments conducted with *Xenopus* eggs using cDNA probes prepared from vegetal pole poly(A)-containing RNA.[210] The possible occurrence of maternal mRNA species coding for alkaline phosphatase in the endodermal cell lineages of *Ciona* embryos[134] is also consistent with the role of mRNA in embryonic determination.

In the third hypothesis proteins, coded for by maternal mRNA synthesized during oogenesis, become localized and serve as morphogenetic determinants (Figure 19C). Although this possibility is inconsistent with the nature of the anuran germ cell determinants and the anterior determinants of *Smittia*, which are probably nucleic acids,[11,181] it was introduced to explain the results of studies on the nature of the o^+ corrective factor of axolotl eggs. The o^+ factor, which is localized primarily in the GV plasm of oocytes and can reverse embryonic defects which arise in o-/o- mutants, appears to be protein.[6] As mentioned previously, the synthesis and localization of proteins during oogenesis also seems to provide the best explanation for the transient occurrence of poly(A)-containing RNA in the polar regions of *Oncopeltus* oocytes.[127]

On the basis of the available data none of the hypotheses depicted in Figure 19 can be favored at this time. If morphogenetic determinants are heterogeneous in nature, which appears to be possible in light of the differences in composition of the germ cell and anterior determinants and the o^+ corrective factor, perhaps localized regulatory RNAs, mRNAs, and proteins are all utilized as spatial regulatory molecules by developing systems.

VI. CONCLUSIONS AND PROSPECTUS

The purpose of this article has been to assess the evidence for the role of localized maternal RNA molecules as morphogenetic determinants. Numerous cases of RNA localization in eggs have been reviewed. Many of these occur in visable cytoplasmic organelles. They leave little doubt that maternal RNA molecules, including those of the poly(A)-containing subclass, are subject to *quantitative* localization and segregation during early development. The nature of the RNA molecules and the mechanisms by which they are localized remain unknown. They could either be regulatory RNA molecules which affect gene action directly or mRNA molecules which code for proteins involved in gene expression. We are left with the uncomfortable feeling that with the exception of only a few cases there is no unequivocal evidence for the *qualitative* localization of maternal RNA in eggs and embryonic cell lineages. Obviously a major goal for the future will be to determine whether specific maternal RNA species are subject to uneven distribution in the embryo. The recombinant DNA technology developed in recent years, which permits the specific amplification of genes coding for maternal RNAs and their utilization as probes for RNA detection, seems to provide a golden opportunity to finally settle this issue.

Although elegant experiments, especially those conducted by Kalthoff and his collaborators,[181] have provided considerable support for the involvement of localized maternal RNA in specific embryonic determinations, further analyses of other developing systems will be needed in order to determine the universality of this phenomenon. It seems crucial at this time to isolate and characterize the RNA molecules which compose the morphogenetic determinants and to develop assay systems to test their developmental potential and mode of action.

ACKNOWLEDGMENTS

The writing of this article was supported by grants from the National Science Foundation (PCM-77-24767) and the National Institutes of Health (GM025119 and HD-13970). I also wish to thank Dr. M. R. Dohmen, University of Utrecht, for providing the photograph used as Figure 9.

REFERENCES

1. **Wilson, E. G.,** *The Cell in Development and Heredity,* 3rd ed., Macmillan, New York, 1925, 1035.
2. **Davidson, E. H.,** *Gene Activity in Early Development,* 2nd ed., Academic Press, New York, 1976, 249.
3. **Hörstadius, S., Josefsson, L., and Runnström, J.,** Morphogenetic agents from unfertilized eggs of the sea urchin *Paracentrotus lividus, Dev. Biol.,* 16, 189, 1967.
4. **Josefsson, L. and Hörstadius, S.,** Morphological substances from sea urchin eggs, *Dev. Biol.,* 20, 481, 1969.
5. **Briggs, R. and Cassens, G.,** Accumulation in the oocyte nucleus of a gene product essential for embryonic development beyond gastrulation, *Proc. Natl. Acad. Sci. U.S.A.,* 55, 1103, 1966.

6. **Briggs, R. and Justus, J. T.,** Partial characterization of the component from normal eggs which corrects the maternal effect of gene O in the Mexican axolotl *(Ambystoma mexicanum), J. Exp. Zool.,* 147, 105, 1968.

7. **Beams, H. W. and Kessel, R. G.,** The problem of germ cell determinants, *Int. Rev. Cytol.,* 39, 413, 1974.

8. **Eddy, E. M.,** Germ plasm and the differentiation of the germ cell line, *Int. Rev. Cytol.,* 43, 229, 1975.

9. **Okada, M., Kleinman, L. A., and Schneiderman, H. A.,** Restoration of fertility in sterilized *Drosophila* eggs by transplantation of polar cytoplasm *Dev. Biol.,* 37, 43, 1974.

10. **Illmensee, K. and Mahowald, A. P.,** The autonomous function of germ plasm in a somatic region of the *Drosophila* egg, *Exp. Cell Res.,* 97, 127, 1976.

11. **Smith, L. D.,** Role of a ''germinal plasm'' in the formation of primordial germ cells in *Rana pipiens, Dev. Biol.,* 14, 330, 1966.

12. **Elpatiewsky, W.,** Die Ungeschlechtszllenbildung bei *Sagitta, Anat. Anz.,* 35, 226, 1909.

13. **Buchner, P.,** Keimbahn and Ovogenese von *Sagitta, Anat. Anz.,* 35, 443, 1910.

14. **Stevens, N. M.,** Further studies on the reproduction of *Sagitta, J. Morphol.,* 21, 279, 1909.

15. **Vasiljev, A.,** La fécondation chez *Spadella cephaloptera* LGRHS et l'origine du corps determinant la voie germinative, *Biol. Gen.,* 1, 249, 1925.

16. **Ghirardelli, E.,** Osservzoini sul determinante germinale (d.g.) e su altre formazioni citoplasmatiche nelle uova di *Spadella cephaloptera* (Chaetognatha), *Pubbls. Staz. Zool. Napoli,* 24, 332, 1953.

17. **Ghirardelli, E.,** Studi sul determinante germinale (e.d.) nei Chaetognati: Ricerche spermentali su *Spadella cephaloptera* Busch, *Pubbls. Staz. Zool. Napoli,* 25, 444, 1954.

18. **Ghirardelli, E.,** Il determinante germinale nell'uovo e nella gastrula di *Spadella cephaloptera* Busch (Chaetognatha). Osservazioni al microscopio elettronico, *Arch. Zool. Ital.,* 51, 841, 1966.

19. **Slyvestri, F.,** Contribuzioni alla conoscenza biologica della Imenotteri parasiti. I. Biologica del *Litomastrix truncatellus* Dalm, *Ann. R. Scuola Sup. Agric. Portici.,* 6, 1, 1906.

20. **Mahowald, A. P.,** The germ plasm of *Drosophila:* an experimental system for the analysis of determination, *Am. Zool.,* 17, 551, 1977.

21. **Mahowald, A. P., Allis, C. D., Karrer, K. M., Underwood, E. M., and Waring, G. L.,** Germ plasm and pole cells of *Drosophila,* in *Determinants of Spatial Organization,* Subtelny, S. and Konigsberg, I. R., Eds., Academic Press, New York, 1979.

22. **Mahowald, A. P.,** Genetic control of oogenesis, in *Mechanisms of Cell Change,* Ebert, J. D. and Okada, T. S., Eds., John Wiley & Sons, New York, 1979, chap. 9.

23. **Mahowald, A. P.,** Polar granules of *Drosophila.* Ultrastructural changes during early embryogenesis, *J. Exp. Zool.,* 167, 237, 1968.

24. **Ullmann, S. L.,** Epsilon granules in *Drosophila* pole cells and oocytes, *J. Embryol. Exp. Morphol.,* 13, 73, 1965.

25. **Wolf, R.,** Der Feinbau des Oosoms Normaler and Zentrifugierter Eier der Gallmücke *Wachtliella persicariae* L. (Diptera), *W. Roux Arch.,* 158, 459, 1967.

26. **Mahowald, A. P.,** Ultrastructural changes in the germ plasm during the life cycle of *Miastor* (Cecidomyidae, Diptera), *W. Roux Arch.,* 176, 223, 1975.

27. **Pasteels, J.,** Recherches sur le cycle germinal chez l'*Ascaris.* Étude cytochimique des acides nucleiques dans l'oogénèse, la spermatogénèse et le développement chez *Parascaris equorum* Goerze, *Arch. Biol.,* 59, 405, 1948.

28. **Stich, V. H.,** Das Vorkommen von Ribonucleinsäure in Kernsaft und spindel sich teilender Kerne von *Cyclops strenuus, Z. Naturforsch.,* B6, 259, 1951.

29. **Mulnard, J.,** Etude morphologique et cytochimique de l'oogénèse chez *Acanthoscelides obtectus* Say (Bruchide, Coleoptera), *Arch. Biol.,* 45, 261, 1954.

30. **Niklas, R. B.,** An experimental and descriptive study of chromosome elimination in *Miastor spec* (Cecidomyidae: Diptera), *Chromosoma,* 10, 301, 1959.

31. **Bantock, C. F.,** Experiments on chromosome elimination in the gall midge, *Mayetiola destructor, J. Embryol. Exp. Morphol.,* 24, 257, 1970.

32. **Bruiyan, N. I. and Shafiq, S. A.,** The differentiation of the posterior poleplasm in the housefly *Musca vicina* Macquart, *Exp. Cell Res.,* 16, 427, 1959.

33. **Panelius, S.,** Germ line and oogenesis during paedogenetic reproduction in *Heteropeza pygmaea* Winnertz (Diptera: Cecidomyridae), *Chromosoma,* 23, 333, 1968.

34. **Counce, S. J.,** Developmental morphology of polar granules in *Drosophila* including observations on pole cell distribution during embryogenesis, *J. Morphol.,* 112, 129, 1963.

35. **Poulson, D. F. and Waterhouse, D. F.,** Experimental studies on pole cells and midgut differentiation in Diptera, *Aus. J. Biol. Sci.,* 13, 541, 196.

36. **Meng, C.,** Autoradiographic Untersuchungen am Oosom in der Oocyte von *Pimpla turionellae* L. (Hymenoptera), *W. Roux Arch.,* 165, 35, 1970.

37. **Gatenby, J. B.,** The cytoplasmic inclusions of germ cells. VI. On the origin and probable constitution of the germ cell determinant of *Apanteles glomeratus* with a note on the secondary nuclei, *Q. J. Microsc. Sci.,* 64, 9, 1920.

38. **Blackler, A. W.,** Contribution to the study of germ cells in the Anura, *J. Embryol. Exp. Morphol.,* 6, 491, 1958.

39. **Czolowska, R.,** Observations on the origin of the "germinal cytoplasm" in *Xenopus laevis, J. Embryol. Exp. Morphol.,* 22, 229, 1969.

40. **Mahowald, A. P. and Hennin, S.,** Ultrastructure of the "germ plasm" in eggs and embryos of *Rana pipiens, Dev. Biol.,* 24, 37, 1971.

41. **Mahowald, A. P.,** Polar granules of *Drosophila.* IV. Cytochemical studies showing loss of RNA from polar granules during early stages of embryogenesis, *J. Exp. Zool.,* 176, 345, 1971.

42. **Watson, M. L. and Aldridge, W. G.,** Methods for the use of indium as an electron strain for nucleic acids, *J. Biophys. Biochem. Cytol.,* 11, 257, 1961.

43. **Allis, C. D., Waring, G. L., and Mahowald, A. P.,** Mass isolation of pole cells from *Drosophila melanogaster, Dev. Biol.,* 56, 372, 1977.

44. **Waring, G. L., Allis, C. D., and Mahowald, A. P.,** Isolation of polar granules and the identification of polar granule specific protein, *Dev. Biol.,* 66, 197, 1978.

45. **Zalokar, M.,** Autoradiographic study of protein and RNA formation during early development of *Drosophila* eggs, *Dev. Biol.,* 49, 425, 1976.

46. **Bounoure, L.,** Recherches sur la ligne'e germinale chez la Grenoville rousse aux premiers stades du development, *Ann. Sci. Nat.,* 17, 67, 1934.

47. **Wakahara, M.,** Partial characterization of "primordial germ cell forming activity" localized in vegetal pole cytoplasm in anuran eggs, *J. Embryol. Exp. Morphol.,* 39, 221, 1977.

48. **Kessel, R. G.,** Cytodifferentiation in the *Rana pipiens* oocyte. II. Intramitochondrial yolk, *Z. Zellforsch.,* 112, 313, 1971.

49. **Williams, M. A. and Smith, L. D.,** Ultrastructure of the "germinal plasm" during maturation and early cleavage in *Rana pipiens, Dev. Biol.,* 25, 568, 1971.

50. **Cambar, R., Delbos, M., and Gipouloux, J.-P.,** Premières observations sur l'infrastructure des cellules germinales à la fin de leur migration dans les crêtes génitales, chez les Amphibiens Anoures, *C. R. Soc. Biol.,* 164, 168b, 1970.

51. **Czolowska, R.,** The fine structure of the "germinal cytoplasm" in the egg of *Xenopus laevis, W. Roux Arch.,* 169, 335, 1972.

52. **Afzelius, B. A.,** Electron microscopy on the basophilic structures of the sea urchin egg, *Z. Zellforsch.,* 45, 660, 1957.

53. **Harris, P.,** Structural changes following fertilization in the sea urchin egg, *Exp. Cell Res.,* 48, 569, 1967.

54. **Verhey, C. A. and Moyer, F. H.,** Fine structure changes during sea urchin oogenesis, *J. Exp. Zool.,* 164, 195, 1967.

55. **Millonig, G., Bosco, M., and Giamberstone, L.,** Fine structure of oogenesis in sea urchins, *J. Exp. Zool.,* 169, 293, 1968.

56. **Anderson, E.,** Oocyte differentiation in the sea urchin, *Arbacia punctulata,* with special reference to the origin of the cortical granules and their participation in the cortical reaction, *J. Cell Biol.,* 37, 514, 1968.

57. **Conway, C. M. and Metz, C. B.,** *In vitro* maturation of *Arbacia punctulata* oocytes and initiation of heavy body formation, *Cell Tiss. Res.,* 150, 271, 1974.

58. **Pasteels, J. J., Castiaux, P., and Vandermeerssche, G.,** Ultrastructure du cytoplasme et distribution de l'acide ribonucléique dans l'eouf fécondé, tant normal que centrifugé de *Paracentrotus lividus, Arch. Biol.,* 69, 627, 1958.

59. **Sanchez, S.,** Effets de l'actinomycine D sur les constituants cellulaires et le metabolise de l'ARN de l'ovocyte d'oursin *(Paracentrotus lividus).* Étude autoradiographique, *Exp. Cell. Res.,* 50, 19, 1968.

60. **Bal, A. K., Jubinville, F., Cousineau, G. H., and Inoue, S.,** Origin and fate of annulate lamellae in *Arbacia punctulata eggs, J. Ultrastruct. Res.,* 25, 15, 1968.

61. **Conway, C. M.,** Evidence for RNA in the heavy bodies of sea urchin eggs, *J. Cell Biol.,* 51, 889, 1971.

62. **Harris, P. J.,** Relation of fine structure to biochemical changes in developing sea urchin eggs and zygotes, in *The Cell Cycle. Gene-Enzyme Interactions,* Padilla, G. M., Whitson, G. L., and Cameron, I. L., Eds., Academic Press, New York, 1969, 315.

63. **Katsura, S. and Inazumi, S.,** Presence of phosphatase activity in heavy bodies of sea urchin eggs, *Dev. Biol.,* 66, 480, 1978.

64. **Kessel, R. G. and Beams, H. W.,** Annulate lamellae and "yolk nuclei" in oocytes of the dragonfly, *Libellula pulchella, J. Cell Biol.,* 42, 185, 1969.

65. **Halkka, L. and Halkka, O.,** Accumulation of gene products in the oocytes of the dragonfly *Cordulia aenea.* II. Induction of annulate lamellae within dense masses during diapause, *J. Cell Sci.,* 26, 217, 1977.

66. **Koulish, S.,** Ultrastructure of differentiating oocytes in the trematode *Gorgoderina attenuata.* I. The "nucleolus-like" cytoplasmic body and some lamellar membrane systems, *Dev. Biol.,* 12, 248, 1965.

67. **Dhainaut, A.,** Étude en microscopie électronique et par autoradiographie à haute résolution des extrusions nucléaires au cours de l'ovogénèse de *Nereis pelagica* (Annélide Polychète), *J. Microsc.,* 9, 99, 1970.
68. **Anderson, E.,** Oocyte-follicle cell differentiation in two species of amphineurans (Mollusca), *Mopalia mucosa* and *Chaetopleura apiculata, J. Morphol.,* 129, 89, 1969.
69. **Rebhun, L. I.,** Electron microscopy of basophilic structures of some invertebrate oocytes. I. Periodic lamellae and the nuclear envelope, *J. Biophys. Biochem. Cytol.,* 2, 93, 1956.
70. **Rebhun, L. I.,** Electron microscopy of basophilic structures of some invertebrate oocytes. II. Fine structure of yolk nuclei, *J. Biophys. Biochem. Cytol.,* 2, 159, 1956.
71. **Bradfield, J. R. G.,** Phosphatase cytochemistry in relation to protein secretion, *Exp. Cell Res.,* 1 (Suppl.), 338, 1949.
72. **Urbani, E.,** Ricerche comparative sui nuclei vitellini di alcune specie animali, *Riv. Biol.,* 41, 331, 1949.
73. **Fautrez, J.,** La distribution de l'acide ribonucleique au cours du grand accroissement de l'oocyte chez *Pholcus* phalanguides (Fuessl.), *C. R. Soc. Biol.,* 144, 1129, 1950.
74. **Hinsch, G.,** Possible role of intracellular membranes in nuclear-cytoplasmic exchange in spider crab oocytes, *J. Cell Biol.,* 47, 531, 1970.
75. **Petit, J.,** Étude morphologique et cytochimique de deux types de groupements mitochondriaux dans les jeunes ovocytes de *Polydesmus angustus* Latz. (myriapode, Diplopode), *J. Microsc.,* 17, 41, 1973.
76. **Palevody, M. C.,** Présénce de noyaux accessores dans l'ovocyte du Collembole *Folsomia candida* Willem (insecte, Aptérygote), *C. R. Acad. Sci.,* 274, 3258, 1972.
77. **Gresson, A. R. and Threadgold, L. T.,** Extrusion of nuclear material during oogenesis in *Blatta orientalis, Q. J. Microsc. Sci.,* 103, 141, 1962.
78. **Anderson, E.,** Oocyte differentiation and vitellogenesis in the roach *Periplaneta americana, J. Cell Biol.,* 20, 131, 1964.
79. **Allen, E. R. and Cave, M. D.,** Formation, transport, and storage of ribonucleic acid containing structures in oocytes of *Acheta domesticus* (Orthoptera), *Z. Zellforsch.,* 92, 477, 1968.
80. **Hopkins, C. F.,** The histochemistry and fine structure of the accessory nuclei in the oocyte of *Bombus terrestris, Q. J. Microsc. Sci.,* 105, 475, 1964.
81. **Cassidy, J. D. and King, R. C.,** Ovarian development in *Habrobracon juglandis* (Ashmead) (Hymenoptera: Braconidae). I. The origin and differentiation of the oocyte-nurse cell complex, *Biol. Bull.,* 143, 483, 1972.
82. **King, P. E. and Fordy, M. R.,** The formation of "accessory nuclei" in the developing oocytes of the parasitoid Hymenopterans *Ophion luteus* (L.) and *Apanteles glomeratus* (L.), *Z. Zellforsch.,* 109, 158, 1970.
83. **Kessel, R. G.,** An electron microscope study of nuclear-cytoplasmic exchange in oocytes of *Ciona intestinalis, J. Ultrastruct. Res.,* 15, 181, 1966.
84. **Jeffery, W. R. and Capco, D. G.,** Differential accumulation and localization of maternal poly(A)-containing RNA during early development of the ascidian, *Styela, Dev. Biol.,* 67, 157, 1978.
85. **Wegmann, I. and Götting, K.-J.,** Untersuchungen zur Dotterbildung in den Oocyten von *Xiphophorus helleri* (Heckel, 1848) (Teleostei, Poeciliidae), *Z. Zellforsch.,* 119, 405, 1971.
86. **Ornstein, L.,** Mitochondrial and nucleoar interaction, *J. Biophys. Biochem. Cytol.,* 2 (Suppl.), 351, 1956.
87. **Wartenberg, H.,** Elektronenmikroskopische and Histochemische Studien uber die Oogenese der Amphibieneizelle, *Z. Zellforsch.,* 58, 427, 1962.
88. **Wittek, M.,** La vitellogénèse chèz les Amphibiens, *Arch. Biol.,* 63, 133, 1952.
89. **Weakley, B. S.,** Basic protein and ribonucleic acid in the cytoplasm of the ovarian oocyte in the golden hamster, *Z. Zellforsch.,* 112, 69, 1971.
90. **Schjeide, O. A., Galey, F., Grellert, E. A., I-San Lin, R., de Vellis, J., and Mead, J. F.,** Macromolecules in oocyte maturation, *Biol. Reproduct.,* 2, 14, 1970.
91. **Chaudry, H. S.,** The yolk nucleus of Balbiani in teleostean fishes, *Z. Zellforsch.,* 37, 455, 1952.
92. **Sotelo, J. R. and Porter, K. R.,** An electron microscope study of the rat ovum, *J. Biophys. Biochem. Cytol.,* 5, 327, 1959.
93. **Brachet, J.,** Localisation de l'acide ribonucléique et des protéines dans l'ovaire de grenoville normal et centrifugé, *Experientia,* 3, 329, 1947.
94. **Schmekel, L. and Fioroni, P.,** The ultrastructure of the yolk nucleus during early cleavage of *Nassarius reticulatus* L. (Gastropoda, Prosobranchia), *Cell Tiss. Res.,* 153, 79, 1975.
95. **Dohmen, M. R. and Verdonk, N. H.,** The structure of a morphogenetic cytoplasm, present in the polar lobe of *Bithynia tentaculata* (Gastropoda, Prosobranchia), *J. Embryol. Exp. Morphol.,* 131, 423, 1974.
96. **Dohmen, R. and Verdonk, N. H.,** Cytoplasmic localization in mosaic eggs, in *Maternal Effects in Development,* Newth, D. R. and Balls, M., Eds., Cambridge University Press, London, 1979, 127.
97. **Clement, A. C.,** Experimental studies on germinal localization in *Ilyanassa.* I. The role of the polar lobe in determination, *J. Exp. Zool.,* 121, 593, 1952.
98. **Dohmen, M. R. and Verdonk, N. H.,** The ultrastructure and role of the polar lobe in development of Molluscs, in *Determinants of Spatial Organization,* Subtelny, S. and Konigsberg, I. R., Eds., Academic Press, New York, 1979, 3.

99. **Cather, J. N.**, personal communication, 1980.
100. **Pasteels, J. and Mulnard, J.**, La métachromasie *in vivo* au bleu de toluidine et son analyse cytochimique dans les oeufs de *Barnea candida*, *Gryphaea angulata* (Lamellibranches) et de *Psammechinus miliaris*, *Arch. Biol.*, 68, 115, 1957.
101. **Raven, Chr. P.**, Further observations on the distribution of cytoplasmic substances among the cleavage cells in *Lymnaea stagnalis*, *J. Embryol. Exp. Morphol.*, 31, 37, 1974.
102. **Wierzejski, A.**, Embryologie von *Physa fontinalis* L., *Z. Wiss. Zool.*, 83, 502, 1905.
103. **van den Biggelaar, J. A. M.**, Development of division synchrony and bilateral symmetry in the first quartet of micromeres in eggs of *Limnaea*, *J. Embryol. Exp. Morphol.*, 26, 393, 1971.
104. **Minganti, A.**, Acidi nucleici e fosfatasi nello sviluppo della *Limnaea*, *Riv. Biol.*, 42, 295, 1950.
105. **Sathananthan, A. H.**, Studies on mitochondria in the early development of the slug *Arion rufus* L., *J. Embryol. Exp. Morphol.*, 24, 555, 1970.
106. **Dohmnen, M. R. and van de Mast, J. M. A.**, Electron microscopical study of RNA-containing cytoplasmic localizations and intercellular contacts in early cleavage stages of the eggs of *Lymnaea stagnalis* (Gastropoda, Pulmonata), *Proc. Konink. Nederland. Akad. Wetenschap. Ser. C*, 8, 403, 1978.
107. **Hadek, R.**, Cytoplasmic whorls in the golden hamster oocyte, *J. Cell Sci.*, 1, 281, 1966.
108. **Szollosi, D.**, Development of "yolky substance" in some rodent eggs, *Anat. Rec.*, 151, 424, 1965.
109. **Weakley, B. S.**, Comparison of cytoplasmic lamellae and membranous elements in the oocytes of five mammalian species, *Z. Zellforsch.*, 85, 109, 1968.
110. **Zamboni, L.**, Ultrastructure of mammalian oocytes and ova, *Biol. Reprod.*, 2 (Suppl.), 44, 1970.
111. **Burkholder, G. D., Comings, D. E., and Okada, T. A.**, A storage form of ribosomes in mouse oocytes, *Exp. Cell Res.*, 69, 361, 1971.
112. **Pucci-Minafra, I., Minafra, S., and Collier, J. R.**, Distribution of ribosomers in the egg of *Ilyanassa obsoleta*, *Exp. Cell Res.*, 57, 167, 1969.
113. **Choi, W. C. and Nagl, W.**, Ribosome crystals in the oocyte of *Gerris najas* (Heteroptera), *Z. Naturforsch.*, 32, 137, 1977.
114. **Taddei, C.**, Ribosome arrangement during oogenesis of *Lacerta sicula*, *Exp. Cell Res.*, 70, 285, 1972.
115. **Pasteels, J. J.**, Comparative cytochemistry of the fertilized egg, in *The Chemical Basis of Development*, McElroy, W. D. and Glass B., Eds., The Johns Hopkins Press, Baltimore, 1958, 381.
116. **Mulnard, J.**, Origine et signification des principaux constituants de l'oocyte chez *Acanthoscelides obtectus* Say (Bruchide-Coléoptère), *Bull. Acad. R. Belg. Cl. Sci.*, 36, 767, 1954.
117. **Afzelius, B. A.**, Basophilic structures in the cytoplasm of the sea urchin egg, in *Proc. Stockholm Conference on Electron Microscopy*, 1956, 147.
118. **Vakaet, L.**, La répartition des acides nucléiques au cours du grand accroissement de l'oeuf de *Libistes reticulatus*, *Bull R. Soc. Belg. Cl. Sci.*, 36, 941, 1950.
119. **Jones-Seaton, A.**, Étude de l'organisation cytoplasmique de l'oeuf des rongers, principalement grand à la basophile ribonucleique, *Arch. Biol.*, 61, 291.
120. **Mancuso, V.**, Gli acidi nucleici nell 'ouvo in sviluppo di *Ciona intestinalis*, *Acta Embryol. Morphol. Exp.*, 2, 151, 1959.
121. **Klag, J. J. and Ubbels, G. A.**, Regional morphological and cytochemical differentiation in the fertilized egg of *Discoglossus pictus* (Anura), *Differentiation*, 3, 15, 1975.
122. **Conklin, E. G.**, Organization and cell lineage of the ascidian egg, *J. Nat. Sci. (Philadelphia)*, 13, 7, 1905.
123. **Inase, M.**, Behavior of the pole plasm in the early development of the aquatic worm, *Tubifex hattai*, *Sci. Rep. Tohoku Univ. IV (Biol.)*, 33, 223, 1967.
124. **Capco, D. G. and Jeffery, W. R.**, Differential distribution of poly(A)-containing RNA in the embryonic cells of *Oncopeltus fasciatus*: analysis by *in situ* hybridization with a (^{3}H)-poly(U) probe, *Dev. Biol.*, 67, 137, 1978.
125. **Jeffery, W. R.**, Polyadenylation of maternal and newly synthesized RNA during starfish oocyte maturation, *Dev. Biol.*, 57, 98, 1977.
126. **Rosbash, M. and Ford, P. J.**, Polyadenylic acid-containing RNA in *Xenopus laevis*, *J. Mol. Biol.*, 85, 87, 1974.
127. **Capco, D. G. and Jeffery, W. R.**, Origin and spatial distribution of maternal messenger RNA during oogenesis of an insect, *Oncopeltus faciatus*, *J. Cell Sci.*, 39, 63, 1979.
128. **Jeffery, W. R. and Wilson, L. J.**, Cortical messenger RNA: localization and segregation in the *Chaetopterus* egg, *J. Embryol. Exp. Morphol.*, in press.
129. **Peterson, J. S. and Jeffery, W. R.**, unpublished data, 1981.
130. **Donohoo, P. and Kafatos, F. C.**, Differences in the proteins synthesized by the progeny of the first two blastomeres of *Ilyanassa*, a "mosaic" embryo, *Dev. Biol.*, 32, 224, 1973.
131. **Newrock, K. M. and Raff, R. A.**, Polar lobe specific regulation of translation in embryos of *Ilyanassa obsoleta*, *Dev. Biol.*, 42, 242, 1975.
132. **Constantini, F. D., Britten, R. J., and Davidson, E. H.**, Message sequences and short repetitive sequences are interspersed in sea urchin egg poly(A)$^{+}$ RNAs, *Nature (London)*, 287, 111, 1980.

133. **Whittaker, J. R.,** Segregation during ascidian embryogenesis of egg cytoplasmic information for tissue specific enzyme development, *Proc. Natl. Acad. Sci. U.S.A.,* 70, 2096, 1973.

134. **Whittaker, J. R.,** Segregation during cleavage of a factor determining endodermal alkaline phosphatase development in ascidian embryos, *J. Exp. Zool.,* 202, 139, 1977.

135. **Whittaker, J. R.,** Cytoplasmic determinants of tissue differentiation in the ascidian egg, in *Determinants of Spatial Organization,* Subtelny, S. and Konigsberg, I. R., Eds., Academic Press, New York, 1979, 29.

136. **Mizuno, S., Lee, Y. R., Whiteley, A. H., and Whiteley, H. R.,** Cellular distribution of RNA populations in 16-cell stage embryos of the sand dollar, *Dendaster excentricus, Dev. Biol.,* 37, 18, 1974.

137. **Rogers, W. H. and Gross, P. R.,** Inhomogeneous distribution of egg RNA sequences in the early embryo, *Cell,* 14, 279, 1978.

138. **Tufaro, F. and Brandhorst, B. P.,** Similarity of proteins synthesized by isolated blastomeres of early sea urchin embryos, *Dev. Biol.,* 72, 390, 1979.

139. **Ernst S., Hough-Evans, B. R., Britten, R. J., and Davidson, E. H.,** Limited complexity of the RNA in micromeres of sixteen-cell sea urchin embryos, *Dev. Biol.,* 79, 119, 1980.

140. **Dan, K., Noguchi, M., and Uemura, I.,** Studies on unequal division in sea urchin embryos: inequality of ribosomal content, in *Mechanisms of Cell Change,* Ebert, J. D. and Okada, T. S., Eds., John Wiley & Sons, New York, 1979, chap. 3.

141. **Ziff, E. B. and Evans, R. M.,** Coincidence of the promotor and capped 5′ terminus of RNA from the adenovirus 2 major late transcription unit, *Cell,* 15, 1463, 1978.

142. **McReynolds, L., O'Malley, B. W., Nisbett, A. D., Fothergill, J. E., Givol, D., Fields, S., Robertson, M., and Brownlee, G. G.,** *Nature (London),* 223, 723, 1978.

143. **Barelle, F. E.,** Complete nucleotide sequence of the 5′ non-coding region of human α- and β-globin mRNA, *Cell,* 12, 1085, 1977.

144. **DeWachter, R.,** Do eukaryotic mRNA 5′ non-coding sequences base pair with the 185 ribosomal RNA 3′ terminus?, *Nuc. Acid Res.,* 7, 2045, 1979.

145. **Milcarek, C. and Penman, S.,** Membrane-bound polyribosomes in HeLa cells: association of polyadenylic acid with membranes, *J. Mol. Biol.,* 89, 327, 1974.

146. **Spirin, A. S.,** On ''masked'' forms of messenger RNA in early embryogenesis and in other differentiating systems, *Curr. Top. Dev. Biol.,* 1, 1, 1966.

147. **Preobrazhensky, A. and Spirin, A. S.,** Informosomes and their protein components: the present state of the knowledge, *Prog. Nuc. Acid Res. Mol. Biol.,* 21, 1, 1978.

148. **Lifton, R. P. and Kedes, L. H.,** Size and sequence homology of masked maternal and embryonic histone messenger RNAs, *Dev. Biol.,* 48, 47, 1976.

149. **Clement, A. C.,** Development of the vegetal half of the *Ilyanassa* egg after removal of most of the yolk by centrifugal force, compared with the development of animal halves of similar visual composition, *Dev. Biol.,* 17, 165, 1968.

150. **Raff, R.,** The molecular determination of morphogenesis, *Biol. Sci.,* 27, 394, 1977.

151. **Lenk, R., Ransom, L., Kaufmann, Y., and Penman, S.,** A cytoskeletal structure with associated polyribosomes obtained from HeLa cells, *Cell,* 10, 67, 1977.

152. **Miller, T. E., Huang, C-Y., and Pogo, A. O.,** Rat liver nuclear skeleton and ribonucleoprotein complexes containing hnRNA, *J. Cell Biol.,* 76, 675, 1978.

153. **Miller, T. E., Huang, C-Y., and Pogo, A. O.,** Rat liver nuclear skeleton and small molecular weight RNA species, *J. Cell Biol.,* 76, 692, 1978.

154. **Herlan, G., Eckert W. A., Jaffenberger, W., and Wunderlich, F.,** Isolation and characterization of an RNA-containing nuclear matrix from *Tetrahymena* macronuclei, *Biochemistry,* 18, 1782, 1979.

155. **Lenk, R. and Penman, S.,** The cytoskeletal framework and poliovirus metabolism, *Cell,* 16, 289, 1979.

156. **Gurdon, J. B.,** Attempts to analyze the biochemical basis of regional differences in animal eggs, in *Cell Patterning,* Porter, R. and Rivers, J., Eds., American Elsevier, New York, 1975, 223.

157. **Froehlich, J. P., Browder, L. W., and Schultz, G. A.,** Translation and distribution of rabbit globin mRNA in separated cell types of *Xenopus laevis* gastrulae, *Dev. Biol.,* 56, 356, 1977.

158. **Capco, D. G. and Jeffery, W. R.,** Regional accumulation of vegetal pole poly(A)$^+$ RNA injected into fertilized *Xenopus* eggs, *Nature (London),* 294, 5838, 1981.

159. **Spurway, H.,** Genetics and cytology of *Drosophila sub obscura.* IV. An extreme example of delay in gene action, causing sterility, *J. Genet.,* 49, 126, 1948.

160. **Felding, C. J.,** Developmental genetics of the mutant *grandchildless* of *Drosophila subobscura, J. Embryol. Exp. Morphol.,* 17, 375, 1967.

161. **Thierry-Mieg, D.,** Study of a temperature sensitive mutant *grandchildless* in *Drosophila melanogaster, J. Microsc. Biol. Cellulaire,* 25, 1, 1976.

162. **Counce, S. J. and Ede, D. A.,** The effect on embryogenesis of a sex-linked female sterility factor in *Drosophila melanogaster, J. Embryol. Exp. Morphol.,* 5, 404, 1967.

163. **Kern, C. H.,** Cytochemical and ultrastructural observations on the polar plasm of eggs produced by the female-sterility mutants *fs(1)N* in *Drosophila melanogaster, Am. Zool.,* 16, 189, 1976.

164. **Bounoure, L., Aubry, R., and Huck, M. L.,** Nouvelles recherches expérimentales sur les origines de la lignée reproductrice chez la grenouille rousse, *J. Embryol. Exp. Morphol.,* 2, 245, 1954.

165. **Padoa, E.,** Le gonadi di girini di *Rana esculenta* da vova irradiate con ultravioletto, *Estratto Monit. Zool. Ital.,* 70-71, 238, 1963.

166. **Ikenishi, K., Kotani, M., and Tanobe, K.,** Ultrastructural changes associated with UV irradiation in the "germinal plasm" of *Xenopus laevis, Dev. Biol.,* 36, 155, 1974.

167. **Ijiri, K-I.,** Existence of ultraviolet-labile germ cell determinant in unfertilized eggs of *Xenopus laevis, Dev. Biol.,* 55, 206, 1977.

168. **McLaren, A. D. and Shugar, D.,** *Photochemistry of Proteins and Nucleic Acids,* Pergamon Press, Oxford, 1964, 320.

169. **Stelow, R. B. and Pollard, E. C.,** *Molecular Biophysics,* Addison-Wesley, Reading, Mass., 1962.

170. **Wakahara, M.,** Induction of supernumerary primordial germ cells by injecting vegetal pole cytoplasm into *Xenopus* eggs, *J. Exp. Zool.,* 203, 159, 1978.

171. **Züste, B. and Dixon, K. E.,** Events in germ cell lineage after entry of the primordial germ cells into the genital ridges in normal and UV-irradiated *Xenopus laevis, J. Embryol. Exp. Morphol.,* 41, 33, 1977.

172. **Smith, L. D. and Williams, M.,** Germinal plasm and germ cell determinants in anuran amphibians, in *Maternal Effects in Development,* Newth, D. R. and Balls, M., Eds., Cambridge University Press, London, 1979, 167.

173. **Geigy, R.,** Action de l'ultraviolet sur le pole germinal dans l'oeuf de *Drosophila melanogaster, Rev. Suisse Zool.,* 38, 187, 1931.

174. **Aboim, A. N.,** Développement embryonnaire et postembryonnaire des gonades normal et agamétiques de *Drosophila melanogaster, Rev. Suisse Zool.,* 52, 53, 1945.

175. **Hathaway, D. S. and Selman, G. G.,** Certain aspects of cell lineage and morphogenesis studied in embryos of *Drosophila melanogaster* with an ultraviolet microbeam, *J. Embryol. Exp. Morphol.,* 9, 310, 1961.

176. **Graziosi, G. and Micali, F.,** Differential responses to ultraviolet irradiation of the polar cytoplasm of the *Drosophila* egg, *W. Roux Arch.,* 175, 1, 1974.

177. **Warn, R.,** Manipulation of the pole plasm of *Drosophila melanogaster, Acta Embryol. Exp.,* 172 (Suppl.), 415, 1972.

178. **Warn, R.,** Restoration of the capacity to form pole cells in UV-irradiated *Drosophila* embryos, *J. Embryol. Exp. Morphol.,* 33, 1033, 1975.

179. **Graziosi, G. and Marzari, R.,** Differential responses to ultraviolet irradiation of the polar cytoplasm of *Drosophila* eggs. II. Response to dose, *W. Roux Arch.,* 179, 294, 1976.

180. **Moritz, K. B.,** Die Blastomerendifferenzierung für Soma und Keinbahn bei *Parascaris equorum.* II. Untersuchungen mittels UV-Bestrahlung und Zentrifugierung, *W. Roux Arch.,* 159, 203, 1967.

181. **Kalthoff, K.,** Analysis of a morphogenetic determinant in an insect embryo (*Smittia* sp., Chironomidae, Diptera), in *Determinants of Spatial Organization,* Subtelny, S. and Konigsberg, I., Eds., Academic Press, New York, 1979, 97.

182. **Kalthoff, K.,** Action spectra for UV induction and photoreversal of a switch in the developmental program of the egg of an insect *(Smitta), Photochem. Photobiol.,* 18, 355, 1973.

183. **Wulff, D. L. and Rupert, C. S.,** Disappearance of thymine photodimer in ultraviolet irradiated DNA upon treatment with a photoreactivating enzyme from Baker's yeast, *Biochem. Biophys. Res. Commun.,* 7, 237, 1962.

184. **Gordon, M. P., Huang, C. W., and Hurter, J.,** in *Photochemistry and Photobiology of Nucleic Acids,* Vol. 2, Academic Press, New York, 1976, 265.

185. **Jäckle, H. and Kalthoff, K.,** Photoreactivation of RNA in UV-irradiated insect eggs (*Smittia* sp., Chironomidae, Diptera). I. Photosensitized production and light-dependent disappearance of pyrimidine dimers, *Photochem. Photobiol.,* 27, 309, 1978.

186. **Kandler-Singer, I. and Kalthoff, K.,** RNase sensitivity of an anterior morphogenetic determinant in an insect egg (*Smittia* sp., Chironomidae, Diptera), *Proc. Natl. Acad. Sci. U.S.A.,* 73, 3739, 1976.

187. **Britten, R. J. and Davidson, E. H.,** Gene regulation for higher cells: theory, *Science,* 165, 349, 1969.

188. **Davidson, E. H. and Britten, R. J.,** Note on the control of gene expression during development, *J. Theor. Biol.,* 32, 123, 1971.

189. **Heywood, S. M., Kennedy, D., and Bester, A. J.,** Separation of specific initiation factors involved in the translation of myosin and myoglobin messenger RNAs and the isolation of a new RNA involved in translation, *Proc. Natl. Acad. Sci. U.S.A.,* 71, 2428, 1974.

190. **Heywood, S. M., Kennedy, D. S., and Bester A. J.,** Studies concerning the mechanism by which translational control RNA regulates protein synthesis in embryonic muscle, *Eur. J. Biochem.,* 58, 587, 1975.

191. **Ro-Choi, T. S. and Busch, H.,** Low molecular weight nuclear RNAs, in *The Cell Nucleus,* Busch, H., Ed., Academic Press, New York, 1974, 152.

192. **Zieve, G. and Penman, S.,** Small RNA species of the HeLa cell: metabolism and sub-cellular localization, *Cell,* 8, 19, 1976.

193. **Lerner, M. R., Boyle, J. A., Mount, S. A., Wolin, S. L., and Steitz, J. A.,** Are snRNPs involved in splicing?, *Nature (London),* 283, 220, 1980.

194. **Mahowald, A. P.,** Fine structure of pole cells and polar granules in *Drosophila melanogaster, J. Exp. Zool.,* 151, 201, 1962.

195. **Enders, A. C. and Schlafke, S. J.,** The fine structure of the blastocyst: some comparative studies, in *Preimplantation Stages of Pregnancy,* Wolstenholme, G. E. W. and O'Connor, M., Eds., Little, Brown, Boston, 1965.

196. **Calarco, P. G. and Brown, E. H.,** An ultrastructural and cytological study of preimplantation development in the mouse, *J. Exp. Zool.,* 171, 253, 1969.

197. **Schultz, R. M. and Wassarman, P. M.,** Biochemical studies of mammalian oogenesis: protein synthesis during oocyte growth and meiotic maturation in the mouse, *J. Cell Sci.,* 24, 167, 1977.

198. **Brower, P. T. and Schultz, R. M.,** Biochemical studies of mammalian oogenesis: possible existence of a ribosomal and poly(A)-containing RNA-protein supramolecular complex in mouse oocytes, *J. Exp. Zool.,* 220, 251, 1982.

199. **Bachvarova, R., DeLeon, V., and Spiegelman, I.,** Mouse egg ribosomes: evidence for storage in lattices, *J. Embryol. Exp. Morphol.,* 62, 153, 1981.

200. **Angerer, L. M. and Angerer, R. C.,** Detection of poly(A)$^+$RNA in sea urchin eggs and embryos by quantitative *in situ* hybridization, *Nuc. Acids Res.,* 9, 2819, 1981.

201. **Capco, D. G. and Jeffery, W. R.,** Transient localizations of messenger RNA in *Xenopus laevis* oocytes, *Dev. Biol.,* 89, 1, 1982.

202. **Wakahara, M.,** Accumulation, spatial distribution and partial characterization of poly(A)$^+$RNA in the developing oocytes of *Xenopus laevis, J. Embryol. Exp. Morphol.,* 66, 127, 1981.

203. **Sagata, N., Okuyama, K., and Yamana, K.,** Localization and segregation of maternal RNA's during early cleavage of *Xenopus laevis* embryos, *Dev. Growth Diff.,* 23, 23, 1981.

204. **Sternlicht, A. L. and Schultz, R. M.,** Biochemical studies of mammalian oogenesis: kinetics of accumulation of total and poly(A)-containing RNA during growth of mouse oocyte, *J. Exp. Zool.,* 215, 191, 1981.

205. **Piko', L. and Clegg, K. B.,** Quantitative changes in total RNA, total poly(A), and ribosomes in early mouse embryos, *Dev. Biol.,* 89, 362, 1982.

206. **Venezsky, D. L., Angerer, L. M., and Angerer, R. C.,** Accumulation of histone repeat transcripts in the sea urchin egg pronucleus, *Cell,* 24, 385, 1981.

207. **Jeffery, W. R., Tomlinson, C., and Brodeur, R.,** Localization of actin mRNA during early development of *Styela,* in press.

208. **Cloney, R. A.,** Cytoplasmic filaments and cell movements: epidermal cells during ascidian metamorphosis, *J. Ultrastruct. Res.,* 14, 300, 1966.

209. **Pucci-Minafra, I. and Ortolani, G.,** Differentiation and tissue interaction during muscle development of ascidian tadpoles. An electron microscope study, *Dev. Biol.,* 17, 692, 1968.

210. **Carpenter, C. D. and Klein, W. H.,** A gradient of poly(A)$^+$RNA sequences in *Xenopus laevis* eggs and embryos, *Dev. Biol.,* 91, 43, 1982.

211. **Brandhorst, B. P. and Newrock, K. M.,** Post-transcriptional regulation of protein synthesis in *Ilyanassa* embryos and isolated polar lobes, *Dev. Biol.,* 83, 250, 1981.

212. **Collier, J. R.,** Protein synthesis in the polar lobe and lobeless egg of *Ilyanassa obsoleta, Biol. Bull.,* 160, 366, 1981.

213. **Collier, J. R. and McCarthy, M. E.,** Regulation of polypeptide synthesis during early embryogenesis of *Ilyanassa obsoleta, Differentiation,* 19, 31, 1981.

214. **Jeffery, W. R. and Meier, S.,** A yellow crescent cytoskeletal complex in ascidian eggs and its role in early development, *Dev. Biol.,* in press.

215. **Moon, R. T., Nicosia, R. F., Olsen, C., Hille, M. B., and Jeffery, W. R.,** The cytoskeletal framework of sea urchin eggs and embryos: developmental changes in the association of messenger RNA, *Dev. Biol.,* in press.

Chapter 4

NUCLEIC ACIDS AND PROTEINS IN THE FIRST STAGES OF EMBRYONIC DEVELOPMENT

Silvio Ranzi and Fiorenza De Bernardi

TABLE OF CONTENTS

I. THE PROTEINS IN THE FIRST STEPS OF DEVELOPMENT

A. The Egg

A process of protein synthesis acts in all eggs. Lampbrush chromosomes are detected in growing oocytes and it is possible to follow in them all the stages of RNA and protein synthesis. It is also possible in many cases to see an action of other cells or organs that collaborate to supply protein to the oocyte.

Giudice[1] presents data suggesting that in the sea urchin the yolk may be synthesized in the cisternae of the endoplasmic reticulum of the oocyte.

The protein synthesis is small in insect oocytes, growing in the panoistic or meroistic ovary. The pinocytotic uptake of hemolymph protein furnishes a material stocked in proteinous yolk platelets.[2]

A considerable amount of RNA is synthesized by the nurse cells of the meroistic ovary and migrates in growing oocytes.

The yolk formation mechanism in large vertebrate eggs starts outside of the ovary, in the liver. The protein, a lipophosphoprotein called vitellogenin, is secreted into extracellular spaces and then removed from circulating blood by the oocyte. This process seems to be related to the follicular epithelium activity, and Lanzavecchia[3] saw an intensive formation of pinocytotic vesicles on the oocyte membrane. Vitellogenin in amphibian oocyte is transformed into yolk. The conversion process appears to involve a proteolytic cleavage of the vittelogenin peptides into those that make up lipovitellenin and phosvitin.[4,5]

Following Parisi and colleagues[6] hemocyanin and ovoverdin, present in the hemolymph of crayfish (*Austropotamobius*), are stored into the growing oocyte.[7]

More specialized structural proteins and enzymes are also synthesized in oocytes. Woodland et al[8,9] showed that early oocytes of *Xenopus* produce histones which are stocked in full-grown oocytes at a sufficient rate up to late cleavage stages.

A list of the enzymes present in unfertilized egg of sea urchin is presented in Giudice's[1] book and for eggs of the other animals in Needham's book.[10]

B. Fertilization

Mirsky[11] extracted unfertilized and fertilized sea urchin eggs in 1 *M* KCl, after having frozen dried and ground up the eggs. The data indicated that 12% of the proteins of the eggs became nonextractable after fertilization. This fact is in complete agreement with Runnström's[12] observation of colloidal organization changes of egg cytoplasm related to fertilization.

Fertilization is related to a decrease in protein solubility observed in sea urchins (*Arbacia*), Anura (*Bufo* and *Rana*) and silkworm (*Bombyx*) (Leonardi Cigada).[13]

It is possible that the increase of proteolytic activity, observed by Lundblad[14] in sea urchin eggs during fertilization, is related to these cytoplasmic changes.

C. Differentiation in Embryonic Development
1. Amphibia

Protein differentiation in normal development of *Rana esculenta, Bufo bufo,* and *Bufo viridis* was studied with salting-out of soluble proteins, with increasing concentration of ammonium sulfate, or salting-in with dilution of the solvent of the soluble proteins. Comparing the optical density (OD) of the protein solution without ammonium sulfate at 280 nm with the OD after addition of various amounts of the precipitating medium, Ranzi et al.[15] obtained the diagrams presented in Figures 1 and 2.

The salting-out diagrams show that after fertilization and during cleavage, the proteins extracted in Weber and Edsall fluid precipitate at an increasingly lower concentration of ammonium sulfate. The maximum absorption corresponds to the fractions which precipitate

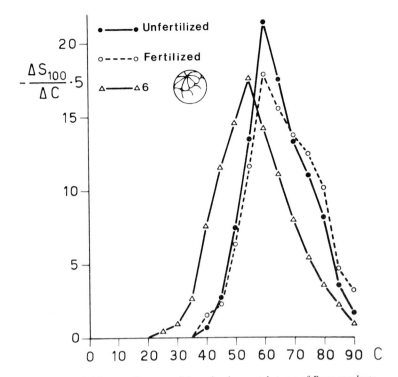

FIGURE 1. Salting-out diagrams of three developmental stages of *Rana esculenta*. The stages are indicated by sketches and stage numbers (following Rugh[16]). On the ordinata the difference between the spectrophotometer reading is given on lower and higher saturation of ammonium sulfate (difference of 5% saturation). (From Ranzi, S., *Adv. Morphog.*, 2, 211, 1962. With permission.)

at 60% saturation in fertilized egg and 50% saturation at the end of the cleavage and during gastrulation (Figure 1). The salting-out diagrams change during neurulation and protein fractions precipitating at low- and high-ammonium sulfate saturation become evident. At the end of neurulation, four fractions precipitating at 25, 40, 55, and 70% (Figure 2) saturation become evident. These four fractions are still in the diagram of the young tadpole when yolk resorption is nearly complete at 20, 40, 60, and 75%. The four peaks are at 10, 35, 55, and 85% ammonium sulfate saturation after the metamorphosis.

Investigations were made to test the appearance of different antigens during embryonic and larval development. The peak at 35% (Figure 2) saturation of the adult corresponds to the peak of adult myosin, prepared according to Szent Gyorgyi. Protein, extracted from gastrula in Weber and Edsall fluid and precipitating between 30 and 70% ammonium sulfate saturation, gives the precipitin test with rabbit antiserum prepared against Szent Gyorgyi myosin. In the same extract the fraction precipitating between 50 and 70% gives precipitin reaction with rabbit serum antiactin. The peak of adult actin is at 60% ammonium sulfate saturation. Sera antired blood cells and antiblood plasma give positive precipitin reaction with fractions precipitating between 50 and 75% ammonium sulfate saturation of *Rana* embryo at stage 15 of Rugh[16] (neural fold in contact); at the same stage, in fractions precipitating between 45 and 75% benzidine positive reactions; during the development it is possible to detect benzidine reactions in the blood island of the more advanced embryo.

The appearance of the different adult proteins was tested by the immunological method in toads (Figure 3). The fraction that precipitates between 50 and 70% ammonium sulfate saturation reacts at the end of the cleavage (stage 8 of Rugh),[15] the fraction between 30 and 50% at the gastrulation (stage 11 of Rugh), the fraction between 70 and 90% at the neural

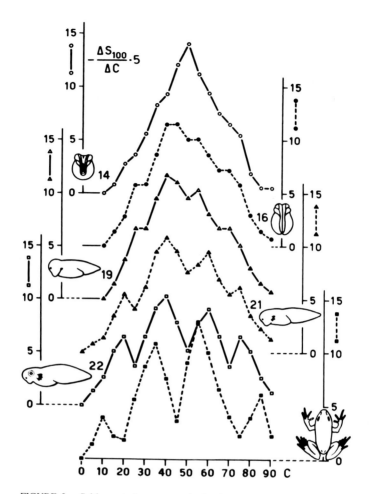

FIGURE 2.   Salting-out diagrams at six developmental stages of *Rana esculenta*. The stages are shown by sketches and stage numbers. (From Ranzi, S., *Adv. Morphog.*, 2, 211, 1962. With permission.)

tube (stage 16 of Rugh), and the fraction that remains in solution at 90% at the tail bud stage (stage 17 of Rugh) (Figure 3).

These data concern separation of some protein and antigen groups during development. But the analysis of soluble proteins by polyacrylamide gel electrophoresis also allowed us to draw a pattern of their evolution during embryonic development in *Pleurodeles*.[17] About 30 bands can be characterized. Most of them are already present in the mature unfertilized oocyte. Protein differentiation is revealed in the strengthening, weakening, or disappearance of the band which seems to correspond to yolk components. The most important changes take place from the tail bud stage onwards. These changes occur more rapidly in the dorsal ectomesodermal part than in the ventral part of the embryo. Ecker and Smith[18] investigated protein synthesis in eggs and first step of development of *Rana pipiens* and were able to show synthesis of some proteins and disappearance of others during cleavage and gastrulation.

The appearance of new proteins during the development can be studied by following the appearance of new antigens. Antigen(s) not present in the egg appear(s) in toad embryo during the cleavage. A second group of antigens appear during the gastrulation, a third during neurulation, and a fourth at tail bud stage. Other antigens are present in adult animals.

These data are in complete agreement with the more recent observations of Spiegel et al.[19] who were able to demonstrate the appearance in *Rana pipiens* proteoinogrammes of a

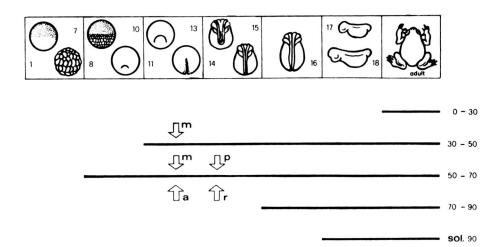

FIGURE 3. Antigen appearance of adult toads *(Bufo bufo)* in embryonic and larval development. The different development stages are shown at the top by sketches and stage numbers (following Rugh[16]). The numbers on the right represent the percentage of ammonium sulfate saturation corresponding to the concentration limits between which the antigens are precipitated. Arrow **a** represents the start of the antiactin precipitation; arrow **m** the start of antimyosin precipitation; and arrows **p** and **r** the start of precipitation with antiplasma and antired cells in our experiments. The horizontal lines represent the presence of the antigens. (From Ranzi, S., *Ist. Lomb. Rend. Sci. B*, 108, 182, 1974. With permission.)

new fraction during the cleavage and of a new fraction during gastrulation. This second fraction of *Rana* is represented by three different bands in *Ambystoma*. The electrophoretic pattern of protein extracted from *Pleurodels* embryos was studied by Denis;[20] a new protein fraction was demonstrated to appear at the beginning of gastrulation.

Our observations on toad antigens can be put in relation also with the data of Clayton,[21] who shows in *Triturus alpestris* at the beginning of gastrulation the appearance of a new antigen, a second antigen between gastrula and neurula, and a third antigen at tail bud stage.

Ballantine et al.[22] confirm the observation about actin demonstrating a large actin synthesis in *Xenopus* embryo during gastrulation. New proteins are synthesized too at this stage and other proteins are synthesized during gastrulation, but it is impossible to identify these proteins, studied by two dimensional gel electrophoresis, with the antigens described by us.[15] The appearance of new protein is evident at the animal pole.[23]

It is possible to collect more data combining this research with crosses and nuclear transplantations. Patternal antigen(s) appear(s) in toad crosses during cleavage, and other patternal antigens appear in more advanced stages.[15] Proteins corresponding to transplanted nucleus appear in axolotl (see Gallien et al.[24]) after tail bud stage when circulation begins.

Another phenomenon must be taken into consideration. Some protein unities show changes during development. This is the case in hemoglobin. Early hemoglobin in the chick between the 48th and 68th hr of incubation shows only two components, both of which show electrophoretic mobility higher than any one of the three components of the adult hemoglobin. On the 11th day of incubation only traces of this embryonic hemoglobin were detected by D'Amelio.[25] Early embryonic hemoglobin has one globin chain different from those of the adult. This globin is, however, no longer evident in hemoglobin from 12-day embryos.[26]

Tadpoles and frogs of *Rana catesbeiana* have different hemoglobins which differ in fingerprinting patterns. A change in hemoglobin is observed at metamorphosis by Baglioni,[27] with gradual replacement of tadpole hemoglobin with frog hemoglobin.

The same is true for the isozyme pattern.[28] Lactate dehydrogenase shows four molecular forms in the early development of *Xenopus* embryo; these bands are already present in oocytes. At stage 27 (muscular response) there is an appearance of a new isozyme with

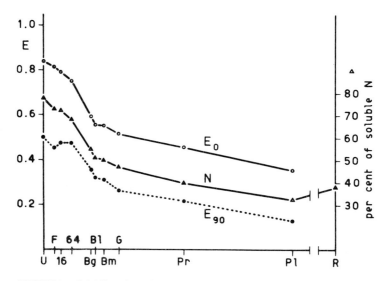

FIGURE 4.   Solubility of proteins of *Arbacia lixula* at different developmental stages of E_O in Weber and Edsall solution (0.5 g in 100 mℓ diluted 1:10) and E_{90} in Weber and Edsall solution with 90% ammonium sulfate solution. The two curves represent absorption values (E). The third curve represents soluble N. U, unfertilized egg; F, fertilized; 16,64 numbers of blastomeres; Bg, young blastula; Bl, swimming blastula; Bm, mesenchyme blastula; G, gastrula; Pr, prism; Pl, pluteus; and R, adult sea urchin. (From Ranzi, S., *Adv. Morphog.*, 2, 211, 1962. With permission.)

slow electrophoretic mobility. At stage 33 to 34 (heart beat) three more new bands appear.[29] The five lactate dehydrogenase bands are present in *Discoglossus pictus* development, but their densitometry changes at different stages of development from growing oocyte up to embryo.[30]

The zymogram of lactate dehydrogenase, of alcohol dehydrogenase,[31] and other zymograms[32] changes during the larval development of *Drosophila melanogaster*.

2. Sea Urchins

The soluble proteins in Weber and Edsall fluid decrease with fertilization and during the development up to pluteus stage[33] of *Arbacia lixula* (Figure 4).

The salting-out diagrams of unfertilized egg show four peaks at 15, 40, 50, and 70% ammonium sulfate saturation (Figure 5).

The diagrams after the fertilization and during cleavage are the same, but it is possible to see a small reduction of the 40% peak and an increase of the 50% peak. The salting-out diagram changes in hatching blastula. In this stage one fraction precipitating between 45 and 60% ammonium sulfate saturation is evident and another fraction, precipitating between 70 and 90% ammonium sulfate saturation can be recognized (Figure 5). The material to build these fractions is supplied by the peak 40% ammonium sulfate saturation, which shows a big decrease. The precipitation diagram changes with the appearance of a fraction precipitating at 55% in swimming blastula and new fractions appear in mesenchyme blastula (Figure 6). These fractions with peaks precipitating at 20 and 45% ammonium sulfate saturation correspond to two fractions present in the adult and there is also an indication of the presence of the fraction with peak at 75% of the adult; consequently, all the fractions of the adult can be recognized at this stage. The adult salting-out pattern is more and more evident in the stages of gastrula, prisma, pluteus, and adult (Figure 7).

The salting-in diagrams also show in the mesenchyme blastula the appearance of a new fraction precipitating at a dilution between 0.45 and 0.35 M KCl concentration and in gastrula the presence of characteristic fractions of the adult (Figure 8).

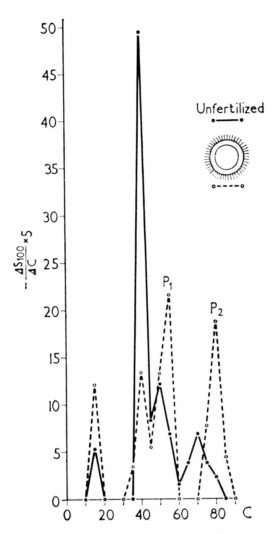

FIGURE 5. Salting-out diagrams for the unfertilized
egg and hatching blastula of *Arbacia lixula*. P_1 and P_2
are two protein fractions synthesized during the development inside the egg envelope.

Spiegel et al.[19,34] were able to show the appearance in *Arbacia* of one fraction of basic
protein during cleavage and a second fraction in the gastrula.

It is possible to collect more information about protein synthesis by studying the isoenzyme
pattern. Differentiation between sea urchin blastomeres is presented by Moore and Villee.[35]
The large blastomeres at the 64-cell stage have two and the small blastomeres have three
bands of L-malate dehydrogenase activity, whereas unfertilized eggs have five.

The disappearance and the appearance of new fractions at the mesenchyme blastula stage
can be shown with immunological research (Figure 9).[36] We separated the fractions of the
egg precipitating with peaks at 15, 40, 50, and 70% ammonium sulfate saturation and the
proteins, which do not precipitate at 90% ammonium sulfate saturation; then, we tested the
fraction with egg antiserum in order to discover the developmental stage in which the
corresponding antigens disappear. The antigen(s) with peak at 15 disappear(s) during cleavage; those not precipitating at 90% ammonium sulfate saturation also disappear during
cleavage; the antigen(s) with peak at 70% (fraction D) disappear(s) at the moment of mesenchyme formation; the fraction with peak at 40% (fraction B) disappears in prisma; and

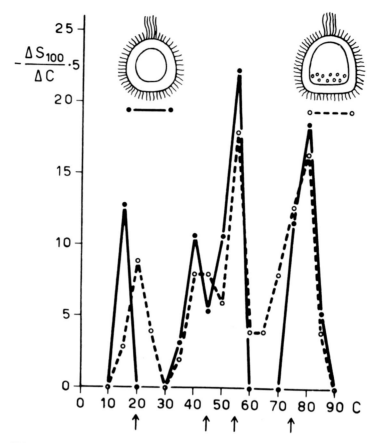

FIGURE 6. Salting-out diagrams for swimming blastula and mesenchyme blastula of sea urchin. The arrows indicate the peaks of the protein fraction of the adult animals. (From Ranzi, S., *Adv. Morphog.*, 2, 211, 1962. With permission.)

the fraction with peak at 50% (fraction C) ammonium sulfate saturation is no longer present in the adult sea urchin.

The appearance of adult proteins has been studied with adult antiserum in different fractions separated following the peaks of the adult 20, 45, 65, and 75% ammonium sulfate saturation and the proteins that remain in solution at 90% ammonium sulfate saturation. The fractions 45, 65, and 75% appear at mesenchyme blastula stage. The fraction that remains in solution at 90% ammonium sulfate saturation reacts with adult antiserum at prisma stage and the fraction precipitating at 20% is present in the adult animal. The cell surface antigens studied by McClay[37] present a difference between blastula and gastrula stages. Hybrid embryos between *Lytechinus variegatus* and *Tripneustes esculentus* at blastula stage adhere to the cells of maternal genotype and, after gastrulation, also to the cells of the paternal genotype. The cell surface changes at the moment of gastrulation; this observation corroborates our observation of the appearance of adult antigens in the mesenchyme blastula stage.

Westin et al.,[38] who in many papers investigated the antigens of sea urchin embryos and found antigens corresponding to the above-mentioned ones, take the protein synthesis into consideration. Eggs and developmental stages of sea urchin *Paracentrotus* were incubated with radioactive [^{14}C]amino acids. Incorporation of label into antigens during different phases of development was studied by means of antisera prepared against sea urchin embryos and larvae. The concentration of antigen-bound radioactivity in extract of proteins increases strongly from unfertilized egg up to the pluteus stage. In unfertilized eggs, less than 10% of antigen-bound label was incorporated into specific antigens, while the rest consisted of

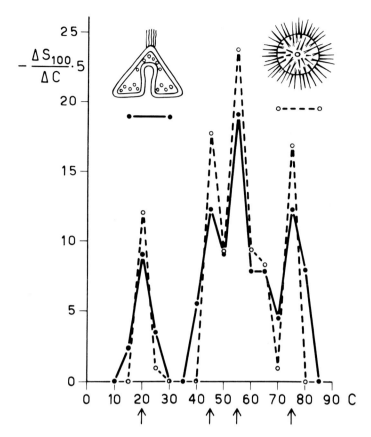

FIGURE 7.   Salting-out diagrams for the prism stage and adult sea urchin. (From Ranzi, S., *Adv. Morphog.*, 2, 211, 1962. With permission.)

antigens common for eggs and older stages. In contrast, in plutei almost 50% of label was in new antigens not present in unfertilized eggs. Extracts of unfertilized eggs contained at least three different antigens. One of these disappears completely before hatching. A second was present throughout development but was unlabeled in older stages; this is perhaps our antigen *C*. A third antigen showed a transient period of more intense labeling during the cleavage stage but disappeared after gastrulation; perhaps, this is our antigen *B*. Other antigens appear during development and, in the plutei, label was found in at least three antigens not detectable in early stages.

3. Hemocyanins and Other Comparative Problems

The hemocyanin synthesis in different animal groups is an interesting case for comparison.

The hemocyanin is already present in the crayfish egg (*Austropotamobius*); this hemocyanin comes from maternal blood synthesized by the cyanocytes in lymphocytogenetic nodules of the gizzard walls.[39] But a synthesis of hemocyanin is also acting during the embryonic development.[6,7]

The hemocyanin of cuttlefish (*Sepia*) is synthesized during the embryonic development.[6] The egg takes the copper from the sea during development and the amount of copper is increased 15-fold (8×10^{-7} mg in egg, 12×10^{-6} mg in embryo at the end of the development)[40] (Figure 10).

Leonardi Cigada,[41] following Banga and Szent Gyorgy, prepared structure proteins II, insoluble in Weber and Edsall fluid with urea; structure proteins I, soluble in Weber and Edsall fluid but precipitated by dilution and neutralization; and globular proteins, soluble and not precipitated by dilution and neutralization.

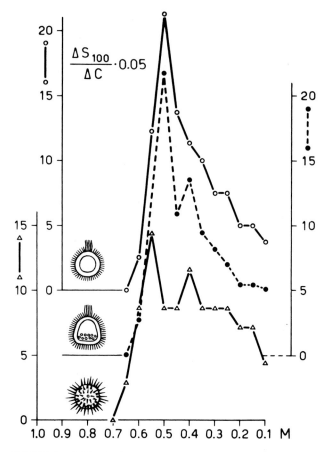

FIGURE 8. Salting-in diagrams at three stages of *Arbacia lixula* as illustrated. The abscissa represent the final concentration of KCl. The ordinate indicates the difference between two successive readings (100 is the solubility of the first sample at 1 *M* KCl). The peak at 0.4 becomes evident at the moment of mesenchyme formation and persists in the adult.

Structural proteins II decrease gradually in the embryonic development of aquatic animals (sea urchin, frog), but increase gradually in the embryonic development of land animals (silkworm, chick, guinea pig) (Figure 11). Globular proteins decrease during development of animals with direct development (chick, guinea pig), and increase during the development of larvae (silkworm, sea urchin, frog) (Figure 12). This behavior can be put in relation to the ecological conditions. The increase in the quantity of structural protein II during development of land animals is related to the protective function of these substances that are in the skin and in the skeleton (the chitin is determined like structural protein II). The increase in quantity of globular protein during larval development is related to the need for a pool of globular protein to use during metamorphosis. During the metamorphosis of the frog and in later stages of the metamorphosis of the silkworm, the soluble protein decreases.

II. TRANSCRIPTION AND DETERMINATION

A. Morphology

One of the main problems of the development and appearance of new proteins is the transcription.

A method for studying it is the blocking of transcription, induced by actinomycin D or

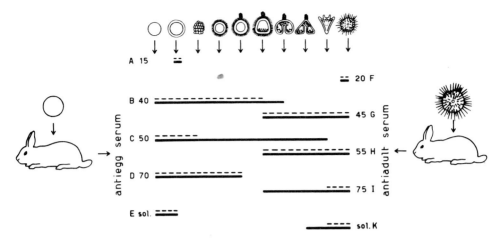

FIGURE 9. Scheme of the immunological investigation on *Arbacia lixula* development. The anti-egg serum induces the precipitation of the fractions A, B, C, D, and E, obtained by fractionating a protein solution in Weber and Edsall fluid extracted from different embryonic stages (A, fraction precipitated between 0 and 30% ammonium sulfate saturation; B, between 30 and 45%; C, between 45 and 60%; D, between 60 and 80%; and E, the fraction which remains in solution at 80% ammonium sulfate saturation). The precipitation induced by the antiserum is indicated by the continuous line. The broken line corresponds to the precipitation induced by the antiserum absorbed on lyophilized powder of different embryonic stages. The anti-adult serum was tested by the same method with the fractions F, G, H, I, and K, obtained by precipitation with ammonium sulfate between 0 and 30% saturation (fraction F); 30 and 50% (fraction G); 50 and 70% (fraction H); 70 and 90% (fraction I); (K) is the fraction remaining in solution at 90% ammonium sulfate saturation.

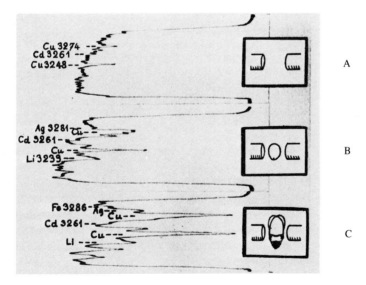

FIGURE 10. Quantitative spectrum analysis of spectra obtained with a graphite arch. (A) The arch; (B) the arch with egg ashes; and (C) the arch with embryo ashes. It is possible to follow the increase of the Cu during the embryonic development (increase of the peaks 324.8 and 327.4 nm).

by daunomycin. Both the substances interact with DNA and stop the RNA synthesis. Puromycin, instead, blocks the translation.

Chick embryos were explanted at primitive streak stages and then cultured on albumen

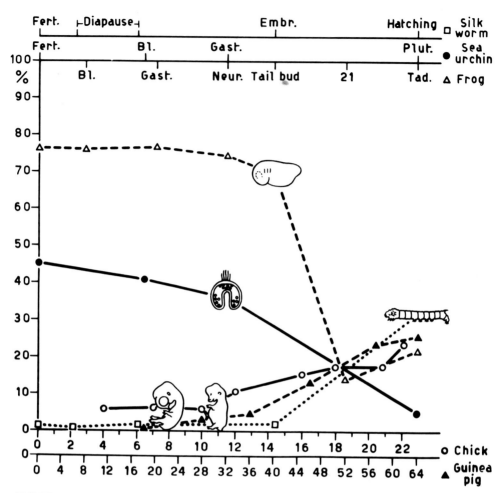

FIGURE 11.   Changes in percentage of the substances isolated as structural protein II in embryonic and larval development of five animal species. The ages of chick and guinea pig are indicated in days on the abscissa. (From Leonardi Cigada, M., *Ist. Lomb. Rend. Sci. B*, 92, 453, 1958. With permission.)

and Pannet Compton solution. Actinomycin D, daunomycin, or puromycin were added to the culture medium.[42]

Embryos treated with actinomycin D or with daunomycin presented quite identical malformations. The less-affected embryos showed a small reduction of the caudal neural tube, the more-affected ones an evident alteration of the nervous system development; in particular, the forebrain was reduced and induction of lens inhibited. In the most-inhibited embryos, the nervous system was strongly reduced or absent. Somites were inhibited. The notochord was present and only seldom did not develop. The heart was always large and well developed. Sometimes it was possible to find embryos with only the heart. Hemoglobin was present in nearly all treated embryos even if the blastoderm expansion was reduced (Figure 13A and B).

Embryos treated with puromycin retarded their development. This retardation was more considerable at the nervous system level; this was very seldom closed and the brain vesicles were reduced, while the length of neural tube was normal. Like the other organs, the heart development was affected by puromycin action. Its volume was generally related to the embryo dimensions and normally smaller than the heart of the embryo treated with actinomycin D and daunomycin. Hemoglobin was frequently absent (Figure 13C).

When actinomycin D or daunomycin act at primitive streak stage, they do not inhibit heart development which, on the contrary, is inhibited by puromycin action. Olivo[43] found

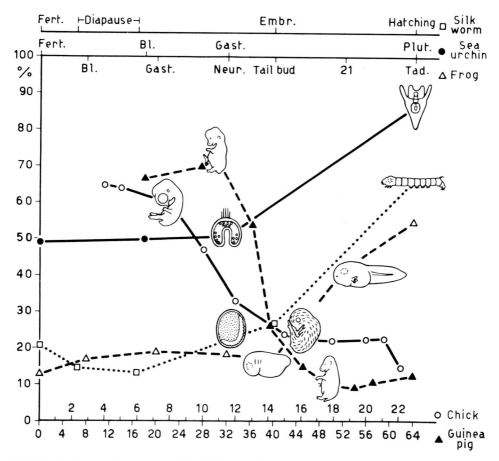

FIGURE 12.   Changes in percentage of the soluble protein in the embryonic and larval development of five animal species. The ages of chick and guinea pig are indicated in days on the abscissa. (From Leonardi Cigada, M., *Ist. Lomb. Rend. Sci. B,* 92, 453, 1958. With permission.)

that myocardial cells are determined before the appearance of the primitive streak. The actinomycin D and daunomycin action show that, at primitive streak stage, the heart transcription has already taken place, but the puromycin shows that the translation does not yet occur. These results agree with the hypothesis that in heart determination a transcription process occurs.

Lanzani Maci[44] observed a great incorporation of uridine in the chick neural rudiment during the earliest stage of neural folds (up to two somites). The synthesis indicated by the uridine incorporation is the same inhibited by actinomycin D.

Our observations agree with the experiments of Denis[45] and of Gallera[46] about the mechanism of nervous rudiment evocation. Denis, in a series of sandwich experiments in Amphibia, treated the reactive ectoderm with actinomycin D for 11 min and the induction did not work. It is necessary to treat the inductor (dorsal lip of the blastopore) for 2 hr to inhibit the evocation of nervous rudiment. Gallera, working on chick embryo, saw that Hensen's node of embryo treated for 150 min with actinomycin D is able to induce neural tube in untreated ectoderm. Instead, a treatment for 30 min of the ectoderm, more or less, inhibits the induction with an untreated Hensen's node. The conclusion is that the evocating power is already present in the dorsal blastoporal lip of Amphibia and in the Hensen's node of the chick. On the contrary, the transformation of ectoderm in neural rudiment is related to a transcription process inhibited by the actinomycin D. Consequently, a transcription process is acting during the determination also in this case.

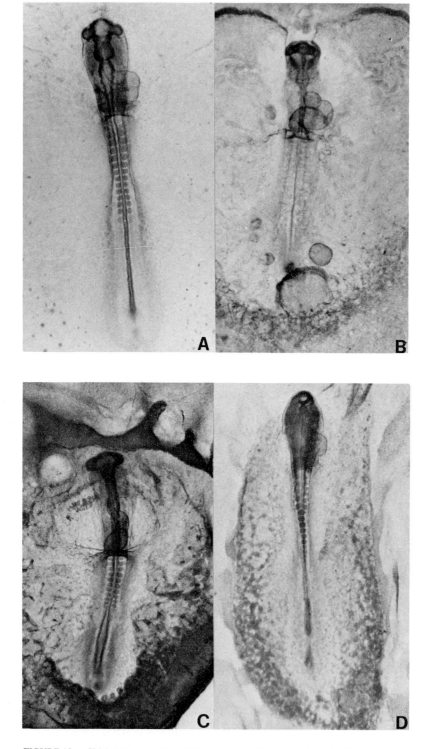

FIGURE 13.   Chick embryos cultured following New,[103] and treated at primitive streak stage for 24 hr with: (B) 0.05 μg/mℓ actinomycin D; (C) 3 μg/mℓ puromycin; and (D) 0.01 *M* LiCl. (A) Control.

Barros and collaborators[47] inhibited the starfish gastrulation with actinomycin D. The transcription is necessary for the development of the entoderm and this datum agrees with

Runnström's[48] observations that actinomycin D inhibits the development of the entoderm of sea urchin gastrula and with Leone's[49] observation that sea urchin larvae do not develop the entoderm in the presence of RNase.

B. Molecular Embryology

1. Amphibia

It is known that during amphibian oogenesis a great quantity of ribosomal RNA and ribosomes is synthesized; these ribosomes are sufficient to bring about the protein synthesis until the gastrula stage. Gurdon's experiments on anucleolate mutants beforehand,[50] and on translation into oocytes and eggs of exogenous mRNA injected afterwards,[51] gave the most convincing proof that these ribosomes were functioning. It was also demonstrated that the RNA sequences synthesized in the oocytes have a great range of complexity; a lot of informational RNA transcribed in this stage can be conserved during first embryogenesis until mid-blastula stage.[52]

The mature amphibian oocyte contains a great variety of gene transcripts, mostly produced during the "lampbrush" stage of diplotenic chromosomes. More recent research has criticized the traditional concept of the lampbrush phase as a continuous period of actively transcribing maternal messages which will be stored in the oocyte. It was calculated that the "lampbrush" phase (1.5 to 2 months in Anura; 7 months in Urodela) would be too long, if we relate it to the quantity of stored transcripts. Otherwise, Rosbash and Ford,[53] using hybridization with [³H]poly(U), found that the rate of synthesis of polyadenylated mRNA is very high in the early oogenesis stage, before the "lampbrush" stage, and then it remains quite constant, despite a greater than 4-fold increase in RNA content per oocyte (it is diluted in ribosomal one) and a 100-fold increase in volume (see also Reference 177). The work of Rosbash and Ford also shows that the poly(A)-containing RNA is mostly contained in ribonucleoprotein particles sedimenting between 25 and 80 S; less than 10% of this RNA cosediments with polyribosomes, is sensitive to EDTA, and may correspond to mRNA which is being actively translated.

In previtellogenic oocytes, on the contrary, about 75% of mRNA could be bound to polyribosomes.[54] The mRNA synthesized in previtellogenic oocytes may have a half-life of more than 2 years and could be stored during the entire oogenesis.[55] Recently Dolecki and Smith,[56] showed that poly(A)⁺RNA is synthesized both at stage 3 (maximal lampbrush chromosome stage) and at stage 6 (completion of oocyte growth): the bulk level could be maintained by continually changing rates of synthesis and degradation.

It is necessary, of course, to maintain this complex steady-state so that the sequences transcribed are the same both at the stage of maximum lampbrush and also in full-grown oocyte: experiments of cross-hybridization between polyadenylated RNA from ovary, immature and mature oocyte, and the corresponding cDNA show quite identical sequences.[57] In this way the idea that a great quantity of maternal information is contained in cytoplasm of eggs before the fertilization is also confirmed for amphibian eggs as well as sea urchins.

Darnbrough and Ford[54] found different electrophoretic patterns for proteins synthesized in vitro by *Xenopus* oocytes at different stages, from early vitellogenesis to mature oocyte; on the contrary, polyadenylated mRNA extracted from oocytes at the same stages and translated by means of a wheat germ cell-free system always gave the same patterns. This suggests, on the one hand, that the mRNAs translated on the few polyribosomes of oocyte are part of a selected population of stored mRNA; on the other hand, some post-transcriptional control in the oocytes may be possible. The estimate of the relative rates of synthesis of different classes of proteins revealed that their accumulation in the oocyte is coupled to and regulated with the accumulation of the RNA with which they interact.[179]

Mature oocyte and unfertilized egg thus represent a very complex system: a single totipotent cell is filled up with items of information such as stable maternal gene transcripts (ribonucleoprotein particles and polyadenylated messengers) and also translational products (histones, DNA polymerases, RNA polymerases) (see Section I.A).

The research carried out on the effects of the actinomycin D on early embryos by Brachet et al.[58] were very important in demonstrating the presence and the utilization of a pool of maternal RNA stored in the eggs.

The synthesis of new embryonic RNA was studied by Brown and Littna in classical research:[59] during the cleavage heterogenous RNA is synthesized which shows a DNA-like base composition and a rapid turnover; Brown and Gurdon,[60] by using anucleolate mutants, were able to show this RNA as nonribosomal but loading the polysomes. The rate of synthesis of this RNA increases in late blastula and during gastrulation, when it is stored; afterwards, the rate is constant. From this research it was already clear that the increase of protein synthesis observed from fertilization in amphibian embryos onwards, although less dramatic than in sea urchin, had been codified by stable mRNA of maternal origin contained in the egg in sufficient amount to program the development of the embryo until gastrula stage. It is well known indeed from the morphological research that actinomycin D cannot inhibit the cleavage, but puromycin, pactamycin, cycloeximide, and all inhibitors of translation can do this.

The ribosomal RNA synthesis, and from here the formation of embryonic ribosome, begins only at early gastrula stage, shortly after the synthesis of new transfer RNA beginning at late blastula.[61]

A still open problem is when the mRNAs directing the embryonic differentiation in postgastrula stage are synthesized. They become the "limitant factors" of protein synthesis accompanying the differentiation, since the entire translational system seems to be ready and fully sufficient, as Gurdon's well-known experiments showed.[51]

Denis identified three forms of informational RNAs during the embryonic development using competition hybridization technique.[62] One of these RNA appears at gastrula stage and it cannot be found in more advanced stages; a second shows a rapid turnover and is also present in the adult; a third one, identified as a stable mRNA, is found from the neurula stage and its appearance accompanies the differentiation period.

The presence of stable mRNA, stored in cytoplasm, explains why the differentiation cannot be blocked by actinomycin D or by daunomycin. In fact, the inhibitors of transcription completely depress the RNA synthesis, but the protein synthesis is nevertheless maintained at the same level as the control or even a little higher.[63] This evidence makes us suppose that some mRNA directing cellular differention is stored in cytoplasmic ribonucleoprotein particles in inactive form and reduced stability; these suggestions derive from evidence of myosin mRNA stored in mRNP sedimenting 70 to 100 S identified in myoblasts before the fusion and differentiation,[64] and from mRNA codifying for globin stored in 20 S particles from hen erythroid cells.[65]

Following recent research the structure of cytoplasmic and polyribosomal RNP seems to be different from the nuclear ones (hnRNP), more resistant to the nucleases.[176]

The available data are not sufficient to clarify many aspects of storage of mRNA in inactive form and of its mobilization and activation. Nevertheless, they suggest that transcription and translation could take place even in widely separated times of cell differentiation, therefore at different developmental stages. Monroy et al.[66] have already found some nontranslated mRNA in unfertilized sea urchin eggs; Gross et al.[67] then identified the messenger for histones and polyadenylated mRNA contained in ribonucleoprotein particles.[68] mRNP particles were furthermore found in a variety of embryonic systems like *Xenopus* oocytes,[53] *Artemia salina* cysts,[69] and *Drosophila* embryos.[70] 5 S RNA and all kinds of RNA are also contained in storage particles with associated enzymes; these particles sediment 42 S with subunits sedimenting 34, 25, and 15 S in amphibian and teleost oocytes.[71]

The lack of homogeneity of amphibian embryo from the gastrula stage suggested the existence of regional differences both in translational activity and in posttranscriptional control of synthesis. Woodland and Gurdon found that synthesis of ribosomal RNA in the endoderm was related to the appearance of nucleoli and to the synthesis of ribosomal proteins and it was starting several hours later than in ectoderm.[72] A RNA called ''heterogeneous'' was found, moreover, quickly synthesized and stored in endodermal cells.

Further research indicated a time sequence of activity of the dorsal region (ectoderm and mesoderm) in respect to the ventral region. Brachet and Malpoix,[73] summarizing cytochemical, autoradiographic, and biochemical data, suggested the well-known diagrammatic distribution of the polysomes (i.e., the ribosomes actively synthesizing proteins) along an anveg gradient in unfertilized egg; this gradient becomes dorso-ventral in the gastrula and cephalo-caudal from tail bud stage on.

The research carried out by Flickinger on neurula embryos of *Rana pipiens* first treated with actinomycin D, then by molecular hybridization with adult DNA, permits the recognition of a greater number of homologous sequences in ectomesodermal cells than in endodermal ones;[74] the heterogeneous nuclear RNA from the ectomesodermal region shows a sedimentation coefficient restricted between 16 and 30 S, thus suggesting a different processing stage of primary transcripts compared with the endodermal region where the mean sedimentation coefficient is higher;[75] previous data also suggested that even RNA in the dorsal region was more stable than in the ventral region.

In previous research we tackled the problem of synthesis of messenger directing the histological differentiation by cutting the embryos in two parts at various development stages, from the gastrula to the neural tube stage, by blocking the RNA synthesis with actinomycin D, and by labeling the embryos with precursors of RNA and protein synthesis.[76] The dorsal region of the embryos cut at gastrula or neural plate stage showed an incorporation reduced by 60% in comparison with the controls. On the contrary the dorsal region of the embryos cut and treated with actinomycin D from the early neurula stage showed an increase in [^{14}C]leucine incorporation in the total TCA-precipitable proteins. Therefore, further research was carried out in order to determine whether the increased incorporation was to be ascribed to the whole translational system or whether there were some polysomal fractions active in translating previously formed RNA even in the presence of actinomycin D.[77]

The sedimentation patterns of polysomes extracted from the dorsal region of the embryos cut at the early neurula stage (st. 15 of Nieuwkoop and Faber[78]) and incubated for 18 hr, up to the muscular response stage (st. 25), show four main classes of polysomes. In almost all fractions, the actinomycin-treated fragments incorporated more than the controls. These results confirmed the data previously obtained by measuring the [^{14}C]leucine incorporated into the total proteins.[76] Therefore, we carried out three series of short incubations (45 min), respectively, at the beginning (st. 15), middle (st. 22), and end (st. 25) of the incubation time in order to see when the various fractions were more active (Figure 14).

This experimental approach enabled us to note that some polysomal classes began to be active in succession. This would correspond to a gradual activation of some mRNAs with the formation of polysomal complexes. The light polysomal fractions seemed to be more active between the neurula and tail bud stages, while at the muscular response stage the greater incorporation was shown by the heavy fractions.

Some polysomal classes show a greater incorporation in the actinomycin-treated embryos than in the controls. The stimulation does not seem to influence the whole polysomal complex, but only some fractions of it, often those which are more active in the controls, too. We cannot imagine a mechanism affecting the translation machinery as a whole. The Tomkins hypothesis for cellular systems subjected to hormonal induction may also fit in with the embryonic differentiation.[79] In those polysomal classes in which the radioactivity is higher in actinomycin-treated embryos than in the controls, the protein synthesis may be supported by mRNAs synthesized before the stage of treatment (early neurula). These

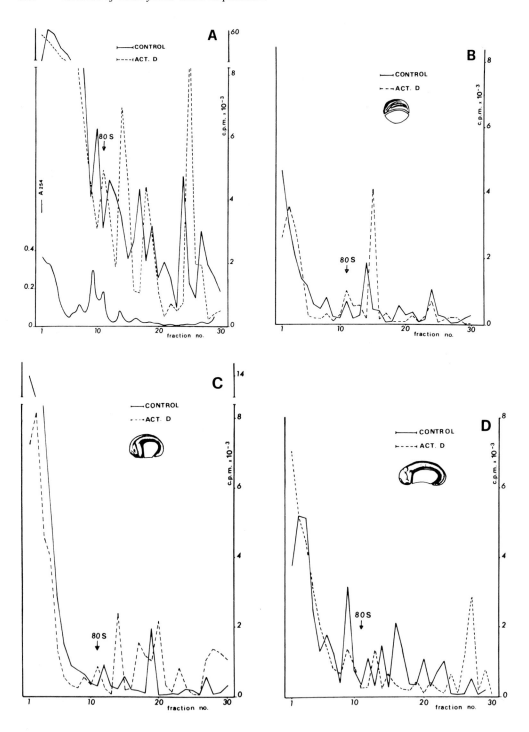

FIGURE 14.   Sedimentation on sucrose gradient of polysomes from the dorsal region of *Xenopus laevis* embryos cut at early neurula stage (st. 15 or N. and F.) and incubated with [^{14}C]leucine for 18 hr (A) and for 45 min (B, C, and D) at neurula, tail bud and muscular response stage, (From De Bernardi, F. and Bolzern, A. M., *Rend. Accad. Naz. Lincei Sci. Fis. 8,* 44, 302, 1978. With permission.)

mRNAs could be controlled by some labile factors, with fast turnover, perhaps some proteins normally produced by the cells. When actinomycin D inhibits the transcription, the production of these repressors is also stopped and the stable mRNAs can be translated without any regulation. Hence the apparent "superinductive" effect of actinomycin D.

The high activity of the polysomes in the dorsal region during the closing of the neural tube corresponds to the differentiation steps of the nervous system and somites. This is a moment of active synthesis of a few kinds of specific proteins generally codified by stable mRNAs which have been present from very early developmental stages. In agreement with the previous hypothesis, we can suppose that these mRNAs, produced before the neurula stage, are controlled by some repressors which are not produced in the actinomycin-treated embryos. The result is the highest incorporation of the precursors in the polysomes loaded by these stable mRNAs.

To try to identify the mRNAs, directing this protein synthesis and to verify when they are synthesized, we extracted them from the heavy polysomal fractions that were also active in actinomycin-treated fragments and we performed a hybridization with [³H]poly(U) in order to show the presence of polyadenylated segments.

The RNA extracted from the same polysomal fractions that we had seen engaged in protein synthesis on evidently performed templates and in which it was possible to show the "superinductive" effect of actinomycin can be hybridized with [³H]-poly(U) in a fraction sedimenting at 26 S, near the ribosomal 28 S RNA. This 26 S RNA was extensively studied by Heywood and Rourke[80] and by Merlie et al.[64] and was identified as the messenger of the heavy chain of myosin. The protein synthesis observed in heavy polysomal classes is clearly supported greatly by this messenger, and it is probably a synthesis of myosin. It is to be noted that at this moment the somites are rapidly isolating and begin the differentiation of muscular cells.

From the data previously obtained it was clear that this mRNA supports protein synthesis in the presence of actinomycin D, too; therefore, it seems to be stable and transcribed before the early neurula stage, when we began the treatment with actinomycin. We tried, then, to identify this 26 S mRNA in earlier stages, beginning from the late blastula.

At each examined stage the embryos were cut into two regions, leaving the ecto- and mesoderm presumptive in the "dorsal" region, and the presumptive endoderm in the "ventral" region. From each group of fragments the total RNAs were extracted and fractionated onto a sucrose gradient; each gradient fraction was then hybridized with labeled poly(U) (Figure 15). The RNA extracted from the animal region of the blastula (st. 9) hybridizes in several fractions but hardly at all in 26 S; hybridization in this fraction increases slightly at the gastrula stage and becomes more evident from the neural plate stage (st. 12-1/2) and at the beginning of neural folds raising (st. 15) (Figures 15C and D). In all the stages³ H-poly (U)-hybridization was not observed in the presence of actinomycin D. If we compare the quantity of poly(U) hybridized in each stage, we can observe that the percentage of poly(U) hybridized in myosin messenger increases about threefold from blastula to gastrula stage, and another threefold from gastrula to neural plate stage (Figure 16). These data suggest that at this moment myosin mRNA is actively transcribed or its own precursor is actively polyadenylated.

In more advanced stages the percentage of hybridization of 26 S decreases, perhaps in connection with differentiation which is also in progress in other cellular types like nervous cells, requiring the presence of a wide range of messengers. At st. 22, when the differentiation of muscular cells and therefore the synthesis of myosin are in progress, the higher proportion of radioactivity bound to the polyadenylate segment of 26 S RNA is restored.

Our results suggest that the protein synthesis in *Xenopus* and particularly the synthesis of myosin during the closing of the neural tube, proceeds on the template of stable mRNA, polyadenylated and transcribed before the raising of neural folds (st. 15). It may be remembered that at late gastrula stage it is possible to detect the presence of myosin with the salting-out method (see Section I.C.1). The paper by Ballantine et al.[22] strengthens our

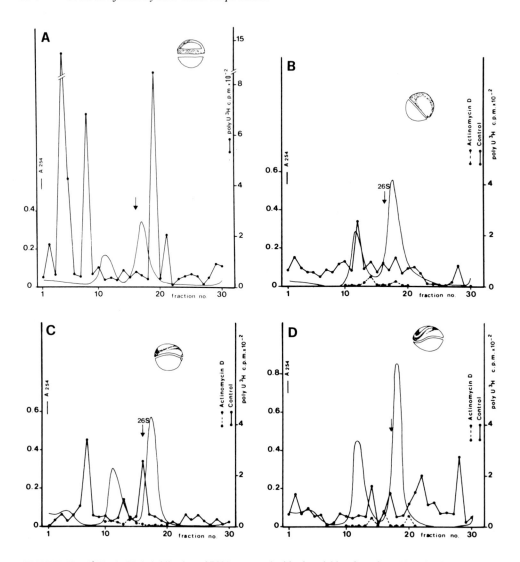

FIGURE 15.   [³H]poly(U) hybridization of RNA extracted with phenolchloroform from (A) animal or (B, C, and D) dorsal regions of *Xenopus laevis* embryos. In (B, C, and D) the embryos were cut at early gastrula stage and were treated with actinomycin D (2 μg/mℓ) up to the indicated stages.

findings showing that another main muscular protein, the α-actin, and its mRNA can be found from the late blastula stage, in the dorsal region.

In the ventral region of the amphibian embryo the histological differentiation happens later, differentiation of the organs beginning at about st. 28 to 30. This fact can explain the very low activity in all polysomal classes we have found at neurula stage. Only at stage 22, corresponding to tail bud formation, could we find some kind of activity, which became particularly intense from the muscular response stage on (Figure 17).

In the ventral region we have not observed a "superinductive" effect in actinomycin-treated fragments. We think that in the ventral region most mRNAs are immediately translated in contrast to what happens in the dorsal region. Most of these mRNAs are probably codifying for "housekeeping" proteins. We can suppose that mRNAs, coding for differentiation of endodermal organs, are produced during the closing of the neural tube and translated at about tail bud stage.

It is interesting to note that the degree of differentiation seems to be correlated to the

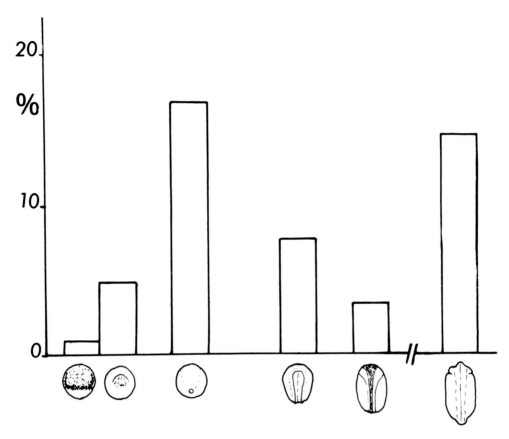

FIGURE 16.   Percentage of [³H]poly(U) hybridized in 26 S RNA compared with the [³H]-labeled poly(U) hybridized in the mRNA at each stage.

degree of inhibition exerted by actinomycin D on the protein synthesis. The comparison of total protein synthesis in fragments of tail bud embryos (st. 22 of N. and F.) cut in three regions (head, dorsal, and ventral regions) and treated with actinomycin D revealed the maximum inhibition in the ventral region, whose differentiation is later, and the highest superinduction in the head region, where the differentiation is more advanced.[76] In this way a correlation between morphological observations and biochemical data could be realized (Figure 18).

2. Sea Urchins

Fertilization induces a 5- to 20-fold increase in the rate of protein synthesis. This stimulation also occurs in the embryos mechanically or chemically enucleated with actinomycin D.[81,82] These embryos are capable of development until the blastula stage, utilizing as template for protein synthesis, the mRNAs synthesized during the oogenesis and stored in the egg. Kaumeyer et al.[68] showed that mRNA polyadenylated is stored in ribonucleoprotein particles of great stability and translatable in vitro only under conditions which do not preserve their stability.

Easily labeled, heterogeneously sized RNA, DNA-like in base composition, was found by several authors synthesized in nuclei of early-cleaving embryos (data reviewed by Davidson).[83] Ninety percent of this RNA remains associated with ribonucleoprotein particles. Giudice et al.[84] showed the presence of large-sized RNA also in the cytoplasm of gastrulae and also in nonnucleated egg portions.

Davidson reported a calculated synthesis and decay rate of mRNA and rRNA in developing

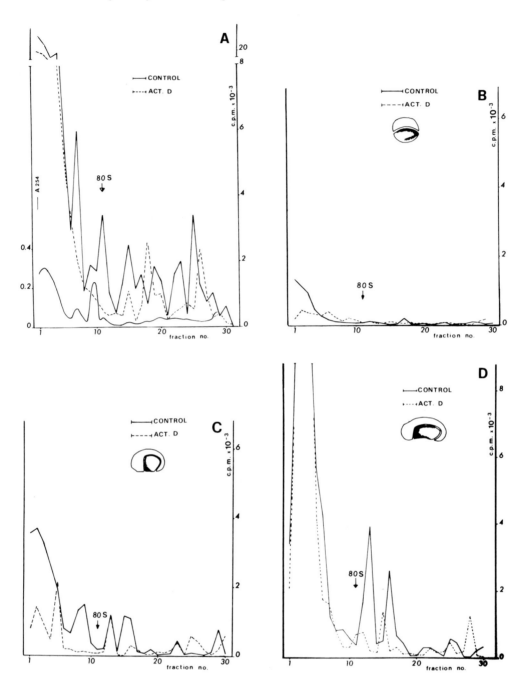

FIGURE 17.   Sedimentation on sucrose gradients of polysomes of ventral region of *Xenopus laevis* embryos cut at the early neurula stage (st. 15 of N. and F.) and incubated with [¹⁴C]leucine for 18 hr (A) and for 45 min (B, C, and D) at the indicated stages. (From De Bernardi, F. and Bolzern, A. M., *Rend. Accad. Naz. Lincei Sci. Fis. 8*, 44, 302, 1978. With permission.)

embryos showing an accumulation of mRNA in postfertilization stages.[83] The mRNA turns over with an average $t_{1/2}$ of 5.3 hr and its synthesis predominates in embryos until the gastrula stage.

The work of Nemer and Surrey distinguishes three classes of mRNA synthesized during the development on separate sets of genes and appearing in the polysomes.[85] Histone mRNA

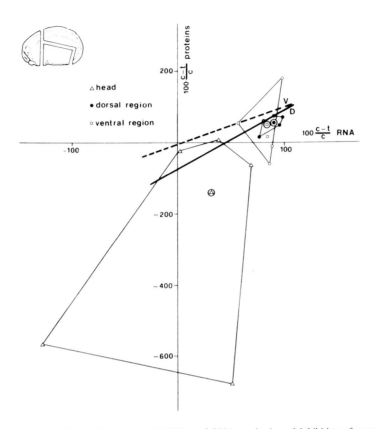

FIGURE 18. Correlation between inhibition of RNA synthesis and inhibition of protein synthesis in *Xenopus laevis* embryos cut in three regions at tail bud stage (st. 22 of N. and F.), treated with actinomycin D and incubated with [³H] uridine and [¹⁴C]leucine. The percentage of inhibition is calculated by the formula 100·c-t/c, where c is the quantity of dpm incorporated by the control and t is the incorporation in the actinomycin-treated fragments. The straight lines represent the correlations between the inhibition obtained in various stages (from gastrula to muscular response) in embryos cut in two regions (D = dorsal region; V = ventral region). The points enclosed in the circles are the mean points of the inhibition in the three regions of the embryo. The distance of the mean points from the straight line representing the dorsal region expresses the amount of histological differentiation, which is maximum in the head region. (From Leonardi Cigada, M., Laria, De Bernardi, F., and Scarpetti Bolzern, A. M., *Rend. Accad. Naz. Lincei Sci. Fis. 8*, 59, 152, 1975. With permission.)

accounts for 60% of the labeled RNA until 10 hr postfertilization; the mRNA polyadenylated increases to more than 50% at gastrula stage, and afterwards it maintains this level. The percentage of nonhistone mRNA poly(A)⁻, decreasing slightly during development, is constantly in approximation to 30 to 40%. A developmental change in initiation activity may account for the translational differences. Young and Raff pointed out that newly transcribed mRNA is stored in ribonucleoprotein particles with little protein content, clearly distinguishable from the protein-rich mRNP containing maternal mRNAs and they can load the polysomes more quickly than the oogenetic mRNAs.[86] These results could support the hypothesis of a posttranscriptional control of gene expression carried out by the proteins contained in RNP.

The synthesis of ribosomal RNA seems to be repressed in the early stage of sea urchin development, as in amphibian embryos. Newly synthesized rRNA becomes identifiable only at the swimming blastula stage with slow labeling kinetics until the feeding stage.[87]

3. Fishes

The very complete research performed by Kafiani[88] on embryos of *Misgurnus fossilis* produced the greatest quantity of available data on the molecular embryology of fishes. As in amphibian and sea urchin embryos, in loach embryo the synthesis of ribosomal RNA is not detectable until the gastrula stage; heterogeneous RNA (a large part of which is presumably mRNA) is already transcribed at the late blastula stage. Moreover, in the loach at the beginning of cellular differentiation new types of RNA are synthesized with a higher degree of diversity. Most of these RNAs seem to be messengers for histone synthesis, which is detectable at late blastula stage on light polysomes associated with nuclear envelope to which the newly transcribed mRNAs may be transferred efficiently.[89] The synthesis of histones, which is different from *Xenopus* and sea urchin, seems to be coupled with DNA replication.

4. Other Organisms

The data available for RNA synthesis in other animals do not permit a careful comparison with those known for the Amphibia and sea urchin molecular embryology. Recent research performed on *Drosophila* and *Ciona intestinalis* embryos indicates that protein synthesis in early embryos proceeds on maternal templates and the expression of embryonal genoma is a later phenomenon.[90-92]

It is more and more likely that the presence of maternal mRNA synthesized during the oogenesis, and stored in mRNP particles, is a general phenomenon involving a lot of problems correlated with the stability of RNA. The only exception seems that of mammalian embryos, where striking changes in RNA and protein synthesis occur during earliest cleavage. This is undoubtedly a consequence of the peculiar process of mammalian development.

The presence of maternal information raises the problem of the control of the translation of the maternal and embryonic message and whether a choice between the two messages exists when, as in the case of histones, similar proteins are codified by both oogenetic and embryonic mRNAs. Although this control is undoubtedly a translational one, its mechanism is still a matter of discussion in the absence of experimental evidence.

Much information on the exchange of developmental interrelationships between nucleus and cytoplasm could be obtained from the unicellular alga *Acetabularia*. The morphogenetic potentialities of the anucleate fragments of this alga are greater than those of the anucleate fragments of the eggs and a close comparison between the two systems is, to a certain extent, available. The classic studies of Haemmerling on the morphology of the cap regenerated after nuclear transplantation between *A. mediterranea* and *A. crenulata* give a very exciting proof of a regeneration directed by the nucleus.[93] Further experiments in biochemical analysis following nuclear transplantation were carried out by Brachet[94] and by Schweiger and Berger.[95] Their analysis of nuclear and cytoplasmic synthesis, by using specific inhibitors in relation to the control of circadian rhythm and of temporal sequence of developmental events, laid the foundations for interesting results on these phenomena genetically determined and controlled by cytoplasmic factors.

C. Developmental Biology

1. Ribonucleic Acids

One of the most important results obtained through molecular embryology was the statement that the differentiation of embryonic cells into differentiated cells depends on the synthesis of specific mRNAs directing the synthesis of specific proteins. While the role of RNA was being clarified, another way to study the problem of the causality of organ differentiation was developed; the technique used was the treatment of various embryonic tissues (which in normal development do not differentiate in any histologically identifiable structure) with informational macromolecules, sometimes capable of directing the synthesis of "luxury" proteins. This point of view became a new research line on the problem of the induction, regarding the formation of specific structures instead of a heterogenic, primary-type induction.

Butros[96] and Benitez et al.[97] tested the effect of heterogeneous DNA and RNA on developing embryos. Niu[98] first recognized the possibility that a ribonucleoprotein extract from calf thymus could induce a sandwich explant of presumptive amphibian ectoderm to differentiate not only in the neural, heterogenetic cell type, but also in the homogenetic, highly specific cell type.

The research carried out by Ranzi et al.[99] was also attempting to show that ribonucleoproteins extracted from a variety of adult organs were able to induce a noncompetent tissue like chorio-allantois of chick embryo to differentiate in cellular elements corresponding to the donor tissue. In these experiments ribonucleoproteins (RNP) from the liver, heart, muscle, kidney, and glandular stomach of adult chick were extracted and a young allantois was filled up with these ribonucleoproteins. It was possible to obtain the formation of glandular tissue, glycogen synthesizing, in the chorio-allantois with heart ribonucleoproteins, multinucleate elements with fibrillar cytoplasm were obtained, whereas using muscle ribonucleoprotein, different elements from those induced by heart RNP were formed (the difference lies in the presence of peripheral nuclei). Sometimes myoblasts forming multinucleate fibers or cross-striated elements were obtained when using muscle RNP. With kidney ribonucleoprotein treatment, spherical, hollow structures covered with a layer of cuboidal cells and other structures, tentatively interpreted as glomeruli, were present. With glandular stomach ribonucleoprotein treatment, epithelial cells of the chorio-allantois showed secretion (Figure 19).

These experiments led to the following hypothesis: the molecules of heart proteins (as well as of liver, skeletal muscles, or kidney) are synthesized on the information given by heart ribonucleic acid or, respectively, by liver, skeletal muscle, or kidney ribonucleic acid. To test this hypothesis other experiments were carried out. A piece of albumin with chick RNP was heated for 10 min in order to denature the ribonucleoprotein before grafting it in the chorio-allantois. The amount of the induced tissue was enormously reduced and few or no cells were transformed.

Ribonucleoprotein extracted from frog heart or liver was able to induce, in chick chorio-allantois, the same kinds of tissues induced by chick heart or liver ribonucleoprotein. The organ differentiation was therefore associated with a ribonucleoprotein differentiation, making similar proteins in the same organ of the chick and of the frog.

During the last 15 years much research, most of which is described in Niu and Segal's book,[100] has accounted for the role of RNA in directing embryonic differentiation. Moreover, a lot of research has identified the messenger ribonucleic acids which can direct the synthesis of typical "luxury" proteins in differentiated tissues. The works of Buckingham and of Heywood have clarified that a 26 S RNA from muscle cells codifies for myosin synthesis in a cell-free system.[64,80]

We have tested the morphogenetic possibility of myosin mRNA in another noncompetent tissue following Niu's experiments on the induction of the heart.[101,175] Fertilized chick eggs were incubated up to primitive streak stage (st. 4 of Hamburger and Hamilton).[102] The anterior part was excised up to 0.6 mm from the node. The remaining part (postnodal piece) was cultured in Pannet Compton solution and albumen following New's method.[103] This method was also improved by replacing the albumen with the egg extract medium of Chauhan and Rao.[104] In the figure and in the table all these explants are called controls. We never observed somites or somitic masses in control explants (Figure 20). Treated postnodal pieces were, on the contrary, cultivated in the presence of RNAs and proteins listed in Table 1.

These results can be summarized in the following conclusion: 26 S mRNA and the pool of mRNA from muscle induce the somites. The action is organospecific because RNA pool

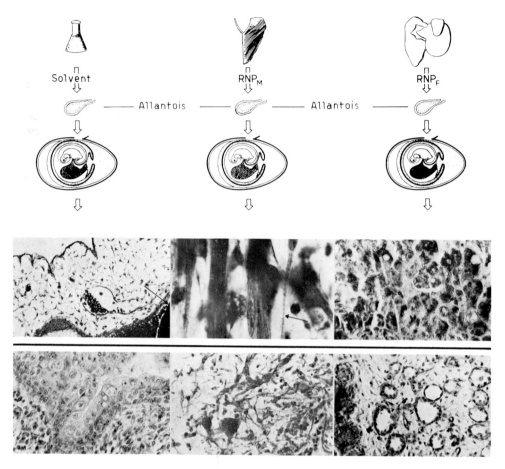

FIGURE 19. Experimental procedure of chorio-allantois grafting experiments. A young allantois, filled with ribonucleoproteins extracted from breast muscle (RNP$_M$) and from liver (RNP$_F$) of adult chick, was implanted on chorio-allantois of 9-day-old chick embryo. The resulting differentiation of chorio-allantoic membrane is shown in the pictures above. Lower row from left: action of RNP of glandular stomach, heart, and kidney.

extracted from liver or kidney does not induce the somites. The action is not related to a transcription process because the actinomycin D does not inhibit the induction.

The induction depends on a translation process; the puromycin added to mRNA inhibits the induction but, if added to the myosin, it is without effect. We can conclude that myosin mRNA may be translated to produce myosin and the myosin itself promotes the differentiation of mesoderm cells to myoblasts which aggregate into somites.

Rabbit or rat mRNA have the same inductive action. The translational possibilities of messenger extracted from a different species is already well known, e.g., the translation of rabbit globin mRNA in *Xenopus* eggs (Gurdon et al.[51]).

The heavy chain of the myosin can also induce somites. Chick myosin denatured with NaSCN or rabbit myosin is unable to induce. We can suppose that rabbit myosin is without action because of the impossibility of penetrating cells of a different zoological species.

The induction obtained with the pool of messengers from muscle raises the question of whether the actin, too, could do it. The explants cultivated in the presence of globular or polymerized actin do not show any mesodermic mass.

The conclusion that we have obtained from the induction experiments can be summarized as follows: the myosin (or perhaps the myosin heavy chain) induces the embryonic mesoderm cells to form some aggregates in the form of somitic masses. Only the homospecific myosin

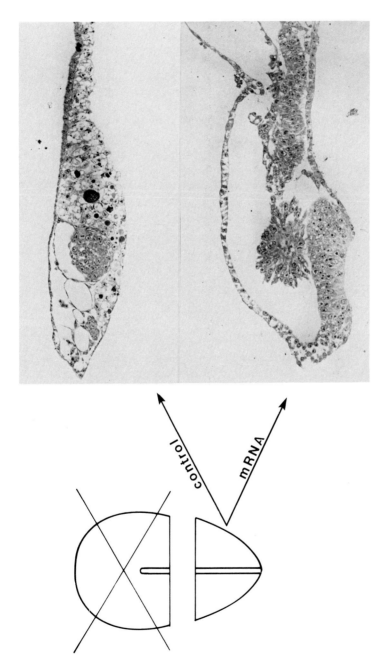

FIGURE 20. Experiment on action of 26 S mRNA or of myosin on the development of postnodal explants of chick embryo.

Table 1
SOMITE INDUCTIONS IN POSTNODAL PIECES OF CHICK EMBRYO

	Explanted embryos	Surviving explants	Inductions	Induction (%)
Controls	160	109	0	0
Myosin mRNA	157	105	42	40.0
mRNA pool: muscle	204	144	76	52.8
mRNA pool: liver	40	30	0	0
mRNA pool: kidney	42	22	0	0
Myosin mRNA + actinomycin D	64	44	9	20.5
Myosin mRNA + puromycin	30	9	0	0
Myosin mRNA ribonuclease-treated	18	12	0	0
Myosin mRNA trypsin-treated	23	15	2	13.3
Myosin mRNA: rabbit	8	6	4	66.7
Myosin mRNA: rat	30	19	2	10.5
Soluble myosin	64	51	17	33.3
Myosin + actinomycin D	6	3	2	66.6
Myosin + puromycin	9	6	3	50.0
Rabbit myosin	15	12	0	0
Heavy chain of chick myosin	34	23	8	30.4
Light chain of chick myosin	8	7	(3)	(42.8)
Myosin + NaSCN	12	7	0	0
Actin F	5	4	0	0
Actin G	18	14	0	0

can enter the cell because the antigenic properties of plasma membrane do not permit rat or rabbit myosin to penetrate the cell; consequently, rat and rabbit myosin fail to induce. The mRNA from heterospecific sources, on the contrary, can always enter the embryonic cells because its antigenic activity is very low. When mRNA is translated inside the cells, the myosin produced overcomes the membrane and it could act in modifying the surface components to a certain extent, thus promoting a specific recognition of the cells and, consequently, their aggregation following the Moscona hypothesis.[105]

As a second hypothesis the myosin itself might, instead, contribute to modifying the cell shape in elongated form. It is possible to discuss whether this "second-order induction" following the classification of Hay[106] that leads the mesodermic cells to round up in somitic masses, is due to an increased amount of cytoplasmic myosin necessary for the shape modification of the cells. These observations suggest testing to see whether the myosin is already present in the blastoderm at the primitive streak stage. We immunized rabbits against chick myosin; antiserum reacts with an extract of whole chick blastoderm at the primitive streak stage (Figure 21). Consequently, we came to the conclusion that myosin is the protein inductor of the somites.

The presence of myosin in the embryonic cells was first recognized by Ebert[107] who found heart myosin diffused in primitive streak chick embryos. Recent research carried out by Pollard et al.[108] found the myosin in a variety of nonmuscular tissues; these studies have more and more emphasized the role of cytoplasmic myosin in cellular structure and motility. The studies carried out on cultured muscular cells demonstrated that prefusion, replicating myoblasts contain a small quantity of myosin, certainly different from skeletal muscle myosin in light chains, and probably in heavy chain, too.[109] It is, moreover, well demonstrated by Merlie et al.[64] that just prior to the fusion the myosin mRNA increases its stability five to six times, and it is accumulated in the cytoplasm. The myosin mRNA transcribed in prefusion myoblast seems to be enriched in specific sequences for terminal differentiation.[110]

The additional quantity of myosin supplied to the mesoderm cells in our experiments both directly, and by translation of mRNA, could achieve the cytoplasmic intracellular level necessary to modify the cell shape in the elongated one, typical of the somitic cells.

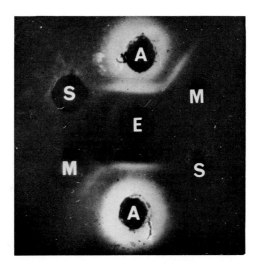

FIGURE 21. Immunodiffusion in Ouchterlony's plate of total extract of chick embryos at primitive streak stage in 0.3 M KCl, pH 6.6 (E). A, myosin antiserum; M, chick myosin; and S, solvent.

The results obtained by us in investigating the mRNAs synthesized in the dorsal region of *Xenopus* embryo also emphasize the induction obtained. In two developmental stages it was indeed shown that a greater proportion of heavy-chain myosin mRNA is detectable; the first when segmentation of somites is beginning, the second when the myotomes are differentiating. Although the second maximum undoubtedly concerns the mRNA coding for muscle myosin, it seems reasonable to connect the first maximum with the synthesis of myosin needed for the formation of the "center of somites". It should be remembered that in detecting the polyadenylate segment by [³H]poly(U) hybridization we cannot distinguish between "muscular" and "nonmuscular" myosin mRNA.

It remains to discuss how the muscular myosin heavy chain can raise the level of cytoplasmic myosin to achieve the effect of changing the form of mesodermal cells; it could be supposed that the heavy chain of cytoplasmic myosin is little different from the muscular one. The differences between the muscular and nonmuscular heavy chain of myosin, in fact, have to be clarified.

The induced somites degenerate after 24 hr and in some cases they remain as undifferentiated epithelial vesicles. This degeneration is related to the fact that the notochord and the ventral part of the neural tube, which are not present in our explants, are necessary for the further development of the somites as Packard and Jacobson have already shown for normally developed somites obtained from excised segmental plate.[111]

The influence of notochord in inducing somites was already well demonstrated in transplant experiments by Nicolet.[112]

Very useful information comes from the observations on the mechanism of somite segmentation, the best of which are made by Bellairs and co-workers.[113,114] A more modern concept recently took the place of the traditional idea that the formation of somites was the result of a specific embryonic induction by another, adjacent tissue; somites could be formed as a result of a programming of the mesoderm at an earlier stage and the role of the notochord may be to help stabilize the somites once they are formed. The recent data have shown that the cells of segmented mesoderm are more adhesive to one another than the cells derived from unsegmented mesoderm, when the adhesiveness of these cells cultured in vitro is measured by the Curtis test. It is also suggested that a higher adhesiveness of clusters of cells, supplemented by the presence of desmosomes,[115] results in the formation of a center of each somite in a fashion similar to those occurring during the neural tube closing.[116] The somite cells are long, rich in microtubules, and anchored by collagen fibers to maintain the internal tension. It seems probable that these thin collagen fibers, connecting the somitic cells to the epithelial cells of the neural tube, exert the stabilizing action observed by Packard and Jacobson.[111]

A prepattern early established before the visible segmentation, was hypothesized for amphibia somites, too, by Cooke.[117]

Some results comparable to ours were obtained also by Niu and Deshpande[118] and by Siddiqui et al.[119,120] who were able to induce a postnodal piece of chick embryo to differentiate a beating heart by treatment with progressively more purified fractions of heart mRNA.

2. Nonhistone Proteins

It is generally accepted that cell development and differentiation result from the derepression of particular sets of genes and from the translation of these gene products. Low-molecular weight RNAs involved in modulation of protein synthesis at a translational level also in cell-free systems have been described in chick embryonic cell muscle (translational control RNA),[121] in *Artemia salina* cysts and developing embryos,[122] and also in heart of chick embryo.[120]

The mechanism of selection and activation of the genes still remains unknown in its molecular detail. Both RNA and proteins were proposed as regulatory elements.[83]

More and more attention has been given in the last few years to the chromatin-associated proteins as playing a role in regulating the pattern of gene expression at a transcriptional level during the embryonic development. In spite of the recently found heterogeneity of histones during the first embryonic stage in Amphibia and sea urchin embryos, the small number of histone classes led to the conclusion that they cannot be responsible for regulating the specific activity of the genes.

The nonhistone (or acidic) protein, instead, seem to be more suitable for a regulatory role. They are heterogeneous, differ considerably in different tissues and in different species, and a great number of them are phosphorylated and can bind specifically to DNA. All these properties have supported the hypothesis that the nonhistone proteins are gene regulators.

The main research carried out on the "reconstituted" chromatin and on the changes in nonhistone protein during the embryonic development and cell differentiation was collected by Stein and Kleinsmith.[123]

An interesting experimental approach in developmental biology was carried out by Duprat et al.[124] by treating some cultured embryonic organs with homogeneous and heterogeneous nonhistone proteins. They were able to show a stimulating effect of the proteins extracted from the same tissue and an inhibitory effect of the proteins extracted from other tissues. Moreover, the nonhistone proteins show a species-specific inhibitory effect when injected in Amphibia blastulae, whereas they have no effect on embryos of different species.[125]

III. THE FIRST EMBRYONIC DETERMINATION

A. Changes in the Embryonic Determination Induced by Salt Solutions

1. Morphology

Runnström and his collaborators, working on sea urchin, showed the LiCl and other substances added in small proportion to sea water inhibit the development of rudiments in the animal half of the embryo by inducing an increased development of the entoderm formed in the vegetal half (vegetalized larvae).[126] Other substances, such as NaSCN, *o*-iodosobenzoate, $ZnCl_2$, instead, inhibit the organs formed in the vegetal half and favor the development of the animal half of the larvae by increasing the apical ciliary tuft (animalized larvae).

The vegetalization, that is, the process giving rise to vegetalized larvae, and the animal-

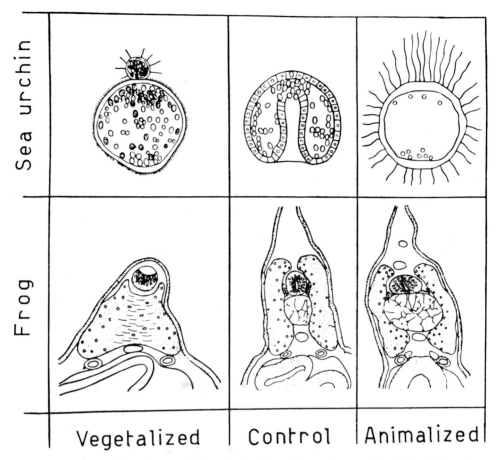

FIGURE 22. Changes induced by vegetalizing and animalizing substances in sea urchin and frog (embryos shown in cross-section).

ization that is the process giving rise to the animalized larvae, have been carefully investigated by the Swedish school. The vegetalization induces a reduction of the apical tuft and an increase of the entoderm inducing an exogastrulation, that is, a protrusion of the large entoderm at the vegetal pole, and an increase of the mesenchyme. The animalization induces an enlargement of ciliary tuft and a reduction of the entoderm and the mesenchyme up to disappearance of the archenteron and of the skeleton (Figure 22).

The vegetalizing substance LiCl induces transformation of ectoderm cells in entoderm in Amphibia, too.[127]

The action of LiCl was investigated in Amphibia and chick development; the LiCl induces a notochord reduction in both cases (Figure 22). This is, in Amphibia, a transformation of notochord presumptive cells into somite cells. A consequence of the notochord reduction is a hypoinduction of nervous rudiment. As a result cyclopic embryos,[128] and in chicks sometimes, omphalocephalic embryos, were obtained.[42]

The first question to be posed is whether LiCl operates like actinomycin D (or daunomycin) or like puromycin. The observations of Leani Collini and Ranzi[42] show large differences in the morphology between chick embryos treated with lithium (Figure 13) and embryos treated with actinomycin D (or daunomycin) or puromycin.

To check the lithium action at molecular level, *Xenopus* embryos were treated with daunomycin, puromycin, or LiCl and labeled with [¹⁴C]protein hydrolysate. The sedimentation pattern of polysomes compared with the sedimentation pattern of the control shows reduction of RNA synthesis with daunomycin, reduction of protein synthesis with puromycin, and reduction of RNA and protein synthesis in LiCl-treated embryos. The action of LiCl at molecular level is different from the action of daunomycin and puromycin just as they are different at morphological level[129] (Figure 23).

The animalizing substance, NaSCN, induces an increase of the notochord in Amphibia[130] and in the chick[131] (Figure 24). The cells for the enlargement of the notochord are taken and transformed from the presumptive somite material. It is also possible to see the appearance of notochord rudiments in the ventral half of the amphibian embryo treated with NaSCN.[132] The embryo with large notochord shows hyperinduction of the nervous system and it is also possible to obtain nervous rudiments in explanted ventral ectoderm.[133]

The actions of LiCl and NaSCN have also been investigated in *Petromyzon,* ascidians, Cephalopoda, and Gastropoda.[128]

A series of research works was carried out to investigate, at a molecular level, the changes related to the salt action on the embryonic determination.

2. Protein Synthesis is Inhibited by LiCl

The *Xenopus* embryos treated with LiCl and labeled with [^{14}C]protein hydrolysate synthesize less protein than the control.[134] The difference is evident in labeled protein and in total protein (Figure 25). LiCl inhibits protein synthesis also in sea urchin *(Lytechinus)* embryo after the cleavage.[135]

3. Breakdown of Proteins is Inhibited by LiCl

Growing *Xenopus* oocytes were labeled with tritiated leucine and afterwards, some developing eggs were treated with lithium. The labeled material disappeared early in untreated controls. The conclusion is that LiCl inhibits the protein breakdown and the utilization of protein stock[136] (Figure 26).

Protein extracted from embryo treated with LiCl is less digested by trypsin or papain than protein extracted from control embryo[33] (Figure 27).

Vegetalized sea urchin embryos are more resistant to urea and trypsin than the control embryos.[33]

Protein extracted from vegetalized embryos is more resistant to the urea-induced denaturation.[33]

Protein solution extracted from adult organisms (serum albumin, myosin, actin, etc.) or from eggs or embryos (lipovitellin, lipovitellenin or other more impure fractions such as euglobulin a, euglobulin b, euglobulin c, pseudoglobulin) is inhibited in denaturation if LiCl, or other substances with a similar effect on embryonic development, are added; i.e., the viscosity of the protein solution increases also for fibrillar proteins (myosin), the reactive groups do not increase, at ultracentrifuge subunities do not appear (Figure 37), and flow birefringence does not change.[33]

Starting from these data, we interpreted the LiCl action as being due to the fact that the stability of a protein molecule is due to the aqueous coat bound to it and the Li ion, bound to the molecules, is highly hydrating in effect. The aqueous coat protects the protein molecules from denaturation, and from the attack of the proteolytic enzymes. This coat preserves the molecules from metabolization, with a process that the old experimental embryologists called developmental inhibition, on the basis of their microscopic observations. Consequently, sea urchin vegetalization, reduction of notochord in Amphibia and chick embryo, reduction of nervous rudiments in Amphibia and chicks appeared, in our research,[33] induced by a stabilization of the protein molecules; this stabilization preserves the molecules from the denaturation.

4. The Synthesis of New Proteins is Increased by NaSCN

The radioactivity of the protein in *Xenopus* embryo, treated with NaSCN at active concentration and labeled with [^{14}C]protein hydrolysate, is stronger than that of the controls[137]

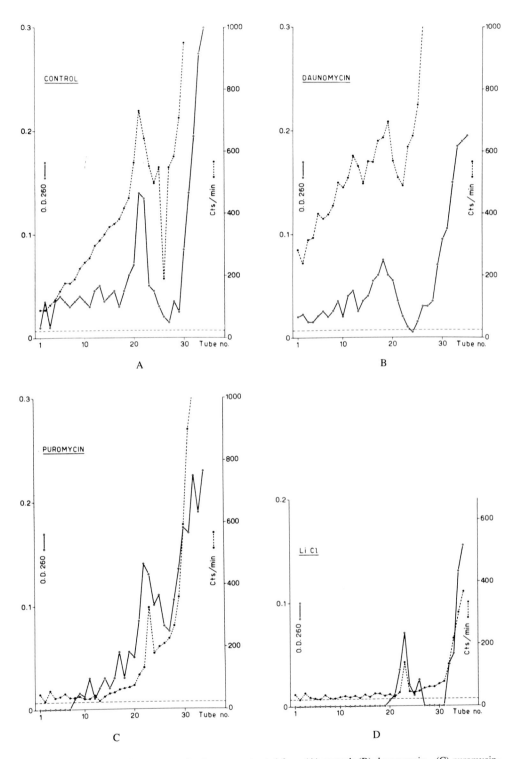

FIGURE 23. Sedimentation pattern of polysomes extracted from (A) control; (B) daunomycin-, (C) puromycin-, (D) LiCl-treated *Xenopus* embryo (labeled with 0.5 mCi of [^{14}]-labeled protein hydrolysate: linear sucrose gradient 15 to 30%). (From Leonardi Cigada, M., Maci, R., and De Bernardi, F., *Ist. Lomb. Rend. Sci. B,* 102, 213, 1968. With permission.)

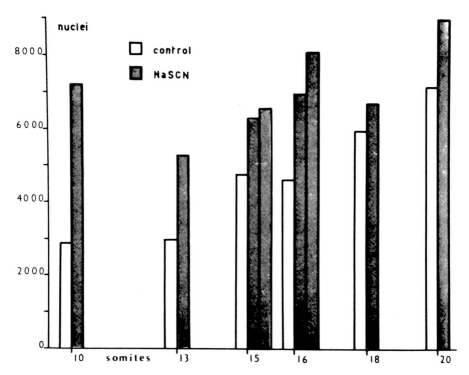

FIGURE 24.   Number of nuclei in notochord of control and NaSCN-treated chick embryos. (From Vailati, G., Vitali, P., and Ranzi, S., *Rend. Accad. Naz. Lincei Sci. Fis. 8,* 53, 594, 1972. With permission.)

(Figure 28). The same phenomenon was observed by Volm et al.[144] from *Tetrahymena* culture treated with 0.01 *M* NaSCN (Figure 30).

5. Protein Breakdown is Increased by NaSCN

Protein extracted from embryos treated with NaSCN is more easily digested by trypsin (Figure 27) or papain than the protein of the control embryos.[33]

The animalized sea urchin embryos are less resistant to the urea or trypsin action than the control embryos. Protein solutions extracted from eggs, embryos, or adult organisms, are denatured by treatment with animalizing substances. The viscosity of the solution changes; more reactive groups are detected. Some proteins such as lipovitellin and lipovitellenin show, at ultracentrifuge, the formation of subunities (Figure 37); if the protein molecule is fibrillar (myosin) the flow birefringence is lost.[33]

The proteins extracted from the embryos showing malformations induced by NaSCN are predisposed to a further denaturation induced by urea.[33]

The conclusion that animalization is related to a protein denaturation suggests testing the action of urea on embryonic development.

Urea, the classic protein denaturing substance, induces animalization if present in seawater where sea urchin eggs are developing.[138] Urea also induces the development of segments of notochord and nervous rudiments in the ventral half of young gastrula of Amphibia.[139] In the ventral half of gastrula, presumptive notochord and nervous system are absent, and these rudiments do not appear in the controls reared in Holtfreter solution.

6. RNA Synthesis is Inhibited by LiCl

Xenopus embryos treated with LiCl at active concentrations and labeled with $^{14}CO_2$ synthesize less RNA than the control. The difference is evident in labeled RNA and in total RNA (Figure 29). These data of Leonardi Cigada et al.[140] are in agreement with Thomason:[141]

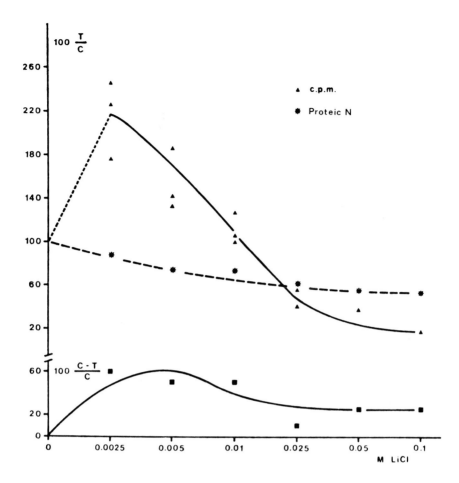

FIGURE 25. Total proteins (broken line) and incorporation of [¹⁴C]-labeled protein hydrolysate (upper unbroken line) in *Xenopus* embryo treated with increasing concentrations of LiCl. The 100 value corresponds to the control. Below, digestion with trypsin: the hydrolysis inhibition induced by LiCl in embryonic protein. (From Leonardi Cigada, M., Laria De Bernardi, F., and Bolzern, A. M., *Rend. Accad. Naz. Lincei Sci. Fis. 8*, 52, 93, 1972. With permission.)

radioactive phosphate incorporation is less active in Li-treated *Xenopus* embryos than in the controls.

In chick embryos treated with LiCl, the incorporation of tritiated uridine was studied with the autoradiographic method. It is thus possible to see the RNA synthesis. In notochord and somites of Li-treated embryos, the incorporation was lower than in the controls.[142] The incorporation in the nervous system rudiment was greater if referred to $100/\mu^2$ than in the controls, but the nervous rudiment of these embryos was smaller than in the controls and so were the cells. Consequently, it is possible to arrive at the same conclusion as Flickinger et al.[143] regarding amphibian embryos — that cells treated with LiCl synthesize less RNA than the control cells. The same phenomenon was observed by Volm et al.[144] for *Tetrahymena* culture treated with 0.01 *M* LiCl (Figure 30). Vegetalized sea urchin larvae synthesize less RNA than the control larvae.[135]

7. RNA Breakdown is Inhibited by LiCl

Yeast RNA treated with LiCl is more resistant to the action of ribonuclease than the control RNA (untreated with LiCl)[140] (Figure 31).

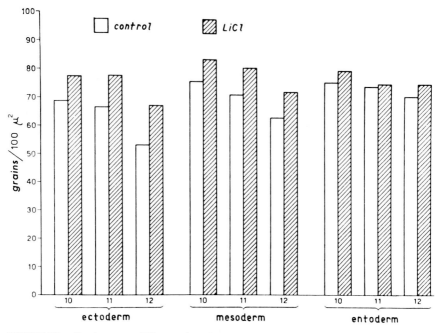

FIGURE 26. Ovarian eggs of *Xenopus* have been labeled with tritiated leucine. Grain numbers in different germ layers: 10, first invagination of dorsal lip; 11, gastrulation; and 12, yolk plug. (From Ranzi, S. and Vailati, G., *Rend. Accad. Naz. Lincei Sci. Fis. 8,* 50, 473, 1971. With permission.)

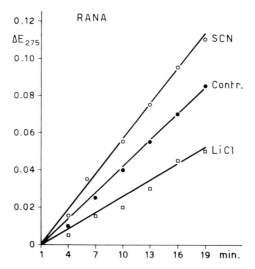

FIGURE 27. Proteins of *Rana esculenta* embryos, treated with NaSCN, LiCl, or untreated (controls) have been digested with trypsin and then precipitated with trichloroacetic acid. The supernatant was read by the spectrophotometer. Abscissa represent the time of trypsin digestion; the values at time 1 and subsequent readings are shown on the ordinate.

8. Embryo Treated with NaSCN Accumulates Neosynthesized RNA

Xenopus embryos treated with NaSCN at active concentration and labeled with $^{14}CO_2$ show a more elevated radioactivity in the RNA than the controls[137] (Figure 32).

Bäckström showed in animalized sea urchin larvae RNA values slightly higher than in

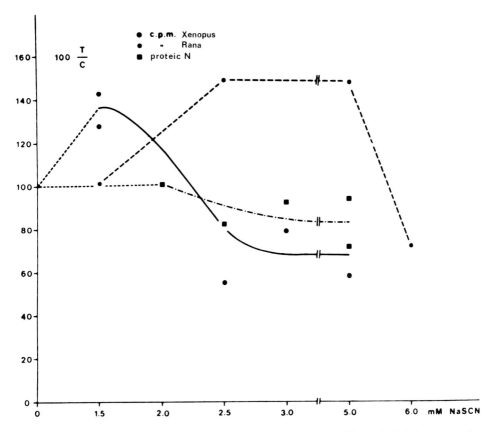

FIGURE 28. Amphibian embryos treated with NaSCN incorporate more [^{14}C]protein hydrolysate than the untreated controls even though the total protein N decreases in treated embryos (the values are expressed in percent of the controls). (From Leonardi Cigada, M., Laria De Bernardi, F., and Scarpetti Bolzern, A. M., *Rend. Accad. Naz. Lincei Sci. Fis. 8,* 55, 755, 1973. With permission.)

the normal larvae.[145] Flickinger et al.[143] and Volm et al.[144] were able to observe an increase of RNA synthesis induced by NaHCO$_3$ or NaSCN in Amphibia or *Tetrahymena* cells.

9. RNA Breakdown is Increased by NaSCN

RNA extracted from yeast and treated with NaSCN is less resistant than the RNA control to ribonuclease digestion[137] (Figure 33).

B. Embryonic Determination: Protein Denaturation and Stabilization

1. Embryonic Determination

All this evidence seems to show that animalization and vegetalization originate from the enlargement or inhibition of the areas in which a process of protein denaturation, necessary for new protein synthesis, is taking place.

If we refer the above to normal development, we can put forward a tentative interpretation. Preexisting protein molecules are metabolized to obtain the protein synthesis related to the formation of sea urchin ectoderm rudiments of the animal pole as well as to the formation of Amphibia and chick notochord and neural tube. This protein breakdown follows a denaturation process. If we add, for example, NaSCN or *o*-iodosobenzoic acid, this denaturation process is enhanced. Consequently, the rudiments in question are larger. If the denaturation is inhibited by an Li ion with its aqueous coat, the development of the rudiments is reduced, and they eventually disappear (loss of ciliar tuft in vegetalized sea urchin, transformation of notochord cells into somite cells and of ectoderm cells into the entoderm in Amphibia).

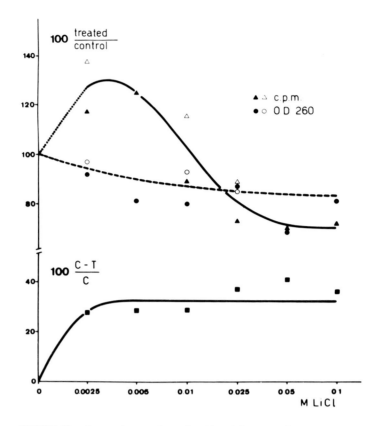

FIGURE 29.   Spectrophotometric reading (dotted line) and [^{14}C] incorporation (upper unbroken line) in total RNA from *Xenopus* embryos treated with various LiCl concentrations. The black symbols denote a 6-hr treatment, the light ones a 20-hr treatment. The value of 100 has been given to the untreated control. Below, yeast digestion by ribonuclease: inhibition by LiCl. (From Leonardi Cigada, M., Laria De Bernardi, F., and Scarpetti Bolzern, A. M., *Rend. Accad. Naz. Lincei Sci. Fis. 8*, 55, 755, 1973. With permission.)

2. Sea Urchins

Fertilization by activating certain enzymic systems according to the data of the school of Runnström[146] produces a higher metabolism at the animal pole where there are more easily digested proteins.

The regions hyperdeveloped by the action of animalizing agents and inhibited by vegetalizing agents show a comparatively high oxidation-reduction potential (see the works of Child,[147] Ranzi,[148] and Hörstadius[149]). We think that this may be interpreted as an indication of a much more active protein breakdown. It is possible to follow the formation of the ciliary tuft and of mitochondria in the cells of the animal pole of the sea urchin embryo,[150] which shows a higher oxidation-reduction potential.

On the other hand, animalization seems in fact to correspond to the breakdown of preexisting protein structures. Proteolytic enzymes (chymotrypsin, ficin, trypsin) are animalizing agents.[151]

Animalization can sometimes be induced under the same conditions as protein denaturation. Hörstadius[152] found that animalization occurs more easily at low temperature, while Jacobsen and Korsgaard Christensen[153] found that denaturation induced by urea is also easier to obtain at low temperatures. Animalization, induced by NaSCN, occurs more easily in calcium-free seawater; Ca prevents many proteins from undergoing denaturation.[154]

The findings of Kavanau[155] also seem important. He found that the content of free amino

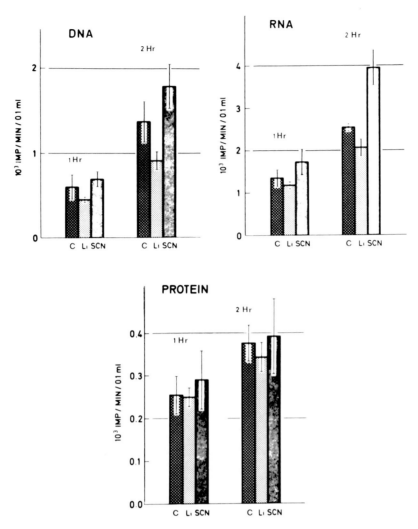

FIGURE 30.   Action of 0.01 *M* lithium and 0.01 *M* thyocyanate on incorporation of [³H]thymidine (DNA), of [³H]uridine (RNA), and [³H]leucine (protein) in *Tetrahymena piriformis:* 1 and 2 hr of experiment. (From Volm, M., Wayss, K., and Schwartz, V., *Roux's Arch.,* 165, 125, 1970. With permission.)

acids was highest in sea urchin embryos exactly at the stage which Bäckström and Gustafson[156] found to be most sensitive to the action of LiCl.

Bearing in mind the fact that several metabolites are animalizing substances, the activation of the enzymes and subsequent formation of metabolites occurring at the animal pole seem to lead to the development of animal hemisphere, while the vegetative pole is stabilized because of failure to form these metabolites. Some regulative processes come into action, e.g., pyruvate in low concentration animalizes and in high concentration vegetalizes. We can tentatively recognize here a phenomenon acting in normal development. Small amounts of pyruvate formed during animal development lead to protein denaturation and increase in the animal structures. After a certain stage in animal development the amounts of pyruvate formed are in excess of that required and vegetal structures are induced.

Feedbacks of this kind may be the basis of the animal-vegetative balance. This balance is changed if artificial animalizing substances such as NaSCN or vegetalizing substances such as LiCl are allowed to act (Figure 34).

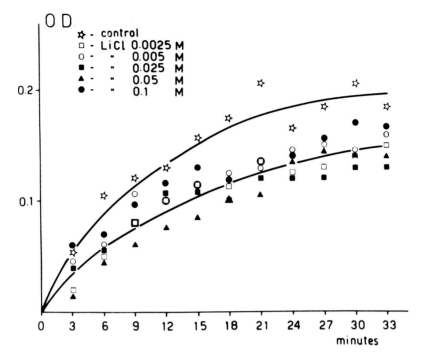

FIGURE 31.   Enzymatic hydrolysis of yeast RNA treated with various LiCl concentrations. Unbroken line, control; dotted line, treated RNA. (From Leonardi Cigada, M., Laria De Bernardi, F., and Scarpetti Bolzern, A. M., *Rend. Accad. Naz. Lincei Sci. Fis. 8,* 55, 755, 1973. With permission.)

3. Amphibia

Animalization and vegetalization in radial symmetric eggs of sea urchins are acting following the animal vegetal axis. The animalization of Amphibia embryo was not observed following the animal vegetal axis, but we must remember that this axis disappears, like a symmetry axis, at the fertilization stage when ventral and dorsal faces of the embryo are determined by the sperm entrance. The formation of the grey crescent, which appears after fertilization, is related to movements of the egg cortex.[157] These movements induce a local center of high metabolism shown by Child[158] which is dependent on the maintenance of a high respiratory rate.[159] This center seems to correspond from a physiological point of view to the animal pole of the sea urchin, and we can imagine it as a center of protein denaturation. The axis, on which the denaturating and stabilizing processes now act, is the dorsoventral axis that corresponds to a ribosome gradient[157] and allows the change of determination between notochord and somites.[130] Direct analyses can also be used to demonstrate protein breakdown. In the frog blastoporal dorsal lip, which is the notochord rudiment, at the time when it is sensitive to animalizing and vegetalizing agents, Deuchar[159] found a higher content of free amino acids than in other areas of the embryo. D'Amelio and Ceas[160] showed that in the same region of the young gastrula there is stronger proteolytic activity. Consequently, notochord determination and neural induction are the effect of the protein denaturation. The experiments leading to this conclusion are summarized in Figure 35, which shows that protein denaturation produces notochord determination or nervous system induction and protein stabilization induces transformation of notochord rudiments into somites.

4. Chick

The research work of Kochav and Eyal-Giladi[161] seems to indicate the possibility of movement of some materials in the cephalic direction induced by egg symmetrization in the

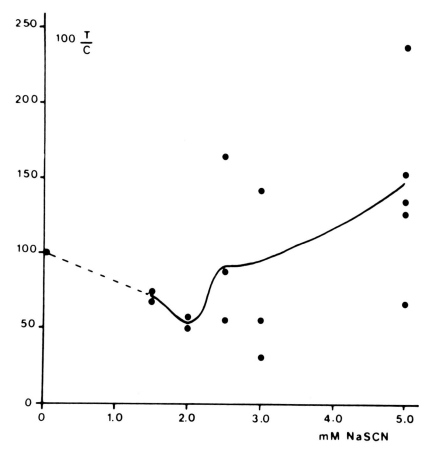

FIGURE 32.   Incorporation of labeled uridine in RNA of *Xenopus* embryos treated with NaSCN. The values are expressed in percent of the controls. (From Leonardi Cigada, M., Laria De Bernardi, F., and Bolzern, A. M., *Rend. Accad. Naz. Lincei Sci. Fis. 8,* 52, 93, 1972. With permission.)

body of the mother. This process can be the equivalent of the amphibian cortex movement related to fertilization. The development of the notochord is related also in chick to a protein denaturation, the notochord increases in dimension in embryos treated with NaSCN,[131] and a small notochord indeed is present after treatment with LiCl.[162]

5. Induction Process in Vertebrata

Neural induction in Vertebrata is related to a process of protein denaturation. NaSCN, at a concentration that can denature the proteins, induces prosencephalic structures from the ventral half of axolotl gastrulae. Holtfreter obtained nervous differentiation from presumptive ectoderm explants of *Triturus* by treating them with solutions whose pH values were lower than 5.0 or higher than 9.2. The explants of presumptive ectoderm develop into epidermis between the two pH values. Embryonic protein extract is denatured at pH value lower than 5.7 or higher than 8.7.[33]

Mineral components of amphibian embryo decrease during the first developmental stages.[163] Na^+, Ca^{2+}, and Mg^{2+} ions are pumped out from the ectoderm cells in the blastocoel, following Barth and Barth. The ectoderm above the archenteron roof pumps out Na^+, Ca^{2+}, and Mg^{2+}, but the presumptive neural plate and archenteron roof are tightly adherent at the end of gastrulation; consequently, the cells of the presumptive neural plate accumulate the ions up to an adequate concentration for the induction of the neural plate. If this hypothesis of Barth and Barth[164] is true, the denaturation induced by ions accumulation is the second step of the induction process.

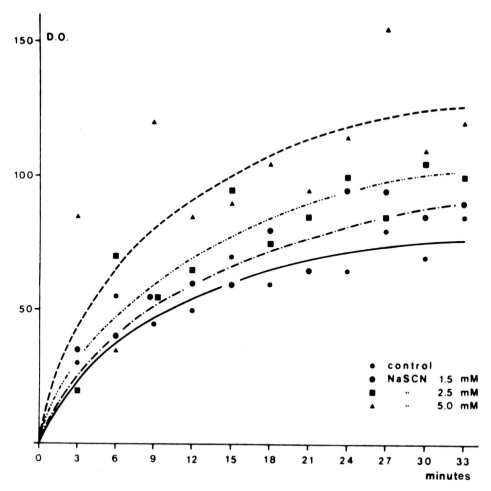

FIGURE 33.   RNase hydrolysis of yeast RNA in presence of NaSCN. Abscissa and ordinata as in Figure 25. (From Leonardi Cigada, M., Laria De Bernardi, F., and Bolzern, A. M., *Rend. Accad. Naz. Lincei Sci. Fis. 8,* 52, 93, 1972. With permission.)

Also, the somite induction is clarified by data presented here. A 26 S mRNA codifying for the myosin appears in amphibian embryo during gastrulation and reaches its maximum at the end of the gastrulation when myosin is present in the embryo (Figure 3). The 26 S mRNA induces the somites in a comparable stage of the chick through a translation into myosin. Consequently, we can conclude that the Spemann organizer promotes the transcription of myosin gene, and the 26 S RNA messenger of myosin is translated in myosin that induces somite differentiation (Figure 36).

6. Molecular Biology of the Embryonic Determination

The question is whether the denaturation and the protein stabilization acting on embryonic determination concern cytoplasmic or nuclear proteins.

The action on the cytoplasm can occur in two processes. The denaturation can put at the disposal of synthetic processes a pool of amino acids allowing more intensive synthesis; the protein stabilization can inhibit the same processes. This interpretation by Ranzi is corroborated by the observation of Moser and Flickinger. These authors observed nervous system induction in ventral explants of *Rana pipiens* cultivated in a rich nutritional medium, containing horse blood serum, vitamins, amino acids, and glutamine.[174]

The observation made by Lindahl et al.[165] in this field is highly significant. From a

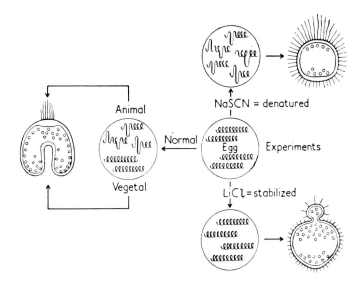

FIGURE 34. Tentative interpretation of the behavior of protein molecules, represented by helixes, during normal development of sea urchin and under experimental conditions. (From Ranzi, S., *Adv. Morphog.*, 2, 211, 1962. With permission.)

population of more easily animalizable sea urchin eggs it is possible to extract, in 0.54 *M* NaI, a greater amount of soluble N (that is to say they contain a lower proportion of insoluble N).

The second process may be the following: the denaturation liberates small components from larger molecules and these components act on DNA as activator of gene transcription[33] (Figure 37). The protein stabilization inhibits the liberation of these active components. Ventral blastopore lip is an inductor of nervous rudiment, if cooked, and Kuusi[166] was able to observe an increase in inducing activity of a protein extract of organizing substances previously denatured by urea.

The action on nuclear protein stimulates or inhibits the gene transcription, denaturing or stabilizing the histones or other proteins that block the transcription. Kaji and Kaji[167] were able to show an inhibition induced by LiCl on interaction between tRNA and polysomes.[129]

The data in hand seem to indicate that protein denaturation and stabilization are acting in enhancing the three kinds of processes: mobilization of small nutritive molecules; liberation of small molecular fragments that can be active in inducing gene transcription; and denaturation of histones and/or other proteins that block the genes.

IV. PHYLOGENETIC QUESTIONS

Haeckel[168] has put forward on a developmental basis the assumption stated by previous authors[169,170] according to which, in an embryo, first the characters of the larger systematic group (today we would say *phylum*) appear and then the less and less comprehensive group characters (today we would say classes and further down till the species). Ontogenesis is a short and rapid recapitulation of phylogenesis. Today, without taking any merit away from Haeckel, everyone is persuaded that the statement should be changed in the sense that the resemblance is not between embryos and ancestors at the adult stage, but between embryos and embryos of other more primitive species. However, substituting for the resemblance between embryos and ancestors, that between embryos and ancestor embryos, the problem stated by Haeckel remains. The question arises spontaneously, whether the recent discoveries in developmental biology can clarify this problem.

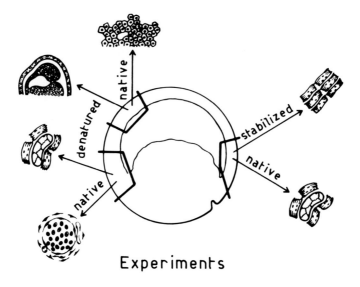

Experiments

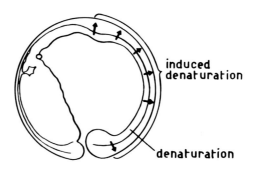

Interpretation

FIGURE 35.   Effects of different experimental conditions on explants of amphibian embryos and tentative interpretations of the determination processes. (From Ranzi, S., *Adv. Morphog.,* 2, 211, 1962. With permission.)

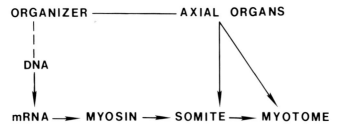

FIGURE 36.   Experimental steps of somite induction in vertebrate embryos. The induction process follows this sequence: Spemann's organizer — myosin gene — mRNA — myosin — somite — differentiation of somite in myotome.

Different hereditary characteristics appear in subsequent stages of development. Lethal characteristics appearing in different stages of *Drosophila* development are well known; they are developmental alterations killing at well-established stages of embryonic or larval de-

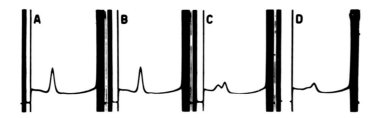

FIGURE 37. Sedimentation patterns of hen egg lipovitellin incubated with different substances. Spinco ultracentrifuge experiments at 59,780 rpm. Centrifugation period shown in brackets. (A) control (45 min); (B) incubated with LiCi (42 min); (C) incubated with NaSCN (39 min); and (D) incubated with *o*-iodosobenzoic acid (45 min).

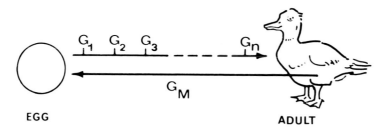

FIGURE 38. Characters corresponding to the different genes (G_1-$G\infty$) appear in the phenotype at subsequent stages of development. Genes ($G\infty$) induce the maternal organism to synthesize substances for the eggs and the embryos. (From Ranzi, S., *Ist. Lomb. Rend. Sci. B,* 108, 182, 1974. With permission.)

velopment or at the metamorphosis. Cleavage, blastoderm formation, gastrulation, germ band extension, and segmentation of the mesoderm are blocked by lethal genes acting at these stages.[171]

The appearance of the different proteins in the embryonic and larval development of the Amphibia and sea urchins presented here causes the appearance of the phenotypes corresponding to the action of the different genes. This appearance occurs in different stages of development and it is the action of genes subjected to the action of natural selection.

Different groups of genes G_1, G_2, G_3 . . . $G\infty$ which act in successive developmental stages can therefore be recognized. This means that in a first developmental period, the embryo will manifest in its phenotype the action G_1, later $G_1 + G_2$, and afterwards $G_1 + G_2 + G_3$ and at the end of development the adult will have a phenotype which is the effect of the action of all the genes $G_1 + G_2 + . . . G\infty$ (Figure 38).

The mother, however, synthesizes substances which are carried to the oocytes and in the case of gestation are carried to the fetus. As far as oogenesis is concerned, among the many examples ovoverdin and hemocyanin should be remembered. These substances in the blood of the shrimp pass into the eggs. Vitellogenin synthesized in liver enters eggs of various Vertebrata to form a yolk.

Moreover, it is now demonstrated that some quantity of ribosomes, protein and mRNA synthesized during the oogenesis onto maternal genes is stored in the oocytes of most animals.

It is clear that G_M genes exist, the translation of which leads to the synthesis of these substances. These genes in the case of vertebrates enter in function in the liver cells where vitellogenin is secreted.

Evolution is the result of gene mutation, and on a first examination of the question we can admit that all genes have an identical mutation frequency. On the basis of this assumption, in an early stage only a few genes can mutate (G_M and G_1 only). A successive stage will

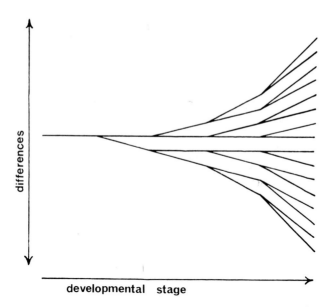

FIGURE 39.   Mutant genes act in subsequent stages of development, and, as a consequence, there are greater differences in the older stages. (From Ranzi, S., *Ist. Lomb. Rend. Sci. B,* 108, 182, 1974. With permission.)

present the possibility of the mutations of a greater number of genes (those of groups G_M, G_1, and G_2) while at the end of development all mutated genes are acting. However, the mutations in effect which occur in earlier stages prejudice the life of the individual more easily as they interact with all the processes as they occur during development. The mutations compatible with the survival of the individual will therefore be fewer for G_M and G_1 than for G_2.

Research on developmental biology, therefore, has shown that the different genes manifest their action on the phenotype during oogenesis or in different moments of development, i.e., in different ages of the egg, of the embryos, of the larvae and young, and of the adult.[173]

Consequently, the embryonic development of the different animals can be presented as a trunk in which the branches, as they are formed, always move further away (Figure 39). The trunk represents the beginning of development, and the branches correspond to the appearance of groups of mutant genes with a progressively later action.

The greater affinity of embryo forms against those of adults, according to Haeckel's statement, has its basis in mutations which are the basis of evolution. The palingenetic characters of the embryo (those that reflect ancestor characteristics) are induced by genes which have not mutated, or the mutation of which have little effect on the embryo phenotype. The cenogenetic characters, which do not reflect ancestor characteristics, can have two origins. Sometimes they appear because of gene mutations, which can bring about phenotypical differences, which are corrected by genes acting in successive stages; such is the origin of larvae, the organization of which is very different from that of the ancestors. On other occasions, the mutations inducing cenogenetic characters can preexist at the beginning of development because they are due to the action of G_M genes. These mutations are represented by structure variations of the egg, but the G_1 genes, which are active at the beginning of development, reproduce the original phenotype in these changed eggs. This is the case of the primitive streak which can be observed in the embryos of birds and mammals, notwithstanding the great difference between the eggs.

REFERENCES

1. **Guidice, G.**, *Developmental Biology of Sea Urchin Embryo,* Academic Press, New York, 1973.
2. **Bier, K.**, Insect oogenesis with special reference to nuclear structure and function, *Acc. Lincei Problemi Sc.,* 104, 61, 1968.
3. **Lanzavecchia, G.**, Problems in the yolk formation in Amphibia, *Acc. Lincei Problemi Sc.,* 104, 101, 1968.
4. **Wallace, R. A. and Dumont, J. N.**, The induced synthesis and transport of yolk proteins and their accumulation by the oocyte in *Xenopus laevis, J. Cell. Physiol.,* 72(Suppl.), 73, 1968.
5. **Wallace, R. A. and Bergink, E. W.**, Amphibian vitellogenein: properties, hormonal regulation of hepatic synthesis and ovarian uptake to yolk proteins, *Am. Zool.,* 14, 1159, 1974.
6. **Parisi, V., Ambrosoli Mognoni, G., and Sozzi, E.**, Studio preliminare delle emocianine estratte da embrioni di *Sepia officinalis* e *Austropotamobius pallipes, Rend. Accad. Naz. Lincei Sci. Fis. 8,* 37, 493, 1964.
7. **Parisi, V., Scagnetti, S., Bozzolati, M., and Faglioni, M.**, Ricerche sulle proteine dell'emolinfa e dell'uovo di *Austropotamobius pallipes* (Ler.), *Rend. Accad. Lincei Sci. Fis. 8,* 46, 209, 1969.
8. **Woodland, H. R.**, The modification of stored histones H3 and H4 during the oogenesis and early development of *Xenopus laevis, Dev. Biol.,* 68, 360, 1979.
9. **Flynn, J. M. and Woodland, H. R.**, The synthesis of histone Hl during early amphibian development, *Dev. Biol.,* 75, 222, 1980.
10. **Needham, J.**, *Chemical Embryology,* Cambridge University Press, London, 1931.
11. **Mirsky, A.**, Protein coagulation as a result of fertilization, *Science,* 84, 333, 1936.
12. **Runnström, J.**, Über die Vandärung der Plasmakolloide bei der Entwicklungserregung des Seeigeleies II, *Protoplasma,* 5, 203, 1928.
13. **Leonardid Cigada, M.**, Le proteine struttura nello sviluppo del *Bombyx mori* (baco da seta), *Ist. Lomb. Rend. Sci.,* 90, 573, 1956.
14. **Lundblad, G.**, *Proteolytic Activity in Sea Urchin Gametes,* Almqvist & Wiksell, Uppsala, 1954.
15. **Ranzi, S., Citterio, P., and Samuelli, C.**, Proteine e antigeni durante l'accrescimento embrionale e larvale di due specie di rospi (*Bufo bufo* e *Bufo viridis*) e dei loro incroci, *Acta Embryol. Morphol. Exp.,* 3, 65, 1960.
16. **Rugh, R.**, *Experimental Embryology,* Burgess, Minneapolis, 1952.
17. **Rakotorivony, J. and Gasser, F.**, Analyse electrophoretique en gel de polyacrylamide des proteines solubles au cours du developpement embryonnaire de l'amphibien urodele Pleurodeles Waltlii Michar, *Ann. Embryol. Morphol.,* 6, 229, 1973.
18. **Ecker, R. E. and Smith, L. D.**, The nature and fate of *Rana pipiens* proteins synthesized during maturation and early cleavage, *Dev. Biol.,* 24, 559, 1971.
19. **Spiegel, M., Spiegel, E. S., and Melzer, P. S.**, Quantitative changes in the basic protein fraction of developing embryos, *Dev. Biol.,* 21, 73, 1970.
20. **Denis, H.**, Recherches sur la differentiation proteique au cours du developpement des amphibiens, *J. Embryol. Exp. Morphol.,* 9, 422, 1961.
21. **Clayton, M. R.**, Distribution of antigens in developing newt embryo, *J. Embryol. Exp. Morphol.,* 1, 25, 1953.
22. **Ballantine, J. E. M., Woodland, H. R., and Sturgess, E. A.**, Changes in protein synthesis during the development of *Xenopus laevis, J. Embryol. Exp. Morphol.,* 51, 137, 1979.
23. **Moen, T. L. and Namenwirth, M.**, The distribution of soluble proteins along the animal-vegetal axis of frog eggs, *Dev. Biol.,* 58, 1, 1977.
24. **Gallien, C. L., Aimar, C., and Guillet, F.**, Nucleocytoplasmic interactions during ontogenesis of individuals obtained by intra- and interspecific nuclear transplantation in the genus Pleudeles (Urodele Amphibian), *Dev. Biol.,* 33, 154, 1973.
25. **D'Amelio, V. and Salvo, A. M.**, Further studies on the embryonic chick hemoglobin, *Acta Embryol. Morphol. Exp.,* 4, 250, 1961.
26. **D'Amelio, V.**, The globins of adult and embryonic chick hemoglobin, *Biochim. Biophys. Acta,* 127, 59, 1966.
27. **Baglioni, C.**, Ontogenesis of erythrocytes and hemoglobin formation, *J. Cell. Physiol.,* 67(Suppl. 1), 169, 1966.
28. **Markert, C. L. and Appella, E.**, Physiochemical nature of isozymes, *Ann. N.Y. Acad. Sci.,* 94, 678, 1961.
29. **Claycomb, W. C. and Villee, C. A.**, Lactate dehydrogenase isozymes of *Xenopus laevis:* factors affecting their appearance during early development, *Dev. Biol.,* 24, 413, 1971.
30. **Bedate, M. A. and Torrellas, A.**, Isozymes of lactic dehydrogenase in *Discoglossus pictus, Acta Embryol. Exp.,* 1973, 313, 1973.
31. **Borner, P. and Chen, P. S.**, Isozymmuster und Enzymaktivität der Lactat und Alkoholdehydrogenase der Letalmutante 1(3) tr und des wildtyps von *Drosophila melanogaster, Rev. Suisse Zool.,* 82, 667, 1975.

32. **Borner, P.**, *Eine vergleichende analyse des Isozymmusters und der Enzymactivität zwischen der Letalmutante 1(3)tr und der wildtyps von Drosophila melanogaster,* Inaug. Diss., Zurich, 1974.

33. **Ranzi, S.**, The proteins in normal and larval development, *Adv. Morphog.*, 2, 211, 1962.

34. **Spiegel, M., Ozaki, H., and Tyler, A.**, Electrophoretic examination of soluble proteins synthesized in early sea urchin development, *Biochem. Biophys. Res. Commun.*, 21, 135, 1965.

35. **Moore, R. O. and Villee, C. A.**, Malate dehydrogenase: multiple forms in separated blastomeres of sea urchin embryos, *Science*, 142, 389, 1963.

36. **Ranzi, S.**, Les changement proteique au cours du developpement embryonnaire et larvaire, *Ann. Biol.*, 33, 523, 1957.

37. **McClay, D. R.**, Surface antigens involved in interactions of embryonic sea urchin cells, *Curr. Top. Dev. Biol.*, 13, 1, 199, 1979.

38. **Westin, M., Perlmann, H., and Perlmann, P.**, Immunological studies of protein synthesis during sea urchin development, *J. Exp. Zool.*, 166, 331, 1967.

39. **Ghiretti Mogaldi, A., Milanesi, C., and Tognon, G.**, Hemopoiesis in crustacea decapoda: origin and evolution of hemocytes and cyanocytes of *Carcinus maenas, Cell Diff.*, 6, 167, 1977.

40. **Ranzi, S.**, Il ricambio minerale dell'uovo durant lo sviluppo della *Sepia officinalis, Arch. Exp. Zellforsch.*, 22, 557, 1939.

41. **Leonardi Cigada, M.**, Ancora sulle varizioni di alcune frazioni proteiche nel corso dello sviluppo del riccio di mare e della rana, *Ist. Lomb. Rend. Sci. B,* 92, 453, 1958.

42. **Leani Collini, R. and Ranzi, S.**, Effetto di actinomicina D, daunomiciana, puromicina e litio cloruro sui primi stadi dello sviluppo embrionale del pollo, *Mem. Accad. Lincei Sci. Fis. 8,* 8(3), 47, 1967.

43. **Olivo, O. M.**, Sulla precoce determinazione del materiale destinato a formare il cuore dell'embrione, *Boll. Soc. Ital. Biol. Sper.*, 3, 18, 1928.

44. **Lanzani Maci, R.**, Sintesi di acido ribonucleico durante la neurulazione dell'embrione di pollo, *Rend. Accad. Naz. Lincei Sci. Fis. 8,* 59, 559, 1975.

45. **Denis, H.**, *Activite des Genes au Cours du Developpement Embryonnaire,* Dosoer, Liege, 1966.

46. **Gallera, J.**, L'action de l'actinomycine D sur le pouvoir inducteur du noeud de Hensen et la competence neurogene de l'ectoblaste de poulet, *J. Embryol. Exp. Morphol.*, 23, 473, 1970.

47. **Barros, C., Hand, G. S., Jr., and Monroy, A.**, Control of gastrulation in the starfish, *Asterias forbesii, Exp. Cell. Res.*, 43, 167, 1966.

48. **Runnström, J. and Markman, B.**, Gene dependency of vegetalization in sea urchin embryos treated with lithium, *Biol. Bull.*, 130, 402, 1966.

49. **Leone, V.**, Effetti del trattamento con ribonucleasi sullo sviluppo embrionale di *Arbacia punctulata* lam., *Acta Embryol. Morphol. Exp.*, 3, 131, 1960.

50. **Brown, D. D. and Gurdon, J. B.**, Absence of ribosomal synthesis in the anucleolate mutant of *Xenopus laevis, Proc. Natl. Acad. Sci. U.S.A.*, 51, 139, 1964.

51. **Gurdon, J. B., Woodland, H. R. and Lingrel, J. B.**, The translation of mammalian globin mRNA injected into fertilized eggs of *Xenopus laevis.* I. Message stability in development, *Dev. Biol.*, 39, 125, 1974.

52. **Crippa, M., Davidson, E. H., and Mirsky, A. E.**, Persistence in early amphibian embryos of informational RNAs from the lamp brush chromosome stage of oogenesis, *Proc. Natl. Acad. Sci. U.S.A.*, 57, 885, 1967.

53. **Rosbash, M. and Ford, P. J.**, Polyadenylic acid containing RNA in *Xenopus laevis* oocytes, *J. Mol. Biol.*, 85, 87, 1974.

54. **Darnbrough, C. and Ford, P. J.**, Cell-free translation of messenger RNA from oocytes of *Xenopus laevis, Dev. Biol.*, 50, 285, 1976.

55. **Ford, P. G., Mathieson, T., and Rosbash, M.**, Very long-lived messenger RNA in ovaries of *Xenopus laevis, Dev. Biol.*, 57, 417, 1977.

56. **Dolecki, G. J. and Smith, D. L.**, Poly(A)$^+$ RNA metabolism during oogenesis in *Xenopus laevis, Dev. Biol.*, 69, 217, 1979.

57. **Perlman, S. and Rosbash, M.**, Analysis of *Xenopus laevis* ovary and somatic cell polyadenylated RNA by molecular hybridization, *Dev. Biol.*, 63, 197, 1978.

58. **Brachet, J., Denis, H., and De Vitry, F.**, The effect of actinomycin D and puromycin on morphogenesis in amphibian eggs and *Acetabularia mediterranea, Dev. Biol.*, 9, 398, 1964.

59. **Brown, D. D. and Littna, E.**, Synthesis and accumulation of DNA-like RNA during embryogenesis of *Xenopus laevis, J. Mol. Biol.*, 20, 81, 1966.

60. **Brown, D. D. and Gurdon, J. B.**, Size distribution and stability of DNA-like RNA synthesized during development of anucleolate embryos of *Xenopus laevis, J. Mol. Biol.*, 19, 399, 1966.

61. **Brown, D. D. and Littna, E.**, Synthesis and accumulation of low molecular weight RNA during embryogenesis of *Xenopus laevis, J. Mol. Biol.*, 20, 95, 1966.

62. **Denis, H.**, Role of messenger ribonucleic acid in embryonic development, *Adv. Morphog.*, 7, 115, 1968.

63. **Leonardi Cigada, M., Maci, R., and De Bernardi, F.**, Azione a livello molecolare di alcune sostanze che interferiscono nei processi di determinazione embrionale, *Ist. Lomb. Rend. Sci. B,* 102, 213, 1968.

64. **Merlie, J. P., Buckingham, M. E., and Whalen, R. G.,** Molecular aspects of myogenesis, *Curr. Top. Dev. Biol.,* 10, 61, 1977.
65. **Civelli, O., Vincent, A., Buri, J. F., and Scherrer, K.,** Evidence for a translational inhibitor linked to globin mRNA in untranslated free cytoplasmic messenger ribonucleoprotein complexes, *FEBS Lett.,* 72, 71, 1976.
66. **Monroy, A., Maggio, R., and Rinaldi, A. M.,** Experimentally induced activation of the ribosomes of the unfertilized sea urchin eggs, *Proc. Natl. Acad. Sci. U.S.A.,* 54, 107, 1965.
67. **Gross, K. W., Jacobs Lorena, M., Baglioni, C., and Gross, P. R.,** Cell-free translation of maternal messenger RNA from sea urchin eggs, *Proc. Natl. Acad. Sci. U.S.A.,* 70, 2614, 1973.
68. **Kaumeyer, J. F., Jenkins, N. A., and Raff, R. A.,** Messenger ribonucleoprotein particles in unfertilized sea urchin eggs, *Dev. Biol.,* 63, 266, 1978.
69. **Nilsson, M. O. and Hultin, T.,** Characteristics and intracellular distribution of messenger-like RNA in encysted embryos of *Artemia salina, Dev. Biol.,* 38, 138, 1974.
70. **Lowett, J. A. and Goldstein, E. B.,** The cytoplasmic distribution and characterization of poly(A)$^+$ RNA in oocytes and embryos of Drosophila, *Dev. Biol.,* 61, 70, 1977.
71. **Wegnez, M. and Denis, H.,** Biochemical research on oogenesis. Transfer RNA is fully charged in the 42-S storage particles of *Xenopus laevis* oocytes, *Eur. J. Biochem.,* 98, 67, 1979.
72. **Woodland, H. R. and Gurdon, J. B.,** The relative rates of synthesis of DNA, sRNA and rRNA in the endodermal region and other parts of *Xenopus laevis* embryos, *J. Embryol. Exp. Morphol.,* 19, 363, 1968.
73. **Brachet, J. and Malpoix, P.,** Macromolecular synthesis and nucleocytoplasmic interactions in early development, *Adv. Morphog.,* 9, 263, 1971.
74. **Flickinger, R. A. and Daniel, J. C.,** Nuclear DNA-like RNA of regions of developing frog embryos, *Roux's Arch.,* 169, 350, 1972.
75. **Sheperd, G. N. and Flickinger, R. A.,** Heterogeneous nuclear RNA (HnRNA) size in developing frog embryos, *Exp. Cell. Res.,* 120, 73, 1979.
76. **Leonardi Cigada, M., Laria De Bernard, F., and Scarpetti Bolzern, A. M.,** Sintesi di acidi ribonucleici e di proteine nelle diverse regioni dell'embrione di *Xenopus laevis, Rend. Accad. Naz. Lincei Sic. Fis. 8,* 59, 152, 1975.
77. **De Bernardi, F. and Bolzern, A. M.,** On protein synthesis in *Xenopus laevis* embryos, *Rend. Accad. Naz. Lincei Sci. Fis. 8,* 44, 302, 1978.
78. **Nieuwkoop, P. D. and Faber, J.,** *Normal Table of Xenopus laevis,* North-Holland, Amsterdam, 1956.
79. **Tomkins, G. M., Gelehrter, T. D., Granner, D., Martin, D., Samuels, H. H., and Thompson, E. B.,** Control of specific gene expression in higher organisms, *Science,* 166, 1474, 1969.
80. **Heywood, S. M. and Rourke, A. W.,** Cell-free synthesis of myosin, *Methods Enzymol.,* 30, 669, 1974.
81. **Epel, D.,** Protein synthesis in sea urchin eggs: a ''late'' response to fertilization, *Proc. Natl. Acad. Sci. U.S.A.,* 57, 899, 1967.
82. **Gross, P. R. and Cousineau, G. H.,** Macromolecules synthesis and the influence of actinomycin on early development, *Exp. Cell Res.,* 33, 368, 1964.
83. **Davidson, E. H.,** *Gene Activity in Early Development,* Academic Press, New York, 1976.
84. **Giudice, G., Sconzo, G., Albanese, I., Ortolani, G., and Cammarata, M.,** Cytoplasmic giant RNA in sea urchin embryos. I. Proof that it is not derived from artifactual nuclear leakage, *Cell Diff.,* 3, 287, 1974.
85. **Nemer, M. and Surrey, S.,** mRNA containing and lacking poly(A) function as separate and distinct classes during embryonic development, *Progr. Nucl. Acid Res.,* 19, 119, 1976.
86. **Young, E. M. and Raff, R. A.,** Messenger ribonucleoproteins particles in developing sea urchin embryos, *Dev. Biol.,* 72, 24, 1979.
87. **Giudice, G. and Mutolo, V.,** Synthesis of ribosomal RNA during sea urchin development, *Biochim. Biophys. Acta,* 138, 276, 1967.
88. **Kafiani, C.,** Genome transcription in fish development, *Adv. Morphog.,* 8, 209, 1970.
89. **Kafiani, C. A., Strelkov, L. A., Chiaureli, N. B., and Davitashvili, A. N.,** Nucleus-associated polyribosomes in early embryos of loach, in *Macromolecules in the Functioning Cell,* Salvatore, F., Marino, G., and Volpe, P., Eds., Plenum Press, New York, 1979, 263.
90. **Mermod, J. J., Schatz, G., and Crippa, M.,** Specific control of messenger translation in Drosophila oocytes and embryos, *Dev. Biol.,* 75, 177, 1980.
91. **Roberts, D. B. and Graziosi, G.,** Protein synthesis in the early Drosophila embryos; analysis of the protein species synthesized, *J. Embryol. Exp. Morphol.,* 41, 101, 1977.
92. **Meedel, T. H. and Whittaker, J. R.,** Messenger RNA synthesis during early ascidian development, *Dev. Biol.,* 66, 410, 1978.
93. **Haemmerling, J.,** Nucleo-cytoplasmic relationships in the development of Acetabularia, *Int. Rev. Cytol.,* 2, 475, 1963.
94. **Brachet, J.,** Synthesis of macromolecules and morphogenesis in Acetabularia, *Curr. Top. Dev. Biol.,* 3, 1, 1968.

95. **Schweiger, H. G. and Berger, S.,** Nucleocytoplasmic interrelationships in Acetabularia and some other Dasycladaceae, *Int. Rev. Cytol.,* 9(Suppl.), 11, 1979.

96. **Butros, J. M.,** Stimulation of cleavage in Arbacia eggs with desoxyribonucleic acid fractions, *Exp. Cell Res.,* 18, 318, 1959.

97. **Benitez, H. H., Murray, M. R., and Chargaff, E.,** Heteromorphic change of adult fibroblasts by ribonucleoproteins, *J. Biophys. Biochem. Cytol.,* 5, 25, 1959.

98. **Niu, M. C.,** Thymus ribonucleic acid and embryonic differentiation, *Proc. Natl. Acad. Sci. U.S.A.,* 44, 1264, 1958.

99. **Ranzi, S., Lucchi Gavarosi, G., Protti Necchi, M., and Citterio, P.,** Ancora sul citodifferenziamento indotto da ribonucleopoteine, *Mon. Zool. Ital.,* 70, 348, 1963.

100. **Niu, M. C. and Segal, S. J., Eds.,** *The Role of RNA in Reproduction and Development,* North-Holland, Amsterdam, 1979.

101. **Bolzern, A. M., Cigada Leonardi, M., De Bernardi, F., Maci, R., and Ranzi, S.,** On the induction of somites in chick embryo, *Rend. Accad. Naz. Lincei Sci. Fis. 8,* 44, 621, 1978.

102. **Hamburger, V. and Hamilton, H. L.,** A series of normal stages in the development of chick embryo, *J. Morphol.,* 58, 49, 1951.

103. **New, D. A. T.,** A new technique for the cultivation of the chick embryo *in vitro, J. Embryol. Exp. Morphol.,* 3, 310, 1955.

104. **Chauhan, S. P. S. and Rao, K. V.,** Chemically stimulated differentiation of post-nodal pieces of chick blastoderms, *J. Embryol. Exp. Morphol.,* 23, 71, 1970.

105. **Moscona, A. A.,** Surface specification of embryonic cells: lectin receptors, cell recognition, and specific cell ligands, in *The Cell Surface in Development,* Moscona, A. A., Ed., John Wiley & Sons, New York, 1974, 67.

106. **Hay, E. D.,** Embryonic induction and tissue interaction during morphogenesis, in *Birth Defects, Excerpta Medica Int. Congr. Ser.* No. 432, Littlefield, J. W. and de Grouchy, J., Eds., Excerpta Medica, Amsterdam, 1977, 126.

107. **Ebert, J. D.,** An analysis of the synthesis and distribution of the contractile protein, myosin, in the development of the heart, *Proc. Natl. Acad. Sci. U.S.A.,* 39, 333, 1953.

108. **Pollard, T. D., Fujiwara, K., Niederman, R., and Maupin-Szamier, P.,** Evidence for the role of cytoplasmic actin and myosin in cellular structure and motility, in *Cell Motility,* Goldman, R., Pollard, T., and Rosenbaum, J., Eds., Cold Spring Harbor Lab., New York, 1976, 689.

109. **Chi, J. C., Fellini, S. A., and Holtzer, H.,** Differences among myosins synthesized in non-myogenic cells, presumptive myoblasts and myoblasts, *Proc. Natl. Acad. Sci. U.S.A.,* 72, 4999, 1975.

110. **Zevin-Sonkin, D. and Yaffe, D.,** Accumulation of muscle-specific RNA during myogenesis, *Dev. Biol.,* 74, 326, 1980.

111. **Packard, D. S. and Jacobson, A. G.,** The influence of axial structures on chick somite formation, *Dev. Biol.,* 53, 36, 1976.

112. **Nicolet, G.,** Is the presumptive notochord responsible for somite genesis in the chick, *J. Embryol. Exp. Morphol.,* 24, 467, 1970.

113. **Bellairs, R., Curtis, A. S. G., and Saunders, E. J.,** Cell adhesiveness and embryonic differentiation, *J. Embryol. Exp. Morphol.,* 46, 207, 1978.

114. **Bellairs, R.,** The mechanism of somite segmentation in the chick embryo, *J. Embryol. Exp. Morphol.,* 51, 227, 1979.

115. **Mestres, P. and Hinrichsen, K.,** Zur histogenese des somiten bein hühnchen, *J. Embryol. Exp. Morphol.,* 36, 669, 1976.

116. **Camatini, M. and Ranzi, S.,** Ultrastructural analysis of the morphogenesis of the neural tube, optic vesicle and optic cup in the chick embryo, *Acta Embryol. Exp.,* 81, 1976.

117. **Cooke, J.,** Cell number in relation to primary pattern formation in the embryo of *Xenopus laevis.* II. Sequential cell recruitment and control of the cell cycle, during mesoderm formation, *J. Embryol. Exp. Morphol.,* 53, 269, 1979.

118. **Niu, M. C. and Deshpande, A. K.,** The development of tubular heart in RNA-treated post-nodal pieces of chick blastoderm, *J. Embryol. Exp. Morphol.,* 29, 485, 1973.

119. **Siddiqui, M. A. Q., Deshpande, A. K., Arnold, H. H., Jakowlew, S. B., and Crawford, P. A.,** RNA-mediated manipulation of early chick embryonic gene functions *in vitro, Brookhaven Symp. Biol.,* 29, 62, 1977.

120. **Arnold, H. H., Innis, M. A., and Siddiqui, M. A. Q.,** Control of embryonic development: effect of an embryonic inducer RNA on *in vitro* translation of mRNA, *Biochemistry,* 17, 2050, 1978.

121. **Heywood, S. M. and Kennedy, D. S.,** Translational control in embryonic muscle, *Proc. Nucl. Acid Res.,* 19, 477, 1976.

122. **Lee-Huang, S., Sierra, J. M., Naranjo, R., Filipowicz, A., and Ochoa, S.,** Eucaryotic oligonucleotides affecting mRNA translation, *Arch. Biochem. Biophys.,* 180, 276, 1977.

123. **Stein, G. S. and Kleinsmith, L. J., Eds.,** *Chromosomal Proteins and Their Role in the Regulation of Gene Expression,* Academic Press, New York, 1975.

124. **Duprat, A. M., Houssaint, E., Suignard, G., and Zalta, J. P.,** Spécificité tissulaire des effets de protéines chromatiniennes non histoniques sur la différenciation d'organes embryonnaires en culture *in vitro, C.R. Acad. Sci. Paris,* 280, 1143, 1975.

125. **Duprat, A. M., Beetschen, J. C., Mathieu, C., Zalta, J. P., and Daguzan, C.,** Effets inhibiteurs de proteins chromatiniennes non histoniques homospecifiques sur la morphogenèse d'embryons d'Amphibiens Urodèles, *Roux's Arch.,* 179, 111, 1976.

126. **Runnström, J.,** Considerations on the control of differentiation in early sea urchin development, *Arch. Zool. Ital.,* 51, 239, 1966.

127. **Johnen, A. G.,** Der Einfluss von Li- und SCN-Ionen auf die Diefferenzierungs-leistungen von Ambystoma-ectoderms und ihre Verhänderung bei Kombinierter Einvirkung beider Ionen, *Roux's Arch.,* 165, 150, 1970.

128. **Ranzi, S.,** Early determination in the development under normal and experimental conditions, in *The Beginning of Embryonic Development,* Vol. 48, Tyler, A., von Borstel, R. C., and Mets, C. B., Eds., American Association for the Advancement of Science, Washington, D.C., 1957, 291.

129. **De Bernardi, F., Leonardi Cigada, M., Maci, R., and Ranzi, S.,** On protein synthesis during the development of lithium treated embryos, *Experientia,* 25, 211, 1969.

130. **Ranzi, S. and Gavarosi, G.,** Dimensions of notochord and somites in embryos of *Xenopus laevis* treated with thiocyanate, *J. Embryol. Exp. Morphol.,* 7, 117, 1959.

131. **Vailati, G., Vitali, P., and Ranzi, S.,** Axione del solfocianuro di sodio nei primi stadi dello sviluppo dell'embrione di pollo, *Rend. Accad. Naz. Lincei Sci. Fis. 8,* 53, 594, 1972.

132. **Ogi, K.,** The effect of sodium thiocyanate on isolates of presumptive ectoderm and medioventral marginal zone in *Triturus gastrulae, J. Embryol. Exp. Morphol.,* 6, 412, 1958.

133. **Ranzi, S. and Tamini, E.,** Azione di NaSCN sullo sviluppo di embrioni di anfibi, *R. Ist. Lomb. Rend. Sci.,* 73, 525, 1940.

134. **Leonardi Cigada, M., Laria De Bernardi, F., and Scarpetti Bolzern, A. M.,** Sintesi di proteine in embrioni di *Xenopus laevis* trattati con LiCl, *Ist. Lomb. Rend. Sci. B,* 107, 117, 1973.

135. **Berg, W. E.,** Effect of lithium on the rate of protein synthesis in sea urchin embryo, *Exp. Cell Res.,* 50, 133, 1968.

136. **Ranzi, S. and Vailati, G.,** Axione del cloruro di litio sul metabolismo proteico dell'embrione di *Xenopus, Rend. Accad. Naz. Lincei Sci. Fis. 8,* 50, 473, 1971.

137. **Leonardi Cigada, M., Laria De Bernardi, F., and Scarpetti Bolzern, A. M.,** Sintesi di acido ribonucleico e di proteine in embrioni di anfibi trattati con solfocianuro, *Rend. Accad. Naz. Lincei Sci. Fis. 8,* 55, 755, 1973.

138. **Pedrazzi, G.,** Sull'azione dell'urea nello sviluppo embrionale dei ricci di mare, *Ist. Lomb. Rend. Sci. Fis.,* 91, 672, 1957.

139. **Leone, V.,** Effetti di trattamento con urea su espianti ventrali di gastrula di *Rana esculenta, Rend. Accad. Naz. Lincei Sci. Fis. 8,* 12, 195, 1952.

140. **Leonardi Cigada, M., Laria De Bernardi, F., and Bolzern, A. M.,** Sintesi di RNA in embrioni di Xenopus trattati con LiCl, *Rend. Accad. Naz. Lincei Sci. Fis. 8,* 52, 93, 1972.

141. **Thomason, D.,** Effect of lithium ions on the penetration of phosphate into Xenopus embryos, *Nature (London),* 179, 823, 1957.

142. **Lanzani Maci, R. and Sotgia, C.,** Cambiamenti della precoce determinazione e sintesi di acido ribonucleico e di proteine nell'embrione di pollo, *Rend Accad. Naz. Lincei Sci. Fis. 8,* 53, 602, 1972.

143. **Flickinger, R. A., Lauth, M. R., and Stambrook, P. J.,** An inverse relation between the rate of cell division and RNA synthesis per cell in developing frog embryos, *J. Embryol. Exp. Morphol.,* 23, 571, 1970.

144. **Volm, M., Wayss, K., and Schwartz, V.,** Virkung von lithium und rhodanidionen auf die nukleinsäure und protein synthese bei *Tetrahymena pyriformis* GL., *Roux's Arch.,* 165, 125, 1970.

145. **Bächström, S.,** Changes in ribonucleic acid content during early sea urchin development, *Ark. Zool.,* (2), 12, 339, 1959.

146. **Runnström, J.,** The mechanism of fertilization in Metazoa, *Adv. Enzymol.,* 9, 241, 1949.

147. **Child, C. M.,** Differential reduction of vital dyes in the early development of echinoderms, *Roux's Arch.,* 135, 426, 1936.

148. **Ranzi, S.,** Ricerche sulle basi fisiologiche della determinazione degli Echinodermi II, *Arch. Zool. Ital.,* 26, 427, 1939.

149. **Hörstadius, S.,** Reduction gradients in animalized and vegetalized sea urchin eggs, *J. Exp. Zool.,* 129, 249, 1955.

150. **Gustafson, T. and Lenique, P.,** Studies of mitochondria in developing sea urchin eggs, *Exp. Cell Res.,* 3, 251, 1952.

151. **Hörstadius, S.,** *Experimental Embryology of Echinoderms,* Claredon Press, Oxford, 1973.

152. **Hörstadius, S.**, *Experimental* researches on the developmental physiology of the sea urchin, *Pubbl. Staz. Zool. Napoli*, 21(Suppl.), 131, 1949.

153. **Jacobsen, C. F. and Korsgaard Christensen, L.**, Influence of the temperature of the urea denaturation of β-lactoglobulin, *Nature (London)*, 161, 30, 1948.

154. **Gorini, L.**, Le role du calcium dans l'activité et la stabilité de quelques protéinases bacteriennes, *Biochim. Biophys. Acta*, 6, 237, 1950.

155. **Kavanau, J. L.**, Amino acid metabolism in the early development of the sea urchin, *Paracentrotus lividus*, *Exp. Cell Res.*, 7, 520, 1954.

156. **Bäckström, S. and Gustafson, T.**, Lithium sensitivity of the sea urchin in relation to the stage of development, *Ark. Zool.*, (2), 6, 185, 1953.

157. **Brachet, J.**, An old enigma: the gray crescent of amphibian eggs, *Curr. Top. Dev. Biol.*, 11, 133, 1977.

158. **Child, C. M.**, Differential dye reduction and reoxidation in Triturus development, *Physiol. Zool.*, 16, 61, 1943.

159. **Deuchar, E. M.**, *Biochemical Aspects of Amphibian Development*, Methuen, London, 1966.

160. **D'Amelio, V. and Ceas, M.**, Distribution of proteinase activity in the blastula and early gastrula of *Discoglossus pictus*, *Experientia*, 13, 152, 1957.

161. **Kochak, S. and Eyal Giladi, H.**, Bilateral symmetry in chick embryo determination by gravity, *Science*, 171, 1027, 1971.

162. **Nicolet, G.**, Action du chlorure de lithium sur la morphogénesè du jeme embryon de poule, *Acta Embryol. Morphol. Exp.*, 8, 32, 1965.

163. **Ranzi, S.**, Ricerche sulla fisiologia dell'embrione degli anfibi. I, *Ist. Lomb. Rend. Sci.*, 75, 3, 1942.

164. **Barth, L. G. and Barth, L. J.**, Ionic regulation of embryonic induction and cell differentiation in *Rana pipiens*, *Dev. Biol.*, 39, 1, 1974.

165. **Lindahl, P. E., Swedmark, B., and Lundin, J.**, Some new observations on the animalization of the unfertilized sea urchin egg, *Exp. Cell Res.*, 2, 39, 1951.

166. **Kuusi, T.**, The effect of urea denaturation of inductor properties of a protein fraction of the bone marrow, *Acta Embryol. Morphol. Exp.*, 4, 18, 1961.

167. **Kaji, A. and Kaji, H.**, Specific interaction of sRNA and polysomes: inhibition by LiCl, *Biochim. Biophys. Acta*, 87, 519, 1964.

168. **Haeckel, E.**, *Naturliche Schöpfungsgeschichte; gemeinverständliche wissenschaftliche Vorträge über die Entwickelungslehre in allgemeinen, und diejenige von Darwin, Goethe und Lamarck in besonderen*, G. Reiner, Berlin, 1868.

169. **v. Baer, K. E.**, *Veber Entwickelungsglschichte der Tiere II*, Teil Bornträgen, Königsberg, 1837.

171. **Darwin, C.**, *The Origin of Species*, 6th ed., Oxford University Press, London, 1956.

172. **Wright, T. R. F.**, The genetics and embryogenesis in Drosophila, *Adv. Genet.*, 15, 262, 1970.

173. **Ranzi, S.**, Biologia dello sviluppo e simiglianze tra embrioni, *Ist. Lomb. Rend. Sci. B*, 108, 182, 1974.

174. **Moser, C. and Flickinger, R. A.**, Effects of nutrient media on gastrula ectoderm of *Rana pipiens* cultured in vitro, *J. Exp. Zool.*, 153, 219, 1963.

175. **Bolzern, A. M., Cigada Leonardi, M., De Bernardi, F., Fascio, U., Maci, R., Ranzi, S., and Sotgia, C.**, Myosin as inductor of somites in chick embryo, *Ist. Lomb. Rend. Sci. B*, 114, 36, 1980.

176. **Pederson, T. and Munroe, S. H.**, Ribonucleoprotein organization of eukaryotic RNA. XV. Different nucleoprotein structures of globin messenger RNA. Sequences in nuclear and polyribosomal ribonucleoprotein particles, *J. Mol. Biol.*, 150, 509, 1981.

177. **Wakahara, M.**, Accumulation, spatial distribution and partial characterization of poly (A)$^+$ RNA in the developing oocytes of *Xenopus laevis*, *J. Embryol. Exp. Morph.*, 66, 127, 1981.

178. **De Bernardi, F.**, Polyadenylated mRNA from various developmental stages of *Xenopus laevis*. Role of 26S mRNA, *Exp. Cell Biol.*, 5, 281, 1982.

179. **Dixon, L. K. and Ford, P. J.**, Regulation of protein synthesis and accumulation during oogenesis in *Xenopus laevis*, *Dev. Biol.*, 93, 478, 1982.

Chapter 5

THE CONTROL OF EARLY DEVELOPMENT IN MOLLUSCS

M. R. Dohmen

TABLE OF CONTENTS

I. INTRODUCTION

The most accepted theory of development states that both self-differentiation of blasto-meres, as a result of localized cytoplasmic determinants, and intercellular interactions are instrumental in development.[1] Eggs in which cytoplasmic localization is rigidly fixed early in development lack field properties and primary determination is the dominant factor at the start. Eggs in which precocious localization is absent or labile rely on positional information in the earliest stages. Mammalian eggs may serve as an extreme example of the latter category as cytoplasmic localization seems to play no role at all in these eggs.

Molluscs are often cited as examples of mosaic eggs. This certainly applies to a certain degree to molluscs showing unequal cleavage or polar lobe formation, but recent work of Morrill et al.[2] and Verdonk,[3] who showed that in eggs of pulmonate gastropods a single blastomere may develop into a normal larva after blastomere deletion at the two- or four-cell stage, demonstrated that equally cleaving molluscan eggs should be considered as regulative eggs. This demonstrates that within one phylum the whole range of devices acting in development may be found. The classical division of the animal kingdom in groups in which certain developmental mechanisms play a dominant role loses much of its signifi-cance. It is becoming clear that in every species at a certain stage or in a certain part, any of these mechanisms may dominate. Molluscs, for instance, are eminently suited for the study of cytoplasmic localization, as will be shown below, but they prove to be equally well suited for studying induction.

It is still impossible to translate the classical embryological concepts, such as morpho-genetic determinants, gradients, fields, and induction into concrete molecular or structural terms, although many efforts are being made to elucidate the mechanisms underlying these concepts. In several systems the identification of localized morphogenetic determinants is being pursued intensively. To bring about primary determination by means of localized morphogens the eggs must have a mechanism that generates cytoplasmic localization and it must be able to cleave precisely in a predetermined manner. The cytoskeleton is probably involved in both processes. Therefore, in the first section a brief review of current ideas of cytoskeletal structures is presented.

II. THE CYTOSKELETON

Cells possess a three-dimensional cytoskeleton consisting of a complex network of at least three major subsystems: microfilaments, microtubules, and intermediate filaments. The latter is a highly polymorphic class of filaments. Five groups have been distinguished in different cell types, and the presence of two different intermediate filament systems in the same cell has been demonstrated by double immunofluorescence.[4] It has been proposed that inter-mediate filaments may function to integrate mechanically the various structures of the cy-toplasmic space.[5] Next to these filaments, the so-called microtrabecular lattice may have a cytoskeletal function.[6] This lattice is not an artifact[7] and represents the nonrandom structure of the cytoplasmic ground substance of living cells. The lattice is confluent with the cortex and interconnects microtubules, endoplasmic reticulum, polysomes, and stress fibers.

The third type of cytoplasmic network that may play a part in bringing about and main-taining a specific spatial structure of the cytoplasm is the endoplasmic reticulum (ER). The ER is a continuous network of cisterns, including the nuclear membrane. In many eggs there is some doubt as to the structure of the ER in the living state. In these eggs the ER appears as separate vesicles in the electron microscope, but this may be an artifact of fixation. The ER is an important vehicle for imparting regional specificity to the various cytoplasmic compartments, in terms of synthetic activity or storage of specific products. An example is found in the ontogeny of the vegetal body of the snail *Bithynia*. This body is a morphogenetic plasm consisting of vesicles derived from the ER.[8] The vesicles originate during oogenesis by budding from the ER. This process takes place in the cytoplasmic compartment which is to become the future vegetal pole region of the egg.

On the basis of centrifugation experiments we may distinguish two different cytoskeletal domains in eggs: the cortex and the endoplasm. In this chapter only the role of the endoplasm will be discussed. The cortex will be treated elsewhere in this volume.[9] The most conspicuous structure in the endoplasmic domain is a concentration of intermediate filaments which are tightly associated with the nucleus, possibly providing it with mechanical support and constraining it in a specific place in the cell, thus serving as a nucleus-anchoring cytoskeleton.[5,10-13] The ER could also play a part in this respect as this network includes the nuclear membrane. In egg cells a nucleus-anchoring system has not yet been demonstrated. As in eggs the position of the nucleus is closely prescribed in relation to the architecture of the cytoplasm, we may also expect that in eggs a cytoskeletal system regulates the movements and positioning of the nucleus. The strictly ordered spatial structure of the egg cytoplasm manifests itself in many ways, most conspicuously in the segregation of cellular components in different compartments. A cytoskeletal basis for these segregation phenomena is presumed but not proven. In order to test the importance of an intact cytoskeleton for bringing about and maintaining spatial structure in eggs and for early developmental processes, a variety of disrupting agents has been used.

Cytochalasin B is known to interfere with microfilaments, although the mechanism of action remains obscure despite its extensive use.[14] A peculiar effect of CB, manifested in various cultured mammalian cells, is that the nucleus is segregated into an evagination of the plasma membrane and eventually expulsed.[15] A similar phenomenon occurs spontaneously in mammalian normoblasts when they differentiate into erythrocytes, and also in food eggs of the snail *Nucella lapillus*.[16] The mechanism is still obscure. Other effects of CB which probably result from breakdown of cytoskeletal elements include the reversible inhibition of polar axis formation in eggs of the brown alga *Fucus*,[17] the prevention of cleavage resulting in multinucleated syncytia, and disturbing effects on morphogenesis after treatment of eggs of the annelid *Sabellaria*[18] and the cephalopod *Loligo*.[19] However, other phenomena in which the functioning of an intact cytoskeleton would seem to be indispensable prove to be unaffected by treatment with CB. For instance, the drug does not disturb ooplasmic segregation in uncleaved ascidian eggs,[20] nor does it prevent the correct segregation of determinants for tissue-specific enzymes in cleavage-arrested ascidian eggs.[21]

Microtubules are affected by drugs such as colchicine, colcemid, vinblastine, and ethylcarbamate. Disruption of the microtubules arrests both nuclear division and cytokinesis, but does not affect segregation in ascidian eggs.[20,21] Centrifugation interferes physically with the cytoskeleton. It is widely used in experimental embryology. Centrifuging eggs at moderate forces (400 to 600 g) easily stratifies the cytoplasmic inclusions such as proteid and lipid yolk and mitochondria. In many eggs these inclusions are found in specific regions as a consequence of ooplasmic segregation. We may suppose that this compartmentation is brought about by the cytoskeleton. At the present moment we do not know whether the stratification of cytoplasmic inclusions involves the breakdown of the cytoskeleton compartments. It may be the organelles are only detached from the cytoskeleton and driven through the skeletal meshes to an equilibrium position, while leaving the skeleton intact. The displacement of the nucleus, however, must at least involve a considerable stretching and deformation of the nucleus-anchoring skeleton. The accumulation of ER vesicles in the hyaline layer in many eggs may result from an extensive breakdown of the ER network, but may also indicate that the ER in these eggs consists of separate cisterns which are not integrated into a continuous network.

Some insight into the dynamic aspects of the cytoskeleton can be gained from studying the further development of centrifuged eggs. Whether development is disturbed by centrifugation or not depends on the egg species and on the stage at which centrifugation was carried out. For example, in insect eggs and in fertilized ascidian eggs, centrifugation may cause severe defects in development.[22,23] In uncleaved molluscan eggs and in many other egg species, however, development is virtually not affected by centrifugation.[24,25] The stage at which centrifugation is carried out is an important factor. For instance, unfertilized ascidian eggs can be centrifuged without affecting further development, whereas centrifugation of fertilized eggs results in abnormal development.[23,26] In molluscs, similar stage-specific sensitivities have been found. Centrifuging at the uncleaved stage affects the development of a small proportion of the eggs, but centrifugation during the early cleavages, particularly just before third cleavage in *Lymnaea,* produces a large proportion of abnormal embryos.[25] These differences between species and stages may reflect both a difference in the timing of segregation of morphogenetic plasms and a difference in localization of these plasms. Plasms located in the endoplasmic domain might be easily disturbed, whereas plasms in the cortical domain might be less sensitive.

Surgical manipulation and compression are other means of disturbing the cytoskeleton. In eggs of *Dentalium,* sucking out parts of the cytoplasm and injecting it in another place have no effect on development.[27] Similarly, in sea urchins half or more of the egg cytoplasm can be removed without impairing normal development.[28] Unfertilized ascidian eggs may be cut into two equal parts and, whatever the plane of section, both parts may give rise to a complete larva.[26] Compression of eggs results in an abnormal orientation of the cleavage spindles and hence in an abnormal orientation of the cleavage plane in relation to the cytoplasmic structure. Nevertheless, normal development after such treatment has been reported for eggs of the snail *Limax.*[29] If, however, the correct distribution of important morphogenetic plasms is affected by the induction of abnormal cleavage planes, abnormal development will result, e.g., in eggs of the bivalves *Pholas* and *Spisula.*[30]

It is hard to draw general conclusions from this wide variety of experiments. It is evident that the integrity of spatial structure of the endoplasmic domain is of vital importance for many egg species. In molluscan eggs the endoplasm seems to be of minor importance. Treatments affecting the endoplasmic structures often do not impair further development. The egg cortex seems to be the most important structure which contains the vital regulative factors. This domain appears to be fairly resistant to experimental interferences of the kind described in this section.

III. CLEAVAGE CHRONOLOGY AND CLEAVAGE PLANES

In eggs with a determinate cleavage pattern the correct timing of cleavage and the correct positioning of the cleavage planes are essential for normal development. However, experimentally induced deviations from the normal pattern can be tolerated to a certain extent in a number of species. This regulative potential is generally explained by assuming that definite determination of blastomeres occurs in later stages in these species and is not the result of precocious localization of morphogenetic determinants.

Cleavage and nuclear division are, under normal circumstances, always closely associated. The coordinated occurrence of nuclear divisions and cleavage probably originates from a functional coupling of these processes. Rappaport[31] has demonstrated that during nuclear division the asters and the spindle provide the stimulus for cleavage to occur. The cleavage stimulus is not provided by a simple increase in the number of aster microtubules at the equatorial cortex in sea urchin eggs,[32] and the stimulus is not of a structural nature as extensive mechanical agitation of the cytoplasm between asters and the equatorial cortex fails to interfere with the formation of the cleavage furrow.[33]

The clock mechanism regulating the cell cycle has not yet been elucidated. The clock probably resides in the cytoplasm or the cortex, and not in the nucleus, as in nuclear transplantation experiments the nucleus adapts its mitotic activity according to the kind of

FIGURE 1. Food egg of the snail *Nucella lapillus*. In this species the nuclei of food eggs are extruded before cleavage and the subsequent cleavage is very irregular. Bar indicates 50 μm.

cytoplasm into which it is implanted. An independent cytoplasmic or cortical clock is inferred from the experiments on *Xenopus* eggs.[34] When eggs are prevented from cleaving by treatment with antimitotic drugs, they undergo a sequence of periodic surface contraction waves timed with the cleavage cycle in untreated eggs. Separated egg fragments, produced by constricting a fertilized egg, still undergo contraction waves with the same period as the cleaving nucleated fragment. Apparently, some expression of the cell cycle persists in the absence of any nuclear material or centrioles. Actual cleavage may occur in the absence of the nucleus. For instance, food eggs of the snail *Nucella lapillus* extrude their nucleus in an early stage.[16] These eggs continue cleaving, though very irregularly (Figure 1). No data are available on the timing of these irregular cleavages. Continued cleavage in the absence of the nucleus or of nuclear activity has also been observed in other species.[35-37] Generally the cleavage pattern is abnormal under these conditions.

The reverse type of uncoupling, in which nuclear division is not impaired but segmentation is inhibited, occurs naturally in early stages of development of insects and triclad turbellarians. Here the egg at first forms a syncytium with many nuclei. Experimental inhibition of segmentation by cytochalasin B has been an important tool in studying early developmental processes, particularly segregation of specific determinants, in ascidia[21] and ctenophores,[38] but not in molluscs.

Retarding nuclear division and cleavage simultaneously can be done by several methods. Antimitotic drugs block both nuclear division and cleavage. After washing the drugs out, mitotic activity and cleavage are resumed. This method has been applied to eggs of *Bithynia* in order to determine whether the morphological changes normally occurring in the structure of the vegetal body and its eventual disappearance are regulated by an inherent autonomous clock. Our results indicate that the vegetal body does not possess an autonomous clock. In insect eggs, on the other hand, structural changes in polar granules occur according to an autonomous timing within the granules themselves, as shown by their behavior after inter-specific transplantation[39] and also in unfertilized eggs.[40] Retarding cell cycles by means of UV irradiation[41] or drugs that inhibit protein synthesis[42] may affect development consider-ably. In *Lymnaea* it has been found that treatment at specific cleavage cycles results in developmental damage, whereas treatment at other stages is relatively harmless. Morpho-genetic abnormalities, viz., head malformations, arose after treatment with puromycin for 30 min at third cleavage, but not after treatment in earlier stages. Specific proteins necessary for normal head formation may be synthesized at third cleavage, but as cleavage cycles are lengthened by this treatment it cannot be ruled out that the cleavage delay itself is the main cause of the abnormalities. The morphogenetic significance of division chronology could reside in the establishment of specific spatial relationships between blastomeres and the resulting interactions could be of vital importance.[43]

The predictable pattern of cleavages in molluscs and other groups with a determinant cleavage type suggests that the cleavage planes play an important role in development. It is not clear which factors determine the position of the cleavage plane. The plane always bisects the spindle perpendicularly, so probably the position of the asters is a determinant, but not the primary one. A prepattern of cleavage furrows in the cortex can be ruled out for several reasons: (1) cortical cytoplasm may be sucked out with a micropipette from the tip of a progressing cleavage furrow and wherever it is injected it will induce a new cleavage furrow;[44] (2) spindles may be displaced in any direction and the position of the cleavage furrows is accordingly adapted; and (3) contractile rings can be induced anywhere by mi-croinjection of calcium ions.[45-47] There is an example, however, of a contractile ring whose position seems to be fixed in a kind of prepattern. This is the ring that causes polar lobe constriction in eggs of *Ilyanassa*. These eggs show large blebs after treatment with ionop-hores, but the response reveals a distinct predisposition of the vegetal hemisphere to form a constriction where the polar lobe constriction normally would form.[48] This ring differs also in other respects from a normal contractile ring. In *Ilyanassa* emetine does not prevent polar lobe formation though it prevents cleavage.[49] In *Dentalium* the formation of the polar lobe can be inhibited by cytochalasin B without interfering with cleavage.[50]

The fact that often the position of the cleavage planes is relatively unaffected by moderate centrifugation[51] indicates that the cortex might serve to position the asters. Meshcheryakov and Veryasova[52-54] have conducted a series of experiments which suggest that the position of the asters depends on the size of the contact zone between the blastomeres and hence on the general form of the adjacent blastomere region. This mechanism evidently cannot explain the determination of the first cleavage plane. In many species it has been observed that the direction of the first cleavage plane is related to the entrance point of the sperm,[55] but the animal-vegetative polarity of the egg seems to be a more universal and more important determinant of the cleavage planes. This is demonstrated by Morgan[56] who found that in centrifuged sea urchin eggs the first three cleavage planes are determined by the stratification of the cytoplasm, but the formation of the micromeres always occurs at the original vegetative pole. It seems therefore that the cleavage planes are primarily determined by cortical factors which act to position the asters. These cortical determinants can be overruled by experimental treatments. A considerable regulative potential ensures normal development after experi-mental disturbance, except in species showing precocious segregation of vital morphogenetic plasms.

IV. LOCALIZATION OF MORPHOGENETIC DETERMINANTS

In the preceding sections a consistent difference in regulative potential was shown to exist between two categories of molluscs. Those showing precocious segregation of morphogenetic

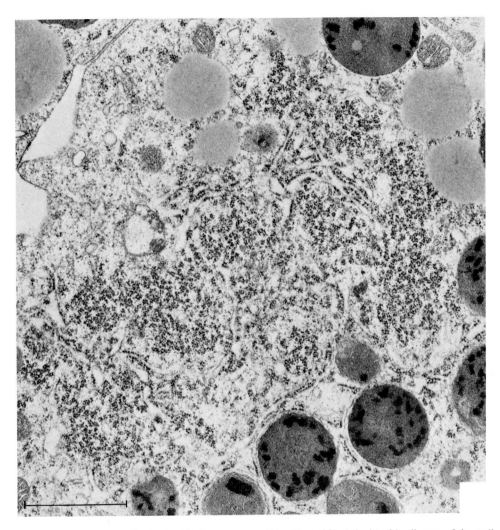

FIGURE 2. Ectosome at the inner end of a macromere (3A, 3B, or 3C) at the late 24-cell stage of the snail *Lymnaea stagnalis*. The ectosome is characterized by the absence of yolk granules, lipid droplets, and mitochondria. It consists of a dense mass of ribosomes, endoplasmic reticulum, and small vesicles with electron-dense contents. (From Dohmen, M. R. and van de Mast, J. M. A., *Proc. K. Ned. Akad. Wet. Ser. C,* 81, 403, 1978.) Bar indicates 1 μm.

determinants, generally manifested by unequal cleavage or polar lobe formation, have poor regulative capacity. On the other hand, equally cleaving eggs show a considerable regulative capacity, most convincingly expressed by experiments in which blastomeres are deleted at the two- or four-cell stage. The surviving cell or cells may develop into normal embryos.[2,3] These equally cleaving eggs do, however, show early segregation phenomena.

In uncleaved eggs of *Lymnaea,* for instance, Raven[51] has demonstrated the existence of a nearly bilaterally symmetrical pattern of subcortical plasms. These plasms probably develop into the ectosomes (Figure 2) found from the eight-cell stage onwards. The ectosomes consist of a complex of ribosomes, endoplasmic reticulum, and electrondense vesicles.[57] They show a differential behavior in the D macromere during the 24-cell stage, when the D macromere is determined to become the dorsal macromere.[51,58] This suggests that the ectosomes play a role in the determination of the D macromere. However, since the available experimental evidence does not support a definite relationship between the developmental capacities of the dorsal macromere and a prelocalization of morphogenetic substances in equally cleaving

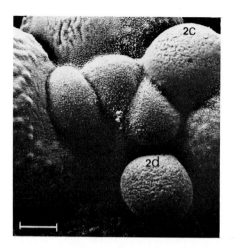

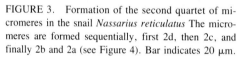

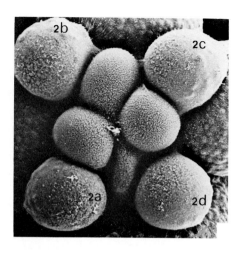

FIGURE 3. Formation of the second quartet of micromeres in the snail *Nassarius reticulatus* The micromeres are formed sequentially, first 2d, then 2c, and finally 2b and 2a (see Figure 4). Bar indicates 20 μm.

FIGURE 4. Formation of the second quartet of micromeres in the snail *Nassarius reticulatus*. The quartet is complete. Bar indicates 20 μm.

molluscan eggs,[3,59] the ectosomes probably function only as an intermediate factor, not as a primary determinant.

Several other types of localizations have been found in equally cleaving molluscan eggs, as reviewed by Raven,[25] but their morphogenetic significance is not clear. The determinative role of cytoplasmic localizations is much more evident in unequally cleaving eggs and polar lobe forming eggs. The latter are the most promising objects for studying localization in molluscs. Two types can be distinguished — large polar lobes, whose size about equals that of a blastomere, and small or intermediate lobes.[60,61] Polar lobes serve the same purpose as unequal cleavage — shunting the vegetal pole area to the D quadrant. The lobes can be removed without immediate damage to the egg. The consequences for further development are dramatic. The most detailed investigations of the effects of lobe deletions have been carried out in the gastropods *Ilyanassa* and *Bithynia* and in the scaphopod *Dentalium*. Deletion of the lobe has essentially the same effects in these species, whether the lobe is small, as in *Bithynia*, or large, as in *Ilyanassa* and *Dentalium*. The effects on terminal differentiation are most conspicuous. Most adult structures, such as shell, foot, operculum, statocysts, eyes, heart, and intestine are lacking in lobeless embryos.

Earlier effects, which are probably causal to the final defects, have been studied in detail in *Ilyanassa* and *Dentalium*. The earliest effect is the determination of cleavage asynchrony. In some species, e.g., *Littorina*,[62] the CD blastomere divides slightly ahead of the AB blastomere. The D-quadrant macromeres continue to divide earlier than the other macromeres in most species.[62,63] The effects of the lobe are not restricted to the cells of the D quadrant. Division chronology is also affected in the other quadrants. In some species, e.g., *Nassarius*, *Crepidula*, and *Littorina*, division asynchrony in the other quadrants becomes manifest already at the third cleavage: the micromere 1d appears first, followed sequentially by 1c, 1b, and 1a. This pattern is maintained in subsequent cleavages (Figures 3 and 4), giving the impression of a mitotic gradient running anticlockwise from D to A.

A pacemaker present in the polar lobe area might initiate the sequence and transmit the message. A mitotic gradient from the vegetal to the animal region has been observed in sea urchin eggs, and a pacemaker in the vegetal region has been proposed by Parisi et al.[64,65] The straight mitotic gradient in sea urchins is easier to grasp than a circular gradient in molluscs. As the lower end of the gradient (the A quadrant) borders the upper end (the D quadrant) this would presuppose a specific directionality in the transmission system of the mitotic stimulus, from D anticlockwise to A. The situation becomes still more complicated when one considers the division of the first micromere quartet. In *Crepidula,* division of 1d lags behind 1a-1c, whereas the other cleavages in the D quadrant are ahead of the other quadrants.[66] Division in *Dentalium* is synchronous until the fifth cleavage. Then 2d and 2D divide ahead of the corresponding cells in the other quadrants.[67] Certain descendants of the cell 1c also divide in advance of the corresponding cells in the A and B quadrant.

The observation that lobe-dependent division asynchrony is also present in the C quadrant may be explained by assuming an inductive stimulus from the D quadrant, but also by assuming that part of the polar lobe determinants has been passed on to the C quadrant at the second cleavage. The latter mechanism is invoked in the case of *Bithynia* to explain the almost equivalent developmental potential of the C and D macromeres.[68] Similarly, in *Dentalium* the apical tuft factor, originally present in the first polar lobe, is supposed to be distributed into C and D at second cleavage.[69] Investigating the significance of division asynchronies for morphogenesis would require the suppression of these asynchronies without removing the polar lobe material. This is not yet possible. Two possible consequences of the effect of the lobe on the cleavage rate can be expected. First, asynchrony may lead temporarily to a specific pattern of cell contacts which may be important for early inductive phenomena. The transient increase in volume of D-quadrant macromeres, observed in *Littorina,*[62] may also contribute to establishing specific cell contacts. Second, a definitive difference in position may be achieved by this method, as suggested by van Dongen and Geilenkirchen[67] for the 1d^{1111} cell in *Dentalium* which is shifted dorsally. Both cleavage asynchrony and deviating cleavage planes undoubtedly contribute to morphogenesis, but they cannot explain the complete absence of such an array of organs as is consistently observed in lobeless embryos. The cells from which lobe-dependent structures are normally derived are still formed after removal of the lobe. Lobeless and normal embryos develop at the same rate for several days and gross changes in the rate of growth occur only after organogenesis has begun. Apparently the D and C quadrant lineages of lobeless embryos lack the potential to give rise to certain larval and adult differentiations as a result of the absence of specific organ-forming factors.

The effects of the polar lobe at the biochemical level have been investigated by measuring the rate of transcription and translation and analyzing general patterns of protein synthesis in lobeless embryos, isolated lobes, and the progeny of separated AB and CD blastomeres. These data are reviewed and discussed by Davidson.[1] Collier and McCarthy[70] recently analyzed protein synthesis in *Ilyanassa* eggs by two-dimensional electrophoresis. Out of 295 polypeptides analyzed, none was unique to either the normal or lobeless embryo. Out of 290 polypeptides synthesized by isolated lobes there was none that did not occur in the normal and lobeless embryo. Differences were observed only in 24-hr-old embryos. These results are not consistent with those obtained by Donohoo and Kafatos[71] who found that in embryos grown for 4 hr from isolated AB and CD blastomeres, distinct sets of proteins were synthesized. An explanation might be that by separating AB and CD a regulative process is being set in train, resulting in a pattern of protein synthesis in one or in both blastomeres that normally does not occur. Comparing the individual proteins in both experiments would be necessary to resolve this question. When protein synthesis after 24 hr of development in the presence of actinomycin D is analyzed in both lobeless and normal embryos, significant differences are still found.[72] These data suggest that the polar lobe exerts its control by regulating the translation of maternal messengers prelocalized in the lobe. Guerrier[73] arrives at the same conclusion on observing that lobe-dependent structures can develop in the presence of actinomycin D in the annelid *Sabellaria.*

From the deletion experiments it is clear that the presence of morphogenetic determinants in polar lobes cannot be questioned. How and where these determinants are localized in the lobes of *Dentalium* and *Ilyanassa* is not yet clear, although these lobes have been extensively investigated.[61] Most probably, they are associated with the cortex as they cannot be displaced

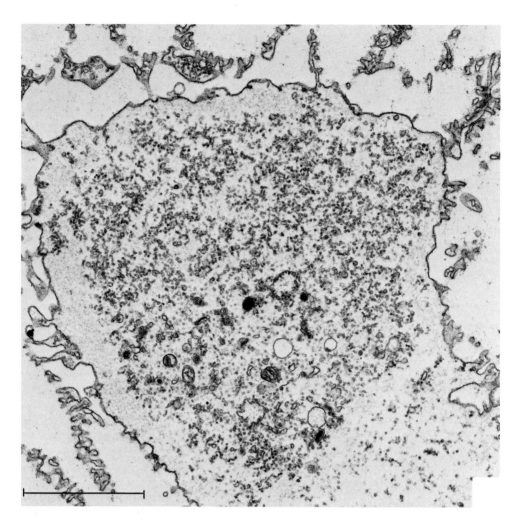

FIGURE 5.   Polar lobe of the gastropod *Crepidula fornicata*. A large number of small vesicles is present in the lobe. These vesicles are thought to contain morphogenetic determinants. The surface of the lobe is extensively folded. Bar indicates 4 μm.

by centrifugation[74,75] nor removed by sucking out the cytoplasm of the lobe.[27] In several species with small polar lobes distinct structures have been observed which might contain the morphogenetic determinants[61] (Figures 5 and 6). The most conspicuous structure is found in *Bithynia*. It is a cup-shaped body (Figures 7 and 8) consisting of a large number of small vesicles with electron-dense contents (Figure 9). This so-called vegetal body is tightly associated with the cortex. It cannot be displaced by centrifugal forces up to 600 g which easily stratify the other cytoplasmic inclusions. Centrifugation at 1400 g detaches the vegetal body from the vegetal pole in about 40% of the eggs. If this is done before first cleavage the polar lobe is formed normally but does not contain the vegetal body. The polar lobe can now be removed without removing the vegetal body. The resulting lobeless egg contains the vegetal body and may develop into a normal embryo, thereby suggesting that the morphogenetic determinants are indeed localized in the vegetal body.[61] Isolation of vegetal bodies in sufficient quantities to permit their fractionation and assaying the components for their morphogenetic potential may be possible as soon as a currently employed technique for isolating polar lobes has been improved. This technique involves the induction of abnormal cleavage by cold treatment. As a result of this treatment the blastomeres are connected

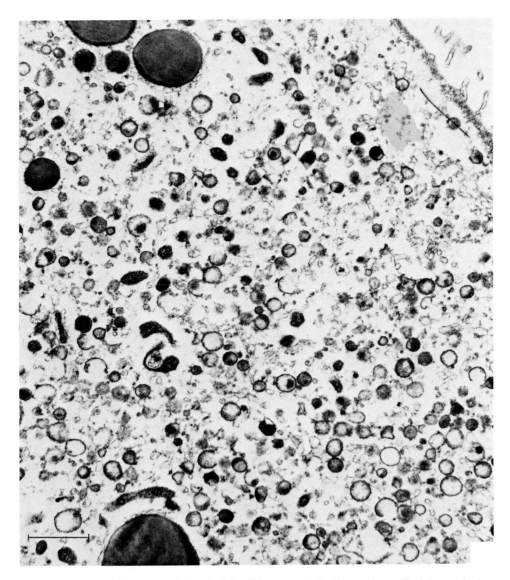

FIGURE 6. Detail of the contents of the polar lobe of the gastropod *Buccinum undatum*. The lobe contains large concentrations of vesicles with electron-dense contents. Bar indicates 1 μm.

by a long, slender strand of cytoplasm from which the polar lobe is suspended. It can thus be easily removed.

The vegetal body contains considerable quantities of RNA, suggesting that RNA may be the morphogenetic determinant.[8,61] *In situ* hybridization experiments with poly(U) gave positive results in the nucleus only, but there is no conclusive evidence that poly(A)-containing RNA is not present in the lobe. The RNA may be masked, so different preparative methods will have to be tried out. Other types of RNA besides messenger RNA are equally probable candidates for the role of determinants. Other chapters in this book deal with these regulatory RNAs. RNA is present in several other plasms with determinative capacity. For instance, in mature eggs of *Drosophila* the polar granules, which contain determinants of the germ cells, stain positively for RNA, but this staining property is lost by the blastoderm stage, before the pole cells are formed.[76] Similarly, the RNA at the vegetal pole of *Bithynia* eggs disappears at a moment, between the 12- and 16-cell stage, when the polar lobe probably

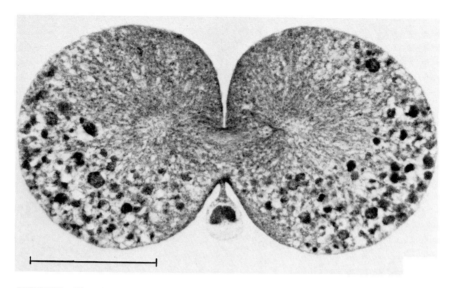

FIGURE 7.   First cleavage stage of the snail *Bithynia tentaculata*. A very small polar lobe is formed and in this lobe a densely staining cup-shaped plasm can be seen. (Hematoxylin-eosin staining.) Bar indicates 50 μm.

has not yet exerted any morphogenetic influence. The chain of events leading from the RNA to morphogenesis is completely obscure, but maybe further study of the origin and effects of the recently isolated polar granule specific protein in *Drosophila* eggs[78] may elucidate part of this process. Germ plasm-like structures have also been observed in polar lobes of the gastropods *Crepidula* and *Buccinum*.[79,80] These structures strikingly resemble the germ plasms of insects and amphibians, but nothing is known about their final destination. In molluscs the gonads develop from descendants of the second somatoblast 4d, so the segregation of these putative germ plasms into the 4d should be demonstrated to start with.

V. CELLULAR INTERACTIONS

Considerable differences exist between equally and unequally cleaving molluscan eggs as regards the importance of cellular interactions in early development. In the nonregulative polar lobe forming eggs development is determined to a large extent by cell-inherent factors. However, cellular interactions are indispensable for normal morphogenesis[81] and a limited degree of regulative capacity is observed. The importance of cellular interactions is clearly shown in *Dentalium*.[82,83] The B and D quadrants follow their inherent developmental pattern, but the A and C quadrants are influenced by the D quadrant. This is demonstrated by lobeless embryos where all four quadrants show the division chronology, the cleavage pattern, and the differentiations characteristic for the B quadrant in normal development. Thus, lobe-dependent structures not only include structures directly derived from the D quadrant, such as heart and intestine, but also structures derived from the other quadrants, such as eyes and tentacles. Inductive as well as inhibitory cellular interactions clearly constitute an important aspect of polar lobe activity.

Equally cleaving eggs have highly regulative capacities, as demonstrated by the ability of single blastomeres, isolated at the two- or four-cell stage, to develop into normal larvae.[2,3] There is no need to invoke vitalistic forces, as Driesch did,[84] to explain the equifinality expressed by isolated blastomeres. This phenomenon illustrates that cell interactions are important already at the two-cell stage. The two blastomeres are not yet committed to a fixed developmental pathway. A continuous signal produced by surface recognition factors may act to inhibit mutually the expression of the totipotency of the blastomeres. Positional

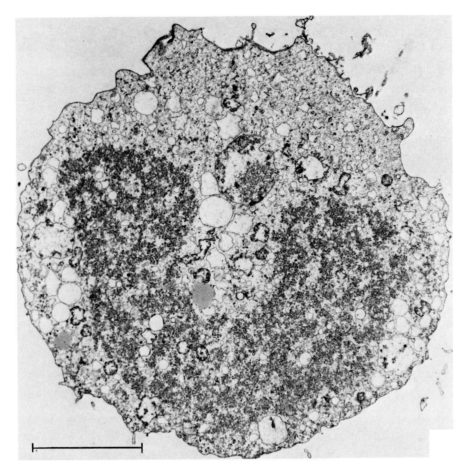

FIGURE 8. Polar lobe of *Bithynia tentaculata*. The cup-shaped "vegetal body" in this lobe consists of small vesicles. Bar indicates 4 μm.

information probably determines to a large extent the early development in equally cleaving eggs. A coordinate system that specifies positional values along the animal-vegetative axis undoubtedly plays an important role, but other systems probably act as well. A most interesting example of cellular interactions occurs during the determination of the dorsal macromere. Van den Biggelaar has made detailed studies of this event.[27,58,59,85] In *Lymnaea* and *Patella* one of the cross-furrow macromeres attains a central position in the embryo: in *Lymnaea* at the 24-cell stage and in *Patella* at the 32-cell stage. This macromere is the dorsal macromere. Its determination is not the result of prelocalized determinants, as deletion of one of the cross-furrow macromeres does not prevent the formation of the mesentoblast. Evidently, one of the remaining macromeres then acquires this specific function.[59] Most probably, interactions with the micromeres constitute the decisive factor. Suppression of the normally occurring contacts between micro- and macromeres, either by deletion of first quartet cells or by dissociation, inhibits normal differentiation of the macromeres.[59]

VI. GENE EXPRESSION IN EARLY DEVELOPMENT

The determinative events described in the previous sections are decisive for further development. They all occur before gastrulation and seem to be independent of the activity of embryonic genes as there are indications that a maternal program guides molluscan development through the cleavage period. Verdonk[87] has shown, by studying the time of effect

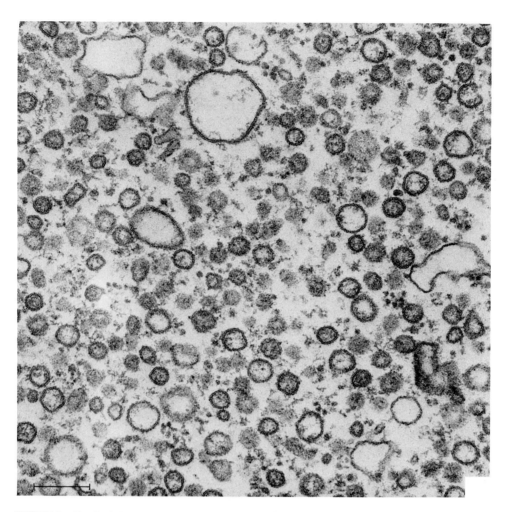

FIGURE 9.   Detail of the vegetal body in the polar lobe of *Bithynia tentaculata*. The body consists of vesicles with electron-dense contents. Bar indicates 0.2 μm.

of recessive lethal genes induced by X-irradiation, that in *Lymnaea* gastrulation requires the activity of embryonic genes. The earliest manifestation of lethal genetic factors was arrest of gastrulation. However, incubation of *Lymnaea* eggs with actinomycin D in concentrations up to 100 μg/mℓ did not suppress gastrulation (Boon-Niermeyer, cited in Reference 87). Morill et al.[88] subjected eggs of *Lymnaea palustris* to actinomycin D concentrations sufficient to reduce uridine incorporation more than 90%. Continuous treatment starting at the two-cell stage slowed development considerably. Gastrulation occurred 6 to 12 hr later than in the controls and the embryos developed into abnormal, arrested trochophores. Thus, actinomycin D appears to exert its first effects at gastrulation and is lethal at the trochophore stage. In *Ilyanassa* eggs treated continuously with actinomycin D at a concentration of 25 μg/mℓ over 90% of high molecular weight RNA synthesis were abolished within 4 hr. This treatment had no effect on gastrulation but further development was arrested.[72] In an actinomycin D experiment carried out on the bivalve *Spisula*,[90] in which uridine incorporation was reduced by 75%, development beyond gastrulation was inhibited. On the other hand, the limpet *Acmaea* seems to be capable of developing into a well-differentiated trochophore larva in the presence of actinomycin,[89] but it is not known to what extent RNA synthesis is affected in this species.

Whatever the exact period proves to be in which development is maternally programmed, it is clear that the cleavage period until gastrulation is predominantly independent of embryonic genome activity. Complete independence is unlikely as transcription has been demonstrated in several molluscan embryos at an early stage.[91] In the bivalves *Spisula*[90] and *Mulinia*[92] early transcription appears to involve only heterogeneous RNA and some transfer RNA. Ribosomal RNA synthesis is not detected until after gastrulation. The bulk of early RNA synthesis is probably histone mRNA, as the most prominent species, both in *Mulinia* and in *Spisula,* sediments around 9S, like histone mRNA from sea urchin or HeLa cells.[93,94] In *Spisula* the only maternal histone mRNA found is the one coding for F1 histone,[95] so the other histone mRNAs are probably newly synthesized during the cleavage period.

In gastropods a different pattern of RNA synthesis is found in the earliest stages. Ribosomal RNA synthesis appears to be activated during the cleavage period. In *Ilyanassa* rRNA synthesis can be detected as early as the four-cell stage.[96] In *Lymnaea* and *Acmaea* rRNA synthesis also occurs in early cleavage stages as judged by the appearance of nucleoli in which incorporation of uridine can be demonstrated by autoradiography.[97,98] Other types of RNA are also synthesized during the cleavage period in gastropods. Collier[99] has demonstrated by MAK chromatographic analysis that high molecular weight nonribosomal RNA, 4S RNA, and 5S RNA are all synthesized from the four-cell stage onwards. The observation that nascent RNA is polyadenylated in early cleavage stages[100] indicates that a substantial proportion of the newly synthesized RNA is messenger RNA.

The question of whether cytoplasmic determinants operate through differential transcription has been studied in *Ilyanassa* by comparing transcription between normal and lobeless embryos. It was found that removal of the polar lobe at first cleavage did not produce any change in the proportion of several size classes of RNA transcribed during the cleavage period. After gastrulation a significant deviation from the normal pattern of RNA synthesis was observed.[101] The lobeless embryo synthesizes less 34S RNA and more 12S and 16S RNA. These data are not sufficient to permit conclusions to be drawn about the mode of operation of the polar lobe determinants. Many more details will be needed. Obviously, differential transcription must occur for organogenesis to take place, but the early determinative events may involve only minute specific transcriptional activity, if any, which can easily go undetected.

VII. CONCLUSION

From the topics discussed in this chapter it is clear that the study of early molluscan embryology touches on every major aspect of developmental biology. An important contribution to the awareness of the intricate pattern of effects exerted by cytoplasmic localizations is furnished by the detailed studies of polar lobe forming eggs. Molluscan eggs in which the morphogenetic plasms are clearly defined structures, e.g., *Bithynia,* seem to be particularly suited for studying the mode of operation of cytoplasmic determinants at the subcellular and biochemical level. The study of the effects of cellular interactions at the level of individual blastomeres is also making rapid progress. The classic example is furnished by mammalian embryology where the position of a cell at the morula stage, inside or outside, determines its fate.[86] In molluscs a similar classic example of cellular interaction is evolving from studies on the determination of the dorsal macromere which plays a key role in the development of molluscs. The decisive events which determine development occur during the cleavage period. This period is maternally programmed and the activity of the embryonic genome is of minor importance, if at all.

REFERENCES

1. **Davidson, E. H.**, *Gene Activity in Early Development*, 2nd ed., Academic Press, New York, 1976, chap. 7.

2. **Morrill, J. B., Blair, C. A., and Larsen, W. J.**, Regulative development in the pulmonate gastropod *Lymnaea palustris*, as determined by deletion experiment, *J. Exp. Zool.*, 183, 47, 1973.

3. **Verdonk, N. H.**, Symmetry and asymmetry in the embryonic development of molluscs, in *Pathways in Malacology*, van der Spoel, S., van Bruggen, A. C., and Lever, J., Eds., Bohn, Scheltema & Helkema, Utrecht, 1979, 25.

4. **Osborn, M., Franke, W., and Weber, K.**, Direct demonstration of the presence of two immunologically distinct intermediate-sized filament systems in the same cell by double immunofluorescence microscopy, *Exp. Cell Res.*, 125, 37, 1980.

5. **Lazarides, E.**, Intermediate filaments as mechanical integrators of cellular space, *Nature (London)*, 283, 249, 1980.

6. **Byers, H. R. and Porter, K. R.**, Transformations in the structure of the cytoplasmic ground substance in erythrophores during pigment aggregation and dispersion. I. A study using whole-cell preparations in stereo high voltage electron microscopy, *J. Cell Biol.*, 75, 541, 1977.

7. **Wolosewick, J. J. and Porter, K. R.**, Microtrabecular lattice of the cytoplasmic ground substance. Artifact or reality, *J. Cell Biol.*, 82, 114, 1979.

8. **Dohmen, M. R. and Verdonk, N. H.**, The structure of a morphogenetic cytoplasm, present in the polar lobe of *Bithynia tentaculata* (Gastropoda, Prosobranchia), *J. Embryol. Exp. Morphol.*, 31, 423, 1974.

9. **Dohmen, M. R.**, The role of the egg surface in development, in *Control of Gene Expression in Early Embryonic Development*, Siddiqui, M. A. Q., Ed., CRC Press, Boca Raton, Fla., 1983, chap. 1.

10. **Lehto, V.-P., Virtanen, I., and Kurki, P.**, Intermediate filaments anchor the nuclei in nuclear monolayers of cultured human fibroblasts, *Nature (London)*, 272, 175, 1978.

11. **Small, J. V. and Celis, J. E.**, Direct visualization of the 10-nm (100-Å)-filament network in whole and enucleated cultured cells, *J. Cell Sci.*, 31, 393, 1978.

12. **Virtanen, I., Kurkinen, M., and Lehto, V.-P.**, Nucleus-anchoring cytoskeleton in chicken red blood cells, *Cell Biol. Int. Rep.*, 3, 157, 1979.

13. **Wang, E., Cross, R. K., and Choppin, P. W.**, Involvement of microtubules and 10-nm filaments in the movement and positioning of nuclei in syncytia, *J. Cell Biol.*, 83, 320, 1979.

14. **Brown, S. S. and Spudich, J. A.**, Cytochalasin inhibits the rate of elongation of actin filament fragments, *J. Cell Biol.*, 83, 657, 1979.

15. **Tsiftsoglou, A. S., Barrnett, R. J., and Sartorelli, A. C.**, Enucleation of differentiated murine erythroleukemia cells in culture, *Proc. Natl. Acad. Sci. U.S.A.*, 76, 6381, 1979.

16. **Pelseneer, P.**, Recherches sur l'embryologie des gastropodes, *Acad. R. Belge Cl. Sci. 2ᵉ Ser.*, 3, 1, 1912.

17. **Quatrano, R. S.**, Separation of processes associated with differentiation of two-celled *Fucus* embryos, *Dev. Biol.*, 30, 209, 1973.

18. **Peaucellier, G., Guerrier, P., and Bergerard, J.**, Effects of cytochalasin B on meiosis and development of fertilized activated eggs of *Sabellaria alveolata* L. (Polychaete Annelid), *J. Embryol. Exp. Morphol.*, 31, 61, 1974.

19. **Arnold, J. M. and Williams-Arnold, L. D.**, Cortical-nuclear interaction in cephalopod development: cytochalasin B effects on the informational pattern in the cell surface, *J. Embryol. Exp. Morphol.*, 31, 1, 1974.

20. **Zalokar, M.**, Effect of colchicine and cytochalasin B on ooplasmic segregation of ascidian eggs, *Wilhelm Roux' Arch. Entwicklungsmech. Org.*, 175, 243, 1974.

21. **Whittaker, J. R.**, Segregation during ascidian embryogenesis of egg cytoplasmic information for tissue-specific enzyme development, *Proc. Natl. Acad. Sci. U.S.A.*, 70, 2096, 1973.

22. **Yajima, H.**, Studies on embryonic determination of the harlequinfly, *Chironomus dorsalis*. I. Effects of centrifugation and of its combination with constriction and puncturing, *J. Embryol. Exp. Morphol.*, 8, 198, 1960.

23. **Conklin, E. G.**, The development of centrifuged eggs of ascidians, *J. Exp. Zool.*, 60, 1, 1931.

24. **Kühn, A.**, *Entwicklungsphysiologie*, 2nd ed., Springer-Verlag, Berlin, 1965.

25. **Raven, Chr. P.**, Mechanisms of determination in the development of gastropods, in *Advances in Morphogenesis*, Vol. 3, Academic Press, New York, 1964, 1.

26. **Reverberi, G.**, Ascidians, in *Experimental Embryology of Marine and Freshwater Invertebrates*, Reverberi, G., Ed., North-Holland, Amsterdam, 1971, 507.

27. **Van den Biggelaar, J. A. M.**, The determinative significance of the geometry of the cell contacts in early molluscan development, *Biol. Cell.*, 32, 155, 1978.

28. **Hörstadius, S., Lorch, J. J., and Danielli, J. F.**, Differentiation of the sea urchin egg following reduction of the interior cytoplasm in relation to the cortex, *Exp. Cell Res.*, 1, 188, 1950.

29. **Guerrier, P.,** Les caractères de la segmentation et la détermination de la polarité dorsoventrale dans le développement de quelques Spiralia. I. Les formes à premier clivage égal, *J. Embryol. Exp. Morphol.,* 23, 611, 1970.

30. **Guerrier, P.,** Les caractères de la segmentation et la détermination de la polarité dorsoventrale dans le développement de quelques Sprialia. III. *Pholas dactylus* et *Spisula subtruncata* (Mollusques Lamellibranches), *J. Embryol. Exp. Morphol.,* 23, 667, 1970.

31. **Rappaport, R.,** Cytokinesis in animal cells, *Int. Rev. Cytol.,* 31, 169, 1971.

32. **Asnes, C. F. and Schroeder, T. E.,** Cell cleavage. Ultrastructural evidence against equtorial stimulation by aster microtubules, *Exp. Cell Res.,* 122, 327, 1979.

33. **Rappaport, R.,** Effects of continual mechanical agitation prior to cleavage in echinoderm eggs, *J. Exp. Zool.,* 206, 1, 1978.

34. **Hara, K., Tydeman, P., and Kirschner, M.,** A cytoplasmic clock with the same period as the division cycle in *Xenopus* eggs, *Proc. Natl. Acad. Sci. U.S.A.,* 77, 462, 1980.

35. **Harvey, E. B.,** Parthenogenetic merogony or cleavage without nuclei in *Arbacia punctulata, Biol. Bull.,* 71, 101, 1936.

36. **Fankhauser, G.,** Cytological studies on egg fragments of the salamander *Triton.* IV. The cleavage of egg fragments without the egg nucleus, *J. Exp. Zool.,* 67, 349, 1934.

37. **Briggs, R., Green, E. U., and King, T. J.,** An investigation of the capacity for cleavage and differentiation in *Rana pipiens* eggs lacking "functional" chromosomes, *J. Exp. Zool.,* 116, 455, 1951.

38. **Freeman, G.,** The role of cleavage in the localization of developmental potential in the ctenophore *Mnemiopsis leidyi, Dev. Biol.,* 49, 143, 1976.

39. **Mahowald, A. P., Illmensee, K., and Turner, F. R.,** Interspecific transplantation of polar plasm between *Drosophila* embryos, *J. Cell Biol.,* 70, 358, 1976.

40. **Schwalm, F. E.,** Autonomous structural changes in polar granules of unfertilized eggs of *Coelopa frigida* (Diptera), *Wilhelm Roux' Arch. Entwicklungsmech. Org.,* 175, 129, 1974.

41. **Labordus, V.,** The effect of ultraviolet light on developing eggs of *Lymnaea stagnalis* (Mollusca, Pulmonata), Thesis, University of Utrecht, 1971.

42. **Boon-Niermeyer, E. K.,** The effect of puromycin on the early cleavage cycles and morphogenesis of the pond snail *Lymnaea stagnalis, Wilhelm Roux' Arch. Entwicklungsmech. Org.,* 177, 29, 1975.

43. **Van den Biggelaar, J. A. M. and Boon-Niermeyer, E. K.,** Origin and prospective significance of division asynchrony during early molluscan development, in *The Cell Cycle in Development and Differentiation,* Balls, M. and Billett, F. S., Eds., Cambridge University Press, London, 1973, 215.

44. **Sawai, T.,** Roles of cortical and subcortical components in cleavage furrows formation in amphibia, *J. Cell Sci.,* 11, 543, 1972.

45. **Conrad, G. W. and Davis, S. E.,** Microiontophoretic injection of calcium ions or of cyclic AMP causes rapid shape changes in fertilized eggs of *Ilyanassa obsoleta, Dev. Biol.,* 61, 184, 1977.

46. **Shimizu, T.,** Deformation movement induced by divalent ionophore A 23187 in the *Tubifex* egg, *Dev. Growth Diff.,* 20, 27, 1978.

47. **Hollinger, T. G. and Schuetz, A. W.,** "Cleavage" and cortical granule breakdown in *Rana pipiens* oocytes induced by direct microinjection of calcium, *J. Cell Biol.,* 71, 395, 1976.

48. **Conrad, G. W. and Davis, S. E.,** Polar lobe formation and cytokinesis in fertilized eggs of *Illyanassa obsoleta.* III. Large bleb formation caused by Sr^{2+}, ionophores X 537A and A 23187, and compound 48/80, *Dev. Biol.,* 74, 152, 1980.

49. **Raff, R. A.,** Polar lobe formation by embryos of *Illyanassa obsoleta:* effects of inhibitors of microtubule and microfilament function, *Exp. Cell Res.,* 71, 455, 1972.

50. **Guerrier, P., Van den Biggelaar, J. A. M., Van dongen, C. A. M., and Verdonk, N. H.,** Significance of the polar lobe for the determination of dorsoventral polarity in *Dentalium vulgare* (da Costa), *Dev. Biol.,* 63, 233, 1978.

51. **Raven, Chr. P.,** The cortical and subcortical cytoplasm of the *Lymnaea* egg, *Int. Rev. Cytol.,* 28, 1, 1970.

52. **Meshcheryakov, V. N.,** Orientation of cleavage spindles in pulmonate molluscs. I. Role of blastomere form in orientation of the second cleavage spindles, *Ontogenez,* 9, 558, 1978.

53. **Meshcheryakov, V. N.,** Orientation of cleavage spindles in pulmonate molluscs. II. Role of architecture of intercellular contacts in orientation of the third and fourth cleavage spindles, *Ontogenez,* 9, 567, 1978.

54. **Mescheryakov, V. N. and Veryasova, G. V.,** Orientation of cleavage spindles in pulmonate molluscs. III. Form and localization of mitotic apparatus in binucleate zygotes and blastomeres, *Ontogenez,* 10, 24, 1979.

55. **Raven, Chr. P.,** in *Morphogenesis: The Analysis of Molluscan Development,* 2nd ed., Macmillan, Pergamon, New York, 1966.

56. **Morgan, T. H.,** *Experimental Embryology,* Columbia University Press, New York, 1927.

57. **Dohmen, M. R. and Van de Mast, J. M. A.,** Electron microscopical study of RNA-containing cytoplasmic localizations and intercellular contacts in early cleavage stages of eggs of *Lymnaea stagnalis* (Gastropoda, Pulmonata), *Proc. K. Ned. Akad. Wet., Ser. C,* 81, 403, 1978.

58. **Van den Biggelaar, J. A. M.,** The fate of maternal RNA-containing ectosomes in relation to the appearance of dorsoventrality in the pond snail *Lymnaea stagnalis, Proc. K. Ned. Akad. Wet., Ser. C.,* 79, 421, 1976.

59. **Van den Biggelaar, J. A. M. and Guerrier, P.,** Dorsoventral polarity and mesentoblast determination as concomitant results of cellular interactions in the mollusk *Patella vulgata, Dev. Biol.,* 68, 462, 1979.

60. **Fioroni, P.,** Zur Struktur der Pollappen und der Dottermakromereneine vergleichende Übersicht, *Zool. Jb. Anat.,* 102, 395, 1979.

61. **Dohmen, M. R. and Verdonk, N. H.,** The ultrastructure and role of the polar lobe in development of molluscs, in *Determinants of Spatial Organization,* Konigsberg, I. R. and Subtelny, S., Eds., Academic Press, New York, 1979, 3.

62. **Moor, B.,** Zur frühen Furchung des Eies von *Littorina littorea* L. (Gastropoda Prosobranchia), *Zool. Jb. Anat.,* 91, 546, 1973.

63. **Clement, A. C.,** Experimental studies on germinal localization in *Illyanassa.* I. The role of the polar lobe in determination of the cleavage pattern and its influence in later development, *J. Exp. Zool.,* 121, 593, 1952.

64. **Parisi, E., Filosa, S., De Petrocellis, B., and Monroy, A.,** The pattern of cell division in the early development of the sea urchin *Paracentrotus lividus, Dev. Biol.,* 65, 38, 1978.

65. **Parisi, E., Filosa, S., and Monroy, A.,** Actinomycin D — disruption of the mitotic gradient in the cleavage stages of the sea urchin embryo, *Dev. Biol.,* 72, 167, 1979.

66. **Conklin, E. G.,** The embryology of *Crepidula, J. Morphol.,* 13, 1, 1897.

67. **Van Dongen, C. A. M. and Geilenkirchen, W. L. M.,** The development of *Dentalium* with special reference to the significance of the polar lobe. I, II, and III. Division chronology and development of the cell pattern in *Dentalium dentale* (Scaphopoda), *Proc. K. Ned. Akad. Wet., Ser. C,* 77, 57, 1974.

68. **Verdonk, N. H. and Cather, J. N.,** The development of isolated blastomeres in *Bithynia tentaculata* (Prosobranchia, Gastropoda), *J. Exp. Zool.,* 186, 47, 1973.

69. **Cather, J. N. and Verdonk, N. H.,** Development of *Dentalium* following removal of D-quadrant blastomeres at successive cleavage stages, *Wilhelm Roux' Arch. Entwicklungsmech. Org.,* 187, 355, 1979.

70. **Collier, J. R. and McCarthy, M. E.,** Protein synthesis in the polar lobe of the *Ilyanassa* egg, *Am. Zool.,* 19, 950, 1979.

71. **Donohoo, P. and Kafatos, F. C.,** Differences in the proteins synthesized by the progeny of the first two blastomeres of *Ilyanassa,* a ''mosaic'' embryo, *Dev. Biol.,* 32, 224, 1973.

72. **Raff, R. A., Newrock, K. M., Secrist, R. D., and Turner, F. R.,** Regulation of protein synthesis in embryos of *Ilyanassa obsoleta, Am. Zool.,* 16, 529, 1976.

73. **Guerrier, P.,** A possible mechanism of control of morphogenesis in the embryo of *Sabellaria alveolata* (Annelide polychaete), *Exp. Cell Res.,* 67, 215, 1971.

74. **Verdonk, N. H.,** The effect of removing the polar lobe in centrifuged eggs of *Dentalium, J. Embryol. Exp. Morphol.,* 19, 33, 1968.

75. **Clement, A. C.,** Development of the vegetal half of the *Ilyanassa* egg after removal of most of the yolk by centrifugal force, compared with the development of animal halves of similar visible composition, *Dev. Biol.,* 17, 165, 1968.

76. **Mahowald, A. P.,** Polar granules of *Drosophila.* IV. Loss of RNA from polar granules during early stages of embryogenesis, *J. Exp. Zool.,* 176, 345, 1971.

77. **Clement, A. C.,** Development of *Ilyanassa* following removal of the D-macromere at successive cleavage stages, *J. Exp. Zool.,* 149, 193, 1962.

78. **Waring, G. L., Allis, C. D., and Mahowald, A. P.,** Isolation of polar granules and the identification of polar granule specific protein, *Dev. Biol.,* 66, 197, 1978.

79. **Dohmen, M. R. and Lok, D.,** The ultrastructure of the polar lobe of *Crepidula fornicata* (Gastropoda, Prosobranchia), *J. Embryol. Exp. Morphol.,* 34, 419, 1975.

80. **Dohmen, M. R. and Verdonk, N. H.,** Cytoplasmic localizations in mosaic eggs, in *Maternal Effects in Development,* Newth, D. R. and Balls, M., Eds., Cambridge University Press, London, 1979, 127.

81. **Cather, J. N.,** Cellular interactions in the regulation of development in annelids and molluscs, in *Advances in Morphogenesis,* Vol. 9, Academic Press, New York, 1971, 67.

82. **Van Dongen, C. A. M.,** The development of *Dentalium* with special reference to the significance of the polar lobe. V and VI. Differentiation of the cell pattern in lobeless embryos of *Dentalium vulgare* (da Costa) during late larval development, *Proc. K. Ned. Akad. Wet. Ser. C,* 79, 245, 1976.

83. **Van Dongen, C. A. M.,** The development of *Dentalium* with special reference to the significance of the polar lobe. VII. Organogenesis and histogenesis in lobeless embryos of *Dentalium vulgare* (da Costa) as compared to normal development, *Proc. K. Ned. Akad. Wet. Ser. C,* 79, 454, 1976.

84. **Driesch, H.,** *Philosophie des Organischen,* 2nd ed., Verlag von Wilhelm Engelmann, Leipzig, 1921.

85. **Van den Biggelaar, J. A. M.,** Development of dorsoventral polarity preceding the formation of the mesentoblast in *Lymnaea stagnalis, Proc. K. Ned. Akad. Wet. Ser. C,* 79, 112, 1976.

86. **Hillman, N., Sherman, M. I., and Graham, C.,** The effect of spatial arrangement on cell determination during mouse development, *J. Embryol. Exp. Morphol.,* 28, 263, 1972.

87. **Verdonk, N. H.,** Gene expression in early development of *Lymnaea stagnalis, Dev. Biol.,* 35, 29, 1973.
88. **Morrill, J. B., Rubin, R. W., and Grandi, M.,** Protein synthesis and differentiation during pulmonate development, *Am. Zool.,* 16, 547, 1976.
89. **Karp, G. C. and Whiteley, A. H.,** DNA-RNA hybridization studies of gene activity during the development of the gastropod *Acmaea scutum, Exp. Cell Res.,* 78, 236, 1973.
90. **Firtel, R. A. and Monroy, A.,** Polysomes and RNA synthesis during early development of the surf clam *Spisula solidissima, Dev. Biol.,* 21, 87, 1970.
91. **Kidder, G. M.,** RNA synthesis and the ribosomal cistrons in early molluscan development, *Am. Zool.,* 16, 501, 1976.
92. **Kidder, G. M.,** Gene transcription in mosaic embryos. I. The pattern of RNA synthesis in early development of the coot clam *Mulinia lateralis, J. Exp. Zool.,* 180, 55, 1972.
93. **Gross, K., Rudermann, J., Jacobs-Lorena, M., Baglioni, C., and Gross, P. R.,** Cell-free synthesis of histones directed by messenger RNA from sea urchin embryos, *Nat. New Biol.,* 241, 272, 1973.
94. **Jacobs-Lorena, M., Baglioni, C., and Borun, T. W.,** Translation of HeLa cells histone mRNA by a mouse ascites cell-free system, *Proc. Natl. Acad. Sci. U.S.A.,* 69, 2095, 1972.
95. **Gabrielli, F. and Baglioni, C.,** Maternal messenger RNA and histone synthesis in embryos of the surf clam *Spisula solidissima, Dev. Biol.,* 43, 254, 1975.
96. **Collier, J. R. and Weinstein, H. M.,** DNA-like RNA synthesis during *Ilyanassa* embryogenesis, *J. Cell Biol.,* 39, 27A, 1968.
97. **Van den Biggelaar, J. A. M.,** RNA synthesis during cleavage of the *Lymnaea* egg, *Exp. Cell Res.,* 67, 207, 1971.
98. **Karp, G. C.,** Autoradiographic pattern of [^{3}H]uridine incorporation during the development of the mollusc *Acmaea scutum, J. Embryol. Exp. Morphol.,* 29, 15, 1973.
99. **Collier, J. R.,** Nucleic acid chemistry of the *Ilyanassa* embryo, *Am. Zool.,* 16, 483, 1976.
100. **Collier, J. R.,** Polyadenylation of nascent RNA during the embryogenesis of *Ilyanassa obsoleta, Exp. Cell Res.,* 95, 263, 1975.
101. **Koser, R. B. and Collier, J. R.,** An electrophoretic analysis of RNA synthesis in normal and lobeless *Ilyanassa* embryo, *Differentiation,* 6, 47, 1976.

Chapter 6

REGULATORY PROCESSES DURING ODONTOGENIC EPITHELIAL-MESENCHYMAL INTERACTIONS

Harold C. Slavkin and Margarita Zeichner-David

TABLE OF CONTENTS

I. "DETERMINATION" DURING EMBRYOGENESIS

The phenomenon of determination during embryogenesis remains one of the most fascinating and as yet poorly understood issues in developmental biology. During the past 3 decades significant attention and progress has been generated regarding the problem of cellular differentiation in multicellular organisms and the mechanisms controlling gene expression in differentiated cells. However, little progress has been made regarding how eukaryotic cells become differentiated. What process or set of processes are involved in selective determination of a particular phenotype? Although the problem of determination in eukaryotic cellular systems was enthusiastically pursued by Spemann,[1] Mangold,[2] and Weiss,[3,4] our collective understanding of how cells become differentiated remains an elusive and as yet a somewhat obscure scientific problem.

Several mechanisms have been suggested which attempt to explain the available data and which have been used rather successfully to foster renewed scientific interest in the problem of how cells become differentiated. These mechanisms include the suggestion of (1) chemical gradients generating morphogenetic fields within the embryo and (2) discrete and localized "inductive" events which translate into cellular differentiation. These suggested mechanisms differ both in their requirement for cellular interactions and in the nature of the cellular interactions involved.[5-8]

Induction has a strict requirement for cell-to-cell interactions. Further, it is assumed that one cell type will "induce" a responsive cell type to acquire a pattern of differential gene expression and a related phenotype which is both original to discrete interaction and the result of a specific interaction. Induction *is not* the amplification of already predetermined gene expression; *it is not* merely a quantitative enhancement of gene expression. Induction must invoke qualitative changes within the responding cell type which are nonreversible and characteristic of a unique cell phenotype.

An interesting concept has recently emerged which attempts to integrate the suggested mechanisms of morphogenetic gradients within the embryo and embryonic induction. In this context, Wolpert has stated that "epithelial-mesenchymal interaction is the transfer of positional information".[9] During embryogenesis clones of dissimilar cells express different and changing quanta of positional information. When a responsive cell responds to an inductive event, according to Wolpert, "positional information is the interpretation of epigenetic signals" from the immediate microenvironment of the responding cell.[9] Despite these challenging and fascinating explanations for the phenomena of determination during embryogenesis, a number of painfully pertinent issues remain as yet unresolved. What are the molecular "epigenetic signals"? What is the mode of transmission for these inductive signals? What mechanism provides for the transduction of positional or inductive information during development? What might the role be for RNA during epithelial-mesenchymal interactions?

II. RNA AS AN EPIGENETIC SIGNAL DURING DEVELOPMENT

The expression of cellular genetic information is a profound process in developmental biology. During embryogenesis, including gastrulation, neurulation, and subsequent organogenesis, cells become determined as a function of spatial and temporal information. Spatial and temporal positional information and the subsequent genesis of pattern and form both constitute a significant linkage between genetics, molecular biology and biochemistry, and morphology. Within this broad and complex problem area is the critical postulate that RNA and/or polynucleotides mediate embryonic induction by representing the transfer of a determinant morphogen between heterotypic cells. A mechanism which has received considerable attention is the possibility that an informational macromolecule is transferred from an inducing cell to a responding cell. RNA has been particularly suggested as a candidate in this regard.[7,10-19]

III. SUGGESTED REGULATORY RNA MOLECULES

A number of RNA molecules have been detected in vivo and in vitro associated with the extracellular microenvironment during cell-to-cell interactions.[20,21] Relatively small molecular weight RNAs, some of which have been found to be methylated, have been found within the extracellular matrix during embryonic epithelial-mesenchymal interactions,[22,23] as well as in the media of cells in culture.[24-26] Evidence has been found which indicates that RNA is associated with outer cell surfaces.[26] Despite this information very little data have been obtained which unequivocally demonstrate (1) the direct transfer of RNA between homotypic and/or heterotypic cells in vivo or in vitro; (2) the isolation and characterization of this "informational RNA"; (3) the direct demonstration of this discrete RNA upon differential gene expression during development; (4) the direct demonstration of "causality" between the transfer of this informational RNA and spatial and temporal positional values for the responding cell; and (5) *direct* evidence indicating that this process actually occurs in vivo during a normal developmental process.

Despite these qualifications, a number of suggested "regulatory RNA molecules" are pertinent to our discussion. Double-stranded RNAs have been isolated from eukaryotic cells.[27-29] These RNAs are low-molecular weight (4 to 14S), appear to originate in the nucleus, and hybridize to cellular DNA. They appear to be transcribed from reiterated DNA sequences. The functional significance of these RNAs is as yet unknown.

A number of low-molecular weight RNAs have been demonstrated (4 to 9S) and suggested to have a regulatory function in eukaryotic cell differentiation.[30-32] Whereas many low-molecular weight RNAs have been found within the nucleus and/or the cytoplasm of cells, and have been shown to be present in amounts equivalent to rRNA, their function(s) remains unknown.

Another approach has been to search for single-stranded RNAs as possible regulator RNAs. A number of such RNAs have been isolated and suggested to function in differential gene activity within eukaryotic cells. The concept of regulatory RNA has recently been supported through the elegant studies which demonstrate a low-molecular weight RNA (7S) from embryonic chick heart capable of inducing a tissue-specific set of changes within progenitor cardiac cells in vitro.[11,14,15] This RNA seems to be heart specific and is not found in other embryonic chick tissues. The RNA sediments at 7S, contains detectable poly(A) sequence, and is not translatable in a cell-free system.[14,15] In addition, two low-molecular weight "regulatory" RNAs have recently been described which are operant during *Artemia salina* embryogenesis.[33] One of these RNAs is a "translational inhibitor", whereas the other RNA is a "translational activator". During embryonic chick myogenesis a translational control RNA (tcRNA) has been identified which might regulate the initiation and translation of mRNAs.[34]

IV. TRANSFER OF RNAs BETWEEN CELLS IN "CONTACT"

Examples of cell-to-cell communication exist throughout metazoan developing systems. The nature of this communication, however, is known to involve a number of different and unique modes of transmission and reception including (1) direct cell-to-cell contacts, (2) short-range cell-to-cell interactions mediated through an interposed extracellular matrix, and (3) long-range intercellular communication mediated by humoral factors such as those associated with polypeptide and/or steroid signals (e.g., the effects of pancreatic insulin on lens cell elongation) (see References 6, 8, and 35 to 39) (Figure 1).

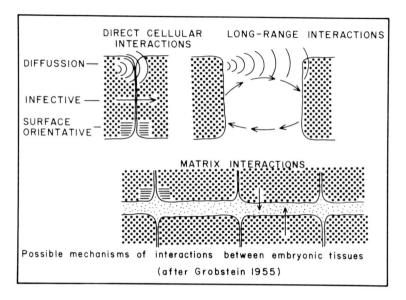

FIGURE 1. Diagrammatic representation of the possible modes for transmission of secondary embryonic induction during epithelial-mesenchymal interactions.[6,37]

Several recent reviews are available which provide a critical assessment of the transfer of the nucleotides and RNAs between cells in contact.[20,21,40,41] The assumed "information" passing between heterotypic cells in vivo and in vitro is further postulated to be highly specific. The specificity might be mediated by the highly specific base sequences and conformational properties of ribonucleic acids or polynucleotides. Several lines of investigation have been very provocative. One approach has been to examine the role of gap junctions in cell-to-cell communication. Available evidence suggests that gap junctions represent ionic coupling between cells achieving metabolic cooperation and a mechanism by which molecules and/or ions can be exchanged during significant phases of development.[41] Gap junctions appear to mediate the flow of ions between similar and dissimilar cells, the intercellular passage of fluorescein molecules, and the autoradiographic demonstration of the transfer of radiolabeled molecules from one cell to another including nucleotides, polynucleotides, and low-molecular weight RNAs.[20,24,41-43] Gap junctions appear in most developing eukaryotic cell systems coincident with major changes in cell differentiation; they are observed within and amongst homotypic cell types as well as between heterotypic cell types. Available information suggests that gap junctions or equivalent forms of intercellular communication might mediate the transfer of RNAs between cells providing both spatial and temporal specificity.

Metabolic cooperation is a form of cell-to-cell communication in which one cell may directly transfer a specific molecule to an immediately adjacent cell. The formation of these direct cell-to-cell contacts, such as gap junctions, occurs in vivo and in vitro.[43] Metabolic cooperation has been classically defined as the phenomenon in which the phenotype of a mutant cell is corrected by intimate cell-to-cell contacts with normal cells. For example, fibroblast cells deficient in hypoxanthine phosphoribosyltransferase (HPRT) incorporate very low levels of tritiated hypoxanthine, guanine, or uridine. When HPRT (−) mutant cells are cocultured with normal cells (HPRT +), gap junctions form and through some form of metabolic cooperation the mutant cells appear to incorporate normal amounts of various labeled purines.[41,42] It is assumed that "informational RNA" or some other regulatory molecule was passed between normal and mutant cells.[20,21] In this context, metabolic cooperation is a phenomenon perhaps comparable to embryonic induction during development.

V. TRANSMISSION OF INDUCTIVE "SIGNALS" DURING EPITHELIAL-MESENCHYMAL INTERACTIONS

The classic experiment of Spemann and Mangold[44] demonstrated that the invaginating chorda-mesoderm during gastrulation is responsible for the induction of the overlaying presumptive medullary plate (neuroectoderm) to become the central nervous system in the amphibian embryo. This experiment established the conditions for "primary embryonic induction".[1-3,44] Recently, direct cell-to-cell contacts have been demonstrated between cells of the inducing chorda-mesoderm and adjacent ectoderm; these intimate contacts between heterotypic cell types may mediate the primary inductive effect.[45]

The classic experiments of Grobstein[6] and Saxen and colleagues[37] have established a number of critical observations regarding secondary embryonic induction or epithelial-mesenchymal interactions. During a number of epithelial-mesenchymal interactions in amphibian, avian, and mammalian developing systems, mesenchymal cells have been found to "induce" adjacent epithelial cells to differentiate into specific phenotypes.[37-40] In each case the mesenchyme-specific induction has been found to be both temporally and spatially specific.[3,6,9,35-40] In examples using heterotypic tissue recombinations, for example embryonic rabbit mammary gland mesenchyme and embryonic chick dorsal skin epithelium, the mesenchyme "induces" the epithelium to express a phenotype which is complementary to that of the inductive mesenchyme; mammalian mammary mesenchyme induces avian skin epithelium to differentiate into branching tubular structures containing acinar cells allegedly synthesizing casein.[46]

What is the mode of transmission for inductive signals during epithelial-mesenchymal interactions? Is the "inductive signal" a diffusible substance or is direct cell-to-cell contact required? A number of transfilter experiments indicate that the inductive signal is a relatively large macromolecule, that transmission of the signal requires nascent RNA and protein synthesis, that the transmission is energy dependent, and that short-range cell-to-cell interactions or direct cell-to-cell contacts are likely modes for transmission.[37] Dialysis tubing of 3,000 or 10,000 molecular weight exclusion inhibits transfilter epithelial-mesenchymal interactions.[37] Nucleopore filters having pore sizes of 0.2 to 0.8 mm diameter support positive inductive interactions.[37] All available information currently argues for direct cell-to-cell contact mediated by particles or cellular processes which are 0.1 μm in diameter or larger.[6,9,19,35-40] It is also appropriate to emphasize that the responding epithelial cells must show DNA replication in order to respond to the mesenchymal inductive signal[37] (Figures 2 and 3).

VI. EPITHELIAL-MESENCHYMAL INTERACTIONS DURING TOOTH MORPHOGENESIS: A MODEL SYSTEM

A. General Embryological Description

The differentiation of epidermal organ systems into functional systems has been found to be dependent upon interactions termed epithelial-mesenchymal interactions. An understanding of such heterotypic tissue interactions (also termed "secondary embryonic induction") is a prerequisite in studies designed to investigate normal and abnormal organogenesis. Common to many different examples of epidermal organogenesis is the critical developmental biology question: What is the nature of and how are inductive signals transmitted over short extracellular distances? Whether one investigates kidney, mammary, salivary, feather, hair, skin, lung, thyroid, or tooth morphogenesis, it becomes readily apparent that several key issues must be resolved. What factors are instructive or inductive toward epidermal morphogenesis and differentiation, and what factors are permissive or supportive of these processes? Can one dissect "instructive" from "permissive" using modern methods of

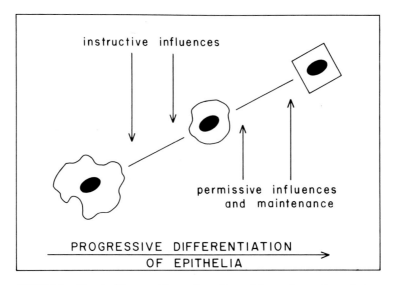

instructive influences

permissive influences
and maintenance

PROGRESSIVE DIFFERENTIATION
OF EPITHELIA

FIGURE 2. The significance of instructive cell-to-cell interactions and permissive factors during epithelial-mesenchymal interactions (e.g., skin, thyroid, salivary gland, mammary gland, thymus, and tooth organ). See extensive discussions by Saxen and colleagues.[37]

developmental biology? One example of epidermal organogenesis which is extremely amenable to the suggested dissection is the developing mouse molar tooth organ[39,40] (Figure 4).

The embryonic murine molar tooth organ provides a unique opportunity to study epithelial-mesenchymal interactions *in situ* as compared to in vitro. In the course of early mouse embryogenesis, immediately following implantation, cranial neural crest cells migrate from the dorsal aspects of the forming neural tube and migrate into the forming craniofacial complex; often in tandem with the opthalmic, maxillary, and mandibular divisions of the trigeminal ganglia. These cranial neural crest cells differentiate into a large number of different phenotypes including most of the mesenchyme of the forming face and oral cavity. The derivatives include a number of ectomesenchymal cell types including the dental papilla mesenchyme and odontoblast cells associated with odontogenesis. Cranial neural crest cell derivatives appear to migrate into the forming oral cavity, within either the first or second branchial arches, and differentiate into an ectomesenchyme which interacts with the oral ectodermal lining of the stomadeum. These initial epithelial-mesenchymal interactions occur during the 9th to 10th days of gestation. During Theiler stage 20, approximately 11.5 to 12.0 days of gestation, the dental lamina forms as a direct consequence of ectomesenchyme-oral ectodermal interactions, which subsequently forms the molar tooth organs. The dental lamina is a band of thickened epithelium which indicates the future position of each posterior dental arch (i.e., "molariform field") (Figure 5A and B).

The embryonic mammalian tooth organ is formed by a series of reciprocal epithelial-mesenchymal interactions which result in the acquisition of highly specific phenotypes as well as unique patterns of cells integrated into tooth form.[39] In the course of early embryogenesis, cranial neural crest cells migrate into the forming craniofacial complex and differentiate into dental papilla mesenchyme as well as other crest cell derivatives. The progenitor tooth mesenchyme differentiates in juxtaposition to an ectodermally derived epithelium that subsequently becomes the enamel organ epithelium. These two tissue types interact throughout the early stages of tooth morphogenesis. The interface between these heterotypic tissues provides a spatial and temporal orientation.

B. Cell-Specific Gene Products Related to Odontogenesis

Prior to overt terminal differentiation, the dental papilla mesenchyme and adjacent enamel

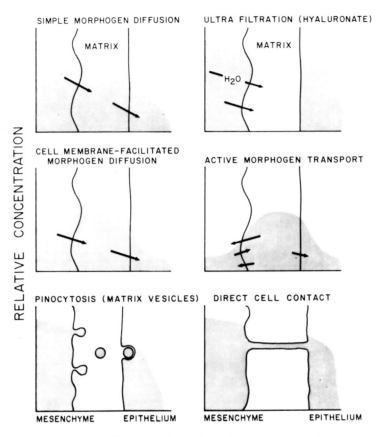

FIGURE 3.   The possible mode of transmission for purine bases, nucleosides, nucleotides, polynucleotides, and ''inductive RNA'' during epithelial-mesenchymal interactions. Diffusion is a physical-chemical process by which molecules move across cell membranes in direct proportion to the electrochemical gradient(s) established across the membrane. Transfer is determined by Fick's law. **Simple diffusion** occurs down a concentration gradient and is not energy dependent. **Ultrafiltration** can result from an influx of hyaluronic acid resulting in a change or shift in hydrostatic pressure. **Cell membrane-facilitated morphogen diffusion** suggests that molecules are transferred via a cell membrane component such as a cell process or matrix vesicle. **Active morphogen transport** implies that energy is required to maintain transmission activity. During **pinocytosis** mesenchymal cell processes form matrix vesicles which transport inductive RNAs to the adjacent epithelia; matrix vesicle contents are then internalized by the epithelia. Direct cell-to-cell contact suggests that instructive molecules can be transmitted from mesenchyme cell process to adjacent epithelial cells via gap-like junctions.

organ epithelia represent dividing and expanding populations of cells. Those cells aligned in juxtaposition to the interface between epithelium and mesenchyme express a set of unique phenotypes. The mesenchymal cells at the interface with epithelium produce type I collagen, cell surface fibronectin, and hyaluronate-rich glycosaminoglycans.[47] The adjacent epithelial cells produce type IV collagen, laminin, and glycosaminoglycans associated with the basal lamina.[47,48] The basal lamina is an epithelial-derived cell surface specialization deposited at the interface between the inner enamel epithelia and the adjacent cranial neural crest-derived mesenchyme[49] (Figure 6A and B).

During subsequent epithelial-mesenchymal interactions, the mesenchymal cells at the interface align themselves into a sheet or ''epithelia'' of mesenchyme to form the progenitor odontoblast cell layer. These cells differentiate into nondividing and highly polarized odontoblast cells which synthesize and secrete extracellular matrix macromolecules including

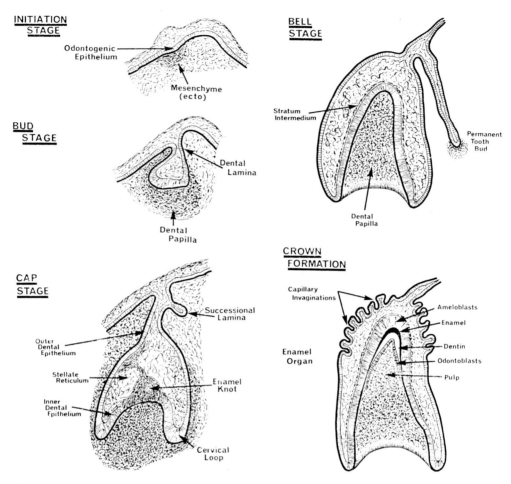

FIGURE 4.   Mammalian tooth morphogenesis consists of a number of developmental stages identified as initiation, bud, cap, bell, and crown formation. During mouse molar odontogenesis, initiation stage occurs at 12-days gestation and reaches the cap stage by 17-days gestation (Theiler stage 25). Secretory amelogenesis is first detected (by histological criteria) at 2-days postnatal age.

types I and III collagen, dentin phosphoprotein, and proteoglycans.[49] Mesenchymal cell differentiation precedes that of the adjacent epithelium by approximately 48 hr.[49] The adjacent inner enamel epithelia differentiate into nondividing and highly polarized ameloblasts which synthesize and secrete extracellular matrix macromolecules of amelogenins (enamel proteins) and glycosaminoglycans.[50-52] Both the dentin and the enamel organic matrices subsequently mineralize.

C. Developmental Comparisons of Murine Odontogenesis In Vivo, as Xenografts on the Chick Chorio-Allantoic Membrane (CAM), and In Vitro

In an attempt to experimentally dissect **instructive** from **permissive** influences upon embryonic tooth morphogenesis and differentiation, our research group has designed a series of investigations which inquire into developmental information which is produced by the tissue interactants **prior to the time of investigations, produced during the periods of observations, or derived directly from humoral factors in the microenvironment.** Modern birds do not express tooth morphogenesis or differentiation. Therefore, mammalian tooth organs explanted as xenografts to the chick chorio-allantoic membrane would obtain simple nutrients and ubiquitous humoral cues from the avian milieu, but perhaps would not receive specific mammalian humoral cues for odontogenesis (see discussion in Reference 49). Suc-

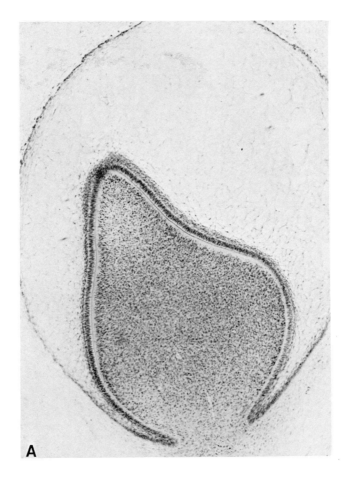

FIGURE 5.   Survey photomicrographs of late cap stage (A) and the initiation
of crown formation with both dentin and enamel extracellular matrix formation
(B).

cessful morphogenesis and differentiation in the avian environment would suggest that
odontogenic tissue interactants produced intrinsic developmental information during tooth
development on the CAM, or that this information had already been produced prior to the
physical isolation of Theiler stage 25 molar tooth organs (see Table 1). In separate investigations,[53,54] our group has developed a chemically defined organ culture medium (Figure
7) which **does not contain** fetal or adult serum, embryonic extracts, and antibiotics; yet is
permissive for tooth morphogenesis and differentiation (Figure 8A, B, and C). This in vitro
approach suggested that external cues derived from exogenous sera or embryonic extracts
were **not required** for epithelial-mesenchymal interactions beyond the cap stage of odontogenesis. Embryonic molar tooth organs were either completely or partially determined for
a specific pattern of morphogenesis and differentiation by the cap stage of development,
and/or subsequent heterotypic tissue interactions generated *de novo* developmental instructions during following stages of odontogenesis. This methodology will be useful in future
studies of epithelial-mesenchymal interactions.

D. Preliminary Characterization of Epithelial-Specific mRNAs during Odontogenesis

Synthesis and secretion of characteristic proteins are essential aspects of merocrine cell
differentiation. Some well-known examples of such differentiation-specific protein synthesis
are the production of ovalbumin by chick oviduct epithelium,[55-57] amylase and insulin by

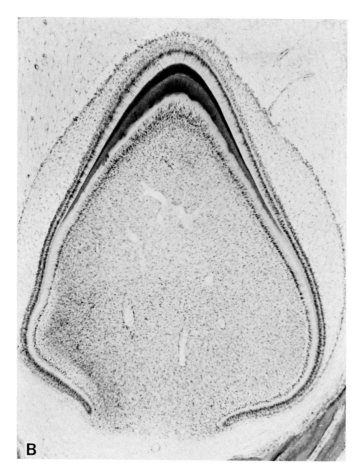

FIGURE 5B.

pancreatic acinar epithelium,[58-61] type II collagen by chondroblast cells, and types I and III collagens by fibroblast cells.[62] Particularly interesting are those factors controlling the synthetic rates of individual proteins during differentiation in epidermal organogenesis (e.g., pancreas, mammary gland, salivary gland, and tooth organs) and the developmental programs for sequential elaboration of several distinct gene products by a single cell type. The study of such processes in epidermal organ systems is often complicated by unavailability of synchronized cells at defined developmental stages, presence of extraneous cell types, and concurrent mitotic activities. In this respect embryonic mammalian tooth organs, which consist of a single sheet of ectodermal-derived ameloblast cells adjacent to a sheet of cranial neural-crest-derived ondotoblast cells, separated by their respective extracellular matrix gene products, developmentally synchronized, and nondividing, have certain advantages.

During embryonic tooth morphogenesis, odontoblast cell differentiation involves sequential elaboration of at least six different extracellular matrix proteins, fibronectin,[47,48] types I and III collagens,[63,64] proteoglycans,[65,66] dentin phosphoproteins,[67,68] and several other noncollagenous glycoproteins.[69] Epithelial ameloblast differentiation involves the production of enamel matrix proteins (i.e., "amelogenins", "enamel proteins", "enameloids"). The dentin and enamel extracellular matrices, therefore, consist of several specific proteins, each produced during a characteristic developmental period.[51,70-73] The timing of synthesis and secretion of these merocrine-type cell proteins is apparently controlled by a series of complex epithelial-mesenchymal interactions.[6,37,39,40,49,74,75]

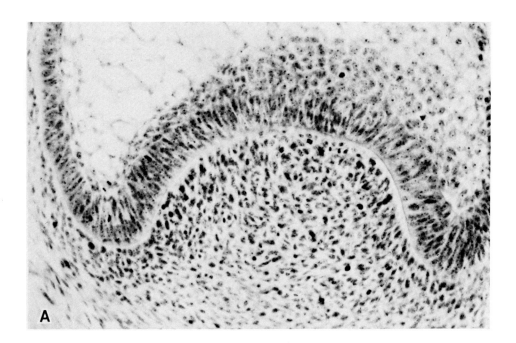

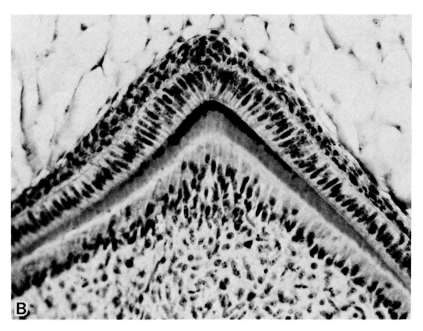

FIGURE 6. A developmental comparison between epithelial and mesenchymal cell differentiation at the cap stage and during early crown formation. (A) During cap stage, inner enamel epithelia and adjacent ectomesenchyme are separated by a metachromatic basement membrane. Both cell types incorporate ^{3}H-thymidine into DNA at this stage of odontogenesis. (B) A gradient of cytodifferentiation is readily apparent within inner enamel epithelial cells which differentiate into secretory ameloblasts, and within adjacent preodontoblast cells which differentiate into odontoblasts. Both odontoblasts and ameloblasts represent nondividing, terminally differentiated, merocrine-type cells producing dentin and enamel extracellular matrices, respectively.

During ameloblast cell differentiation, the synthesis of unique epithelial-derived extracellular matrix proteins can be used in experiments related to isolation and characterization of epithelial-specific mRNA molecules. We have previously identified four extracellular

Table 1
CHRONOLOGICAL COMPARATIVE DESCRIPTIONS OF MURINE
MOLAR TOOTH ORGAN DEVELOPMENT IN VIVO, ON THE CAM,
AND IN VITRO

	Developmental staging			
Descriptive characteristics[a]	Theiler stage	In vivo time gestation	CAM time	In vitro time
Dental lamina initiation	20	12 days	—	—
Tooth bud stage	22	14 days	—	—
Enamel organ formation	24	16 days	—	—
Cap stage	25	17 days	0	0
Bell stage	26	18 days	ND[b]	2 days
Dentinogenesis initiated	27	20 days	ND	5 days
Crown pattern completed	27	Birth	5	7 days
Amelogenesis initiated		2 days postnatal	7	10 days
Enamel formation completed		10 days postnatal	ND	ND

[a] Chronology is a summary of data describing in vivo odontogenesis.[39,40,49] Chorioallantoic membrane (CAM) xenograft data based on 33 explants. In vitro morphogenesis and cytodifferentiation data based on 17 different experiments and 160 mandibular first molar tooth organ (cap stage) explants from Theiler stage 25 C57BL/6 embryos.

[b] ND, not determined.

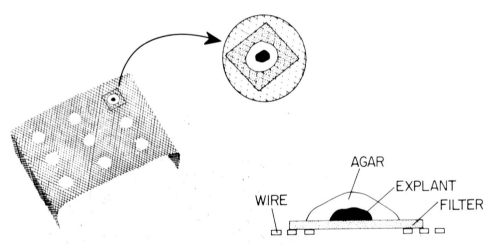

FIGURE 7. Chemically defined organ culture for embryonic mouse molar tooth organs. This diagram indicates the assembly for long-term (0 to 21 days) organ culture of cap stage molar organs **without** supplementation with serum, exogenous embryonic extracts, or antibiotics.[53,54]

matrix enamel proteins of 65,000, 58,000, 22,000, and 20,000 molecular weight which characterize ameloblast differentiation in the embryonic New Zealand White rabbit.[52] We have also reported preliminary evidence supporting the feasibility of isolating biologically active mRNA species from embryonic rabbit molar tooth organs.[6] Techniques have been reported for the solubilization and partial characterization of the tooth epithelial-derived proteins, which are unique and distinct from the odontoblast-derived dentin matrix proteins.[51]

To characterize tissue-specific gene products, those that best resembled the nascent translated gene products, we designed studies to isolate the recently synthesized proteins contained in the enamel organ epithelia from organ cultures. The intact molar organs were labeled with (^{35}S) methionine, and then analyzed as such or mechanically separated into mesenchymal

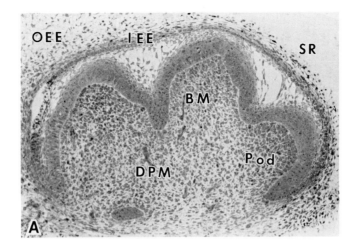

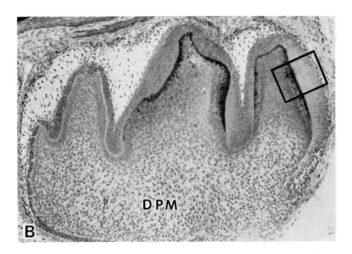

FIGURE 8. Chemically defined organ cultures permissive for tooth morphogenesis and both mesenchymal and epithelial cytodifferentiation. (A) Late cap stage mouse molar tooth organ illustrating the inner (IEE) and outer enamel epithelia (OEE), stratum reticulum (SR), basement membrane (BM), preodontoblasts (Pod), and dental papilla mesenchymal cells (DPM). (B) After 10 days in vitro tooth morphogenesis and cytodifferentiation is evident. (C) The rectangle indicated in (B) is shown here to demonstrate the relative degree of cytodifferentiation of odontoblasts (Od), dentin (D), enamel (arrows), and ameloblasts (Am).

or epithelial components. Under these incubation conditions, only the secretory ameloblast-cell layer and the forming enamel matrix were labeled (Figures 9 to 11). Although the odontoblasts were also actively synthesizing proteins, very few silver grains resulting from light microscopic autoradiography were localized either over the odontoblast cells or over the dentin matrix. The absence of silver grains from the odontoblast cells or dentin matrix was possibly the result of a physical barrier from the surrounding epithelia which did not allow diffusion of the (^{35}S) methionine into the dental pulp mesenchyme. This observation was extremely fortuitous; we could use the intact molar organ for culture studies in vitro, having confidence that the newly synthesized isotopically labeled proteins were epithelial-specific gene products. This conclusion has proven to be valid; for example, when the fluorographs obtained from intact molar organs or isolated enamel organ epithelia samples were critically compared, they were identical.

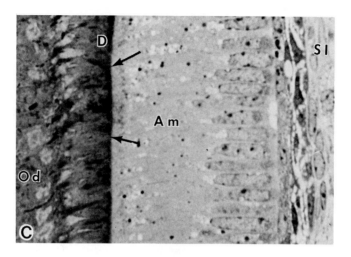

FIGURE 8C

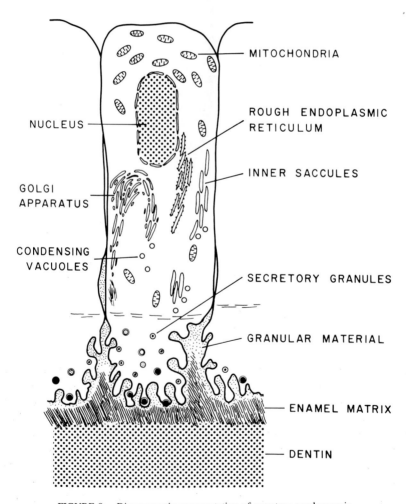

FIGURE 9. Diagrammatic representation of secretory amelogenesis.

Although many of the proteins present in the molar organs were labeled with the amino acid precursor, most labeled proteins were of relatively high molecular weight (43,000 and

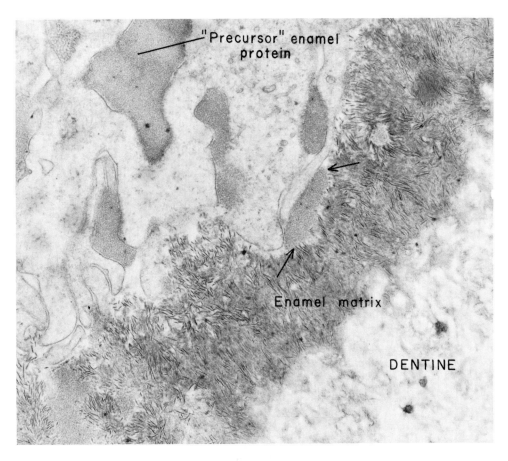

FIGURE 10. Transmission electron microscopy evidence has suggested a "precursor" form of enamel matrix material (arrows). The precursor form appears as a "stippled" or "granular" material in contrast to the enamel matrix associated with forming calcium hydroxyapatite crystals.

higher). Only one protein of lower molecular weight (approximately 16,000) was routinely labeled. In contrast, the Coomassie Blue-stained gels, representing the total number of extracted proteins, also contained several low molecular weight proteins assumed to be derived from larger proteins.

Secreted extracellular matrix proteins are often synthesized in a precursor form which are subsequently processed into modified forms by sequential enzymatic cleavage and posttranslational processing during intracellular transport and secretion. Using the paradigm preprocollagen → procollagen → collagen, we have assumed that "preproenamel" is also posttranslationally processed into increasingly smaller polypeptides concomitant with increased extracellular-matrix-enamel mineralization.[73] After secretion from the ameloblasts (Figures 9 and 10), the enamel proteins are further processed by an enamel peptidase.[77,78] The fact that most of the newly synthesized proteins present were of high-molecular weight favors the concept of a precursor(s) for enamel proteins (Figures 9 and 10).

Our results indicated that there were at least six proteins (A, B, C, D, E, and F) possibly derived from epithelial cells; all six polypeptides were routinely labeled in vitro (Figures 11A and B and 12). Two of these proteins (E and F) were of low-molecular weight, whereas proteins A, B, C, and D ranged from 72,000 to 43,000 molecular weight (Figure 12). These results agree with previous reports from our laboratory[51,52] and other laboratories,[71,72] which indicate that "enamelin" has a molecular weight of 70,000 and "amelogenins" have molecular weights of 25,000 or less. However, it still remains to be determined if each of these

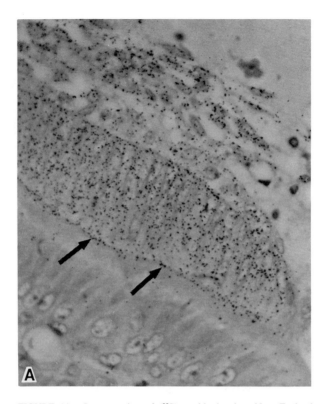

FIGURE 11. Incorporation of (^{35}S) methionine into New Zealand White embryonic rabbit molar tooth organs in vitro. High-resolution, light microscopic autoradiography after 4 hr incubation indicated that the majority of silver grains were localized over the enamel organ epithelia. (A) Localization of silver grains over stellate reticulum, stratum intermedium, secretory ameloblasts, and newly secreted enamel matrix proteins. (B) Dark field microscopy of region comparable to that shown in (A). Enamel, arrows.

proteins is an individual gene product. To address this question (Figure 13), we isolated mRNA fractions large enough to contain mRNA species that could code for these proteins (Table 2). Assuming that the major epithelial-derived gene products were in a range from 72,000 to 43,000 daltons, we selected mRNA molecules sedimenting between 16S and 26S (assuming that a protein of 43,000 daltons needs 1,290 bases for coding and a protein of 72,000 daltons requires 2,100 bases; these mRNA species should be contained in these fractions). The translated tooth epithelial mRNA fraction in a cell-free system produced only three of these proteins, (those of 65,000, 58,000, and 43,000 daltons) which corresponded to proteins B, C, and D, respectively.[79] Gene products B and C were definitely epithelial-specific proteins, as established by the fact that the dental pulp mesenchyme does not contain these specific mRNA species (Figure 12). The origin of protein C (43,000 daltons) was more difficult to establish, since both epithelial and mesenchymal mRNA species synthesized a protein having this relative electrophoretic mobility on SDS polyacrylamide slab gels.

In additional studies we have determined the location of embryonic enamel proteins in mouse and rabbit molar tooth organs by indirect immunofluorescent microscopy using a specific immunoglobulin raised against mouse enamel-containing extracellular matrix.[80,81] Antigenic determinants were localized within secretory ameloblast cells and newly secreted enamel extracellular matrix (Figure 14A and B). In separate studies the antibody has been found to show specificity for five of the polypeptides shown in Figure 12 (^{35}S-methionine

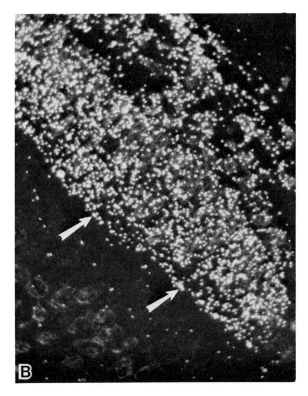

FIGURE 11B.

labeled enamel organ epithelial gene products in organ culture experiments) and three of
the translation reaction products obtained in the cell-free system.

At present we cannot discard the possibility that the lower molecular weight proteins (E
and F) are also nascent gene products (Figure 12). Lower sedimentation coefficient mRNA
fractions are now being tested in the cell-free system to obtain this information. Also, further
characterization of the proteins extracted from the organ cultures and the proteins synthesized
in the cell-free system using amino acid composition and peptide maps is in progress (Figure
15).

E. Heterologous Tissue Interactions: Epithelial Specificity

The classic studies of Spemann[1] have provided convincing evidence regarding the importance of heterotypic tissue interactions during vertebrate embryogenesis. Numerous molecular biology approaches have suggested that in many of the previous studies of "primary
and secondary embryonic induction", the expression of the genome had already been pre-
selected before the experimental manipulations; inadequate assays and overzealous interpretations have perhaps created an illusion regarding the actual mechanism(s) of heterologous
tissue interactions. For example, might heterotypic tissue interactions serve to enhance
previously determined transcriptive decisions? Or, might these tissue interactions actually
mediate *de novo* transcription and translational processes? Mesenchymal specification of
epithelial cytodifferentiation and morphogenesis has been considered to be a general feature
of various epithelial-mesenchymal interacting systems (e.g., salivary gland, mammary gland,
feather, hair, and tooth morphogenesis). In contrast, we have demonstrated that a mesenchyme can be induced by a heterologous epithelium to synthesize in quantity a specific gene
product(s) unorthodox to the organ from which the mesenchyme was taken (Figure 16).
Stage 22-23 avian limb bud epithelium induced 17-day embryonic mouse tooth mesenchyme

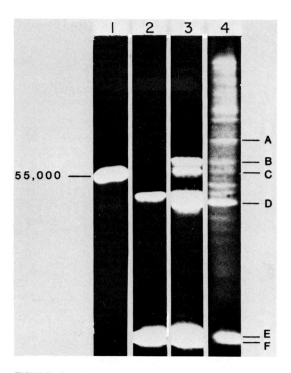

FIGURE 12.   Comparison between the translational reaction products obtained from epithelial-specific 16-26 S mRNAs and isotopically labeled, epithelial-specific gene products extracted from organ cultures of embryonic rabbit molar organs. (1) (^{14}C)-labeled amylase molecular weight marker; (2) rabbit reticulocyte lysate endogenous cell-free system (without exogenous mRNA added); (3) the translation reaction products using 16-26 S mRNAs from enamel organ epithelia; and (4) (^{35}S) methionine-labeled epithelial gene products extracted from short-term organ cultures of molar tooth organs. All proteins were extracted and separated in 10% SDS polyacrylamide slab gels and then prepared for fluorography. Epithelial-specific gene products assumed to be enamel proteins are designated A through F.

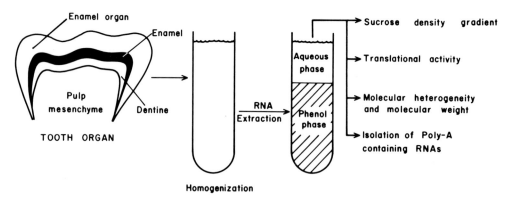

FIGURE 13.   Diagrammatic representation of experimental strategy to obtain enamel mRNAs.

to differentiate into cartilage (Figure 17).[82] Peptide analysis (cyanogen bromide cleavage after purification of extracted collagen chains) demonstrated that heterologous tissue recombinations produced type II collagen [α(II)]$_3$ (i.e., cartilage type) in addition to type I collagen

Table 2
ENRICHMENT OF mRNA TRANSLATIONAL
ACTIVITY THROUGH PURIFICATION

Sample	Intact molars		Dental pulp mesenchyme	
	Specific activity	Enrichment	Specific activity	Enrichment
Total RNA	19521	1	01732	1
Poly(A)$^+$ RNA	130166	6.6	138353	12.8
16-26S mRNA	244037	13	278011	25.8

$[\alpha(I)]_2\alpha_2$ (Figure 18). Intact or reconstituted mouse molar tooth organs synthesized type I collagen and type I trimer $[\alpha(I)]_3$ collagen. Immunohistochemical criteria using anti-type II collagen antibodies identified type II collagen in cartilage-like matrix within the mesenchymal component of heterologous tissue recombinants. Cartilage has never been described during in vivo or in vitro tooth tissue differentiation or associated with the pathology of dental papilla mesenchyme. These results support the hypothesis that epithelial-mesenchymal interactions during embryonic development can selectively induce *de novo* synthesis of unique gene products.

F. Heterologous Tissue Interactions: Mesenchymal Specificity

One very interesting question regarding the nature of epithelial-mesenchymal interactions during tooth morphogenesis is the issue of **determination.** When during development are the mesenchymal and epithelial phenotypes determined? When during development is the mesenchyme inductive? Using inbred mouse strains, it is well known that the dental papilla mesenchyme is highly inductive from Theiler stage 23 through stage 25 (e.g., the so-called "cap stage" of tooth development).[74,75] For example, Theiler stage 25 dental papilla mesenchyme obtained from a molar tooth organ by enzymatic dissociation of the tooth analagen has been shown to induce embryonic mouse footpad epidermis to differentiate into secretory ameloblasts.[82]

Embryonic induction has been shown to experimentally mediate epithelial-mesenchymal tissue recombinations when the sources of each tissue type represent diverse phylogenetic origin (Table 3). A number of fascinating heterochronic (diverse chronology of development) and heterotypic (diverse tissue sources such as chick/mouse chimeras) have demonstrated the conservation of mesenchyme-specific embryonic induction.[1,46,83-85] Recently, Kollar and Fisher demonstrated that dental papilla mesenchyme from Theiler stage 25 mouse embryos (circa days 16 to 18 gestation) induced 5-day-old chick embryonic pharyngeal arch epithelium to differentiate into secretory ameloblasts.[86] These investigators have suggested that the loss of teeth in Aves did not result from a loss of genetic information coding for enamel polypeptides in the avian epithelium but rather the epithelial-mesenchymal interactions in chick embryos are unable to express odontogenesis (Figure 19).

Is there reason to believe that the unique sequences for enamel proteins are present in avian chromatin? No teeth are present in lower chordates or fossil ostracoderms! In all jawed vertebrates teeth are universally present except where secondarily lost as in turtles and modern birds.[87] Teeth have been identified within the fossils of certain Jurassic and Cretaceous birds, however, most experts agree that modern birds are **toothless** and do not contain dental vestiges. Therefore, during evolution the reptilian genes for enamel proteins may have been deleted during the genesis of modern birds, or modern birds do contain unique sequences for enamel proteins but do not express amelogenesis due to an alteration in epithelial-mesenchymal interactions required for tooth formation. The recent findings by Kollar and

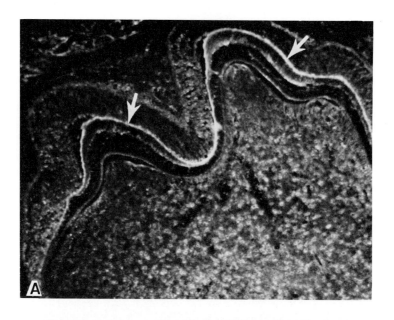

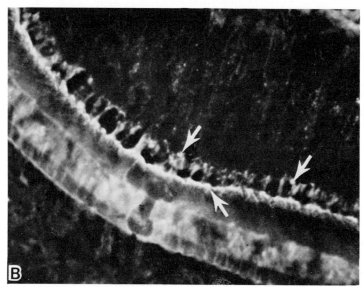

FIGURE 14.   Indirect immunofluorescent localization of newly secreted enamel proteins. The rabbit antimouse enamel matrix antibody stains intracellular antigens within the ameloblast as well as newly secreted enamel antigens (arrows). (A) Low magnification; (B) higher magnification.

Fisher[86] would indicate that modern birds (5-day-old chick embryonic pharyngeal arch epithelium) do contain the structural genes for enamel-matrix proteins. However, if present, these highly conserved sequences are usually repressed during embryogenesis.

In contrast, Karcher-Djuricic and Ruch[88] reported negative results from organ cultures in which interspecies heterologous tissue associations between embryonic chick mandibular epithelium (5-day-old) and embryonic mouse dental mesenchyme (11- to 12-days gestation); under these experimental conditions tooth morphogenesis, dentinogenesis, and amelogenesis were not detected within avian epithelium/mouse mesenchyme recombinants during 2 weeks in vitro cultures. However, mouse mesenchyme differentiated into cartilage when recombined with chick mandibular epithelium.[88]

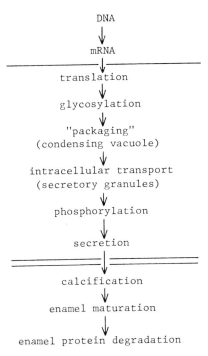

FIGURE 15. Summary of the processing of enamel mRNAs during transcription, translation, and posttranslational modifications.

Yet another interpretation is that during these chick/mouse heterologous tissue recombinations, dental papilla mesenchyme from the mouse transferred "information RNA" which specifically induced amelogenesis within a responding avian ectodermal derivative. In either case, each postulate can be tested using modern methodology currently available in molecular biology.

VII. DIRECT CELL-TO-CELL CONTACT DURING EPITHELIAL-MESENCHYMAL INTERACTIONS

Available evidence suggests that epithelial-mesenchymal interactions during tooth development require direct cell-to-cell contact between heterotypic tissues for some duration (circa 24 to 48 hr) in order to accommodate the instructive phases of differentiation and subsequent morphogenesis.[6,35,37,39,49,89] Direct cell-to-cell contacts have been demonstrated during tooth development in vivo and in vitro in positions of orientation prior to ameloblast differentiation.[37,39,49,89,90] In addition to mesenchymal cell processes positioned within epithelial clefts following the degradation of the epithelial-derived basal lamina,[90] discrete mesenchyme-derived matrix vesicles (circa 0.1 μm in diameter) are also deposited along the undersurface of the preameloblast cells and in contact with the progenitor ameloblast cells.[19,39,40,49,90,91] Using ultracytochemical methods[40] and biochemical analysis we have determined that some of these matrix vesicles contain methylated, low-molecular weight RNAs[19,22,23,92] (Figure 20).

VIII. TRANSMISSION OF "INDUCTIVE RNA" DURING EPITHELIAL-MESENCHYMAL INTERACTIONS

During embryonic tooth morphogenesis RNA has been detected within the extracellular

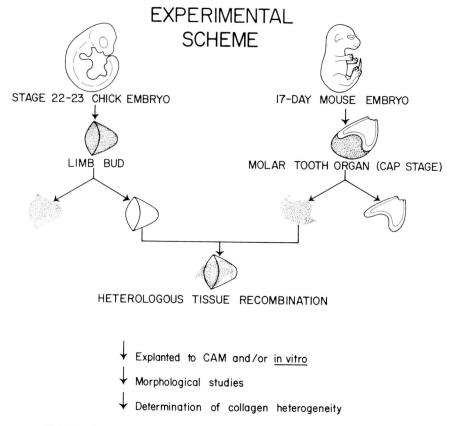

FIGURE 16. Experimental scheme of interspecies heterologous tissue interactions.

matrix interposed between epithelium and adjacent mesenchyme.[19,22,23,40] Approximately 4 to 7 μg low molecular weight, methylated RNA per 24 mg (dry weight) extracellular matrix was obtained.[23] More recently, the development of techniques to physically isolate matrix vesicles from the extracellular matrix enabled the subsequent identification of RNA within these vesicles.[19] The RNA within the matrix vesicle is resistant to RNAase A and T_1 digestion.

Organ culture experiments suggested that extracellular matrix RNAs were critically dependent upon nascent transcription. Preincubation of physically isolated tooth organs with dactinomycin, followed by pulse-chase additions of a number of different labeled RNA precursors, prevented silver grain density from being detected by light microscopic autoradiography localized over the extracellular matrix.[22,23] Descriptive studies using tritiated methyl-methionine, cytosine, uridine, adenosine, or guanosine produced qualitatively similar results.[19]

More recently, we have attempted to demonstrate the incorporation of tritiated uridine into molecules localized within mesenchymal cell processes and matrix vesicles in discrete positions relative to those regions related to embryonic induction during tooth morphogenesis. The in vivo incorporation of tritiated uridine by progenitor odontoblast cells (mesenchyme) and into matrix vesicles derived from these cells, has been analyzed by high-resolution electron microscopic autoradiography. Special attention has been focused on the unique extracellular matrix organelle, the matrix vesicle derived from the mesenchymal cell processes. Experiments were designed to determine whether, during the phases of epithelial differentiation prior to cessation of DNA replication, matrix vesicles associated with the mesenchyme incorporate (5, 6-³H) uridine. Moreover, in studies ranging from 5 min to 24 hr, following the intraperitoneal injection of tritiated uridine to newborn mice, we assessed the relative distribution of silver grains relative to the epithelial-mesenchymal interface.

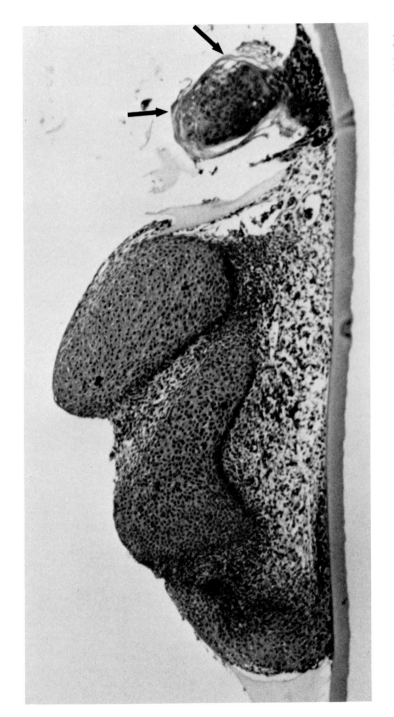

FIGURE 17. Stage 22-23 chick limb epithelium "induced" chondrogenesis within a responding Theiler stage 25 dental papilla mesenchyme within 5 days in vitro. Avian epithellium, arrows.

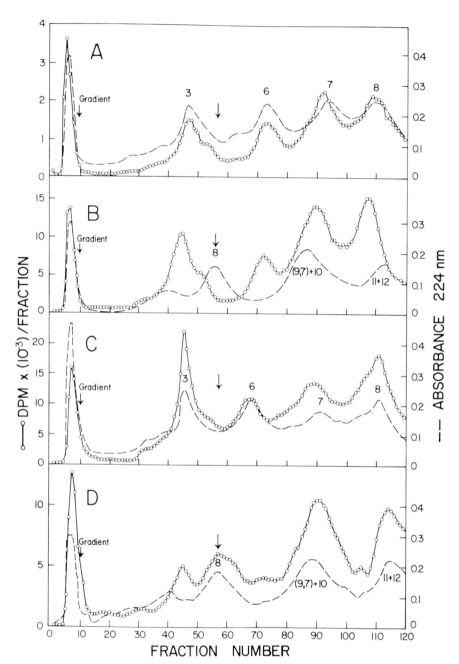

FIGURE 18. Analyses of collagen types after CNBr cleavage of differentially salt separated (³H) proline-containing material purified from tooth organs and from heterologous tissue recombinants (open circles). (A) Radioactive material synthesized by tooth organs and precipitated with 2.6 M NaCl coincided with CNBr-cleaved type I [α₁(II)] carrier collagen chains (broken line). (B) Tooth organ material that remained in the supernatant compared with the elution profile of CNBr-cleaved type II [α₁(II)] carrier collagen chains (broken line). (C) Heterologous recombinants synthesized (³H) proline-labeled material that precipitated with 2.6 M salt is compared with CNBr-cleaved type I [α₁(I)] carrier chains (broken line). (D) Major peaks of the CNBr-cleaved radioactive supernatant collagen produced from heterologous tissue recombinants coincided with the general profile of the CNBr-digested type II carrier collagen (broken lines).

Initial observations showed that silver grains were immediately localized over mesenchymal cell nuclei as well as mesenchymal cell processes and matrix vesicles adjacent to the extra-

Table 3
SOURCES OF MESENCHYMAL
SPECIFICITY DURING SECONDARY
EMBRYONIC INDUCTION

Mesenchyme-derived type I collagen
Mesenchyme—derived type III collagen
Mesenchyme-derived proteoglycans
Mesenchymal cell surface fibronectin
Mesenchymal cell surface alloantigens
Mesenchyme-derived matrix vesicles
Mesenchyme-derived ''extracellular matrix RNAs''

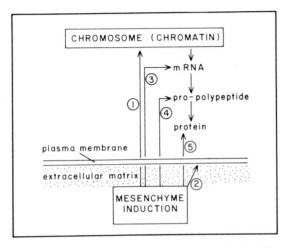

FIGURE 19.   Possible sites of mesenchyme ''inductive influences'' upon a responsive epithelium during heterotypic tissue interactions.

cellular matrix interface. Within 15 min following the administration of the tritiated uridine, silver grains were localized over matrix vesicles and mesenchymal cell processes immediately adjacent to inner enamel epithelial cells (Figure 21).

Caution must be exercised when considering these observations. The matrix vesicles are 0.1 μm in diameter and the silver grains per se are 0.1 μm in diameter. The resolution of this technique is limited and it is very difficult to localize the specific origin of the radioactivity as detected by autoradiography of such small structures. In addition, it is not known that the tritiated uridine is incorporated into RNA as visualized in these preparations. However, analysis of the number of significantly labeled matrix vesicles and mesenchyme cell processes observed at various times after (^{3}H) uridine injection indicates that incorporation takes place very rapidly and within critical regions of cell-to-cell interactions. In regions of overt cytodifferentiation of odontoblast and ameloblast cells, silver grains *were not observed* within the extracellular matrix.

The progression of epithelial-mesenchymal interactions, including those associated with tooth morphogenesis, is arrested by puromycin, cycloheximide, and dactinomycin.[37,39] During epithelial-mesenchymal interactions associated with tooth formation, electron microscopic cytochemistry revealed matrix vesicles which contain ribonucleoprotein particles.[40] In the chordamesoderm-neuroectoderm interface in amphibia embryos electron microscopy has revealed extracellular matrix ribonucleoproteins.[93] A number of investigators have identified extracellular ''organelles'' containing ribonucleoproteins in locations associated with

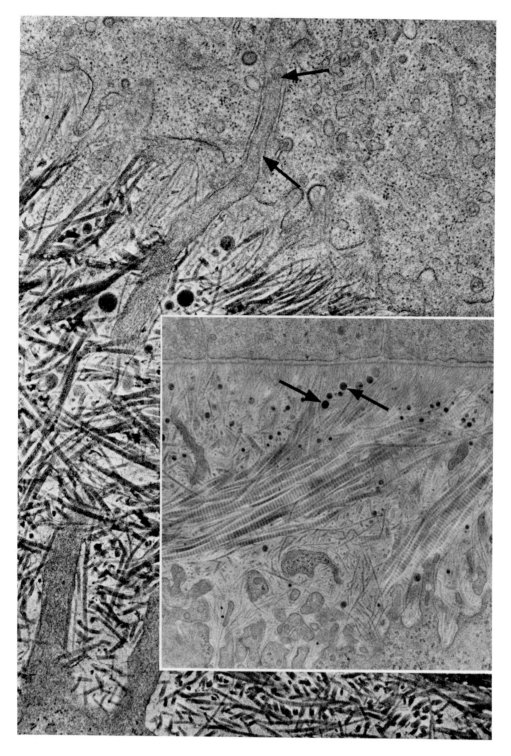

FIGURE 20.   Direct cell-to-cell contact is characteristic in both *in situ* and in vitro epithelial-mesenchymal interactions associated with odontogenesis. Sites of mesenchyme contact with epithelium following basal lamina degradation (arrows). **Insert** illustrates the accumulation of matrix vesicles (arrows) adjacent to intact basal lamina.

secondary embryonic induction.[19,23,25,40,49,91-94] RNA may be transmitted from the inducing tissue to the responding tissue type during direct cell-to-cell interactions.

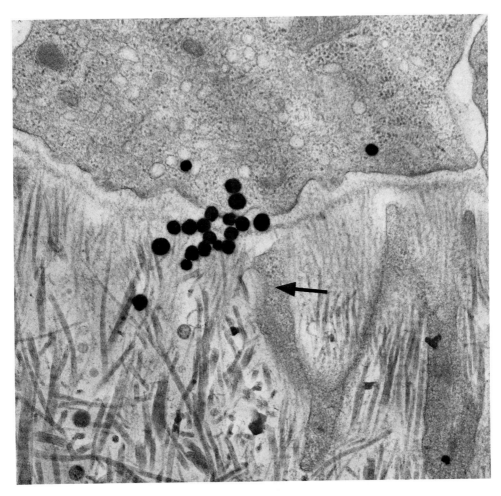

FIGURE 21. High-resolution, electron microscope autoradiograph. This electron photomicrograph of high-resolution autoradiography represents 20 min following the intraperitoneal administration of 25 μCi (5, 6⁻³H) uridine (specific activity 55 Ci/mmol). Silver grains were often localized in discrete regions along the interface between mesenchyme and epithelium. Pulse/chase studies, ranging from 5 min to 24 hr in vivo suggest that silver grains appear to be transmitted from mesenchyme cell processes (arrows) and matrix vesicles to the adjacent epithelial cell surfaces.

However, extreme caution must be exercised regarding the interpretation of these as yet descriptive findings. The notion that RNAs may be transferred between tissues during organogenesis often infers that the RNA has a unique specificity, has inductive effects, and that transmission and reception of the RNA molecule is time and space specific. Actual evidence for each of these criteria is not available within one developing system. Cause and effect relationships have not been established under critical experimental conditions. In an attempt to meet some of these criteria, Grainger and Wessels[95] designed a series of experiments which asked if RNA is transferred from mesenchyme to epithelium during embryonic heterologous tissue interactions associated with lung morphogenesis. Eleven-day embryonic mouse lung corresponds to 5-day-old chick embryonic lung development.[96] Using interspecies heterologous tissue recombinations, mouse epithelia, and chick mesenchyme, epithelia responded to lung mesenchyme instructions by forming budding and branching patterns of epithelial morphogenesis characteristic of early morphogenesis *in situ*.[95] Appropriate controls, including transfilter studies, indicated that dissociation methods and isotopic labeling procedures did not inhibit positive heterotypic tissue interactions.[95] Under these experimental conditions, following preincubation of chick lung mesenchyme in vitro with density labeled nucleotides (^{13}C-^{15}N-labeled nucleotides), approximately 2×10^6 cpm were detected in density labeled mesenchyme RNA. During transfilter heterologous tissue interactions, previously identified as being critical toward epithelial differentiation, no transfer of intact RNA (assuming 1500 bases per molecule) was detected.[95] These investigators suggested that the level of sensitivity for such experiments was such that less than 0.01% of the labeled RNA molecules in the mesenchyme could have been detected in the epithelium, corresponding to less than 75 labeled RNA molecules transferred to each epithelial cell.[95]

IX. LOCALIZATION OF "TRANSCRIPTIVELY ACTIVE" CELLS DURING ODONTOGENESIS

An extremely critical problem in developmental biology is understanding the regulation of cellular differentiation. How are epigenetic influences translated into sequential activation of structural genes? How does gene activation become expressed as the synthesis of specific messenger RNAs specifying proteins characteristic of a particular stage in the genesis of a unique phenotype? As is evident from the previous discussions, the obvious technical difficulties are twofold: (1) to detect in a specific cell type during differentiation extremely small amounts of specific mRNAs and (2) to be able to correlate very small quantitative changes of specific mRNAs with subtle changes in phenotype. Precise methods are essential toward the elucidation of these problems.

An electron microscopic technique has been reported which enabled visualization of the binding sites of acridine orange to DNA within fixed individual cells.[97] The acridine orange was found to bind selectively to the transcriptively active extended euchromatin portion of nuclear DNA and enabled localization within individual bone marrow, blood, and lymph node cells.[98,99]

Acridine orange (AO) was shown to be an effective probe for the ultrastructural localization of gene expression in embryonic epithelial and ectomesenchymal cells during tooth organogenesis.[100] A concentration of 10^{-2} M AO provided high-resolution localization of intranuclear binding sites indicative of "transcriptively active" euchromatin portions of the cell nucleus. This method was combined with an analysis of the distribution of silver grains resulting from the incorporation of labeled uridine into ribonucleic acid molecules as evaluated by RNAse T_1 and actinomycin D sensitivity. Comparisons of the spatial distribution patterns of (1) electron-dense reaction products resulting from AO binding to euchromatin and (2) silver grains resulting from the incorporation of uridine into newly synthesized RNA molecules, localized several discrete "transcriptively active" regions in the cervical loop of progenitor ectomesenchymal and adjacent epithelial cell prior to overt cell differentiation. The results justify further investigations of acridine orange and other related molecules as ultrastructural probes for gene expression during embryonic epithelial-mesenchymal interactions associated with epidermal organogenesis (Figure 22).

Another approach which promises to provide exquisite specificity for detection of the specific mRNAs in cytological preparations is *in situ* hybridization. The presence of polyadenylic acid (poly A) sequence in many examples of specific eucaryotic mRNA molecules[101] provides a potentially superior means for the detection of poly(A)-containing mRNA in histological preparations by *in situ* hybridization.[102] One unique application of this principle has been the recent report of an *in situ* hybridization technique which employs a (^{3}H) poly(U) probe to assay the relative localization and patterns of distribution of poly(A)-RNA molecules within histological preparations.[103] These investigators have suggested that the optima sensitivity of (^{3}H)poly(U) *in situ* hybridization, assuming a 10% efficiency for a 14-day exposure autoradiograph, might be that the detection of one silver grain would represent approximately 288 putative mRNA molecules containing poly(A) sequences of about 100 nucleotides.[103]

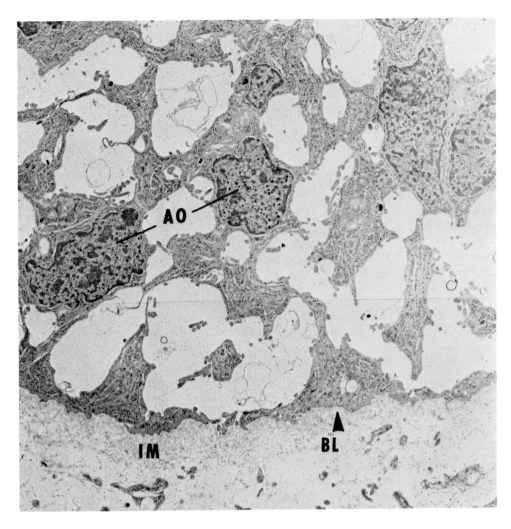

FIGURE 22. Ultracytochemical localization of acridine orange (AO) assocated with euchromatin in specific cell nuclei. IM, intercellular matrix; and BL, basal lamina associated with inner enamel epithelia.

Recent advances in molecular biology now provide an additional and as yet formidable task in the form of producing specific complementary cDNA probes for specific mRNAs. A number of important technical problems require as yet further investigation before such sensitive probes can effectively localize unique mRNAs within embryonic cells.[104]

X. PROSPECTUS

Recent studies of epithelial-mesenchymal interactions indicate that direct cell-to-cell contact is a requirement for organogenesis. In a number of developing organ systems it has been shown that mesenchyme "induces" epithelium to differentiate into a cell type complementary to that of the mesenchyme. Heterochronic and heterotypic tissue recombinations across vertebrate species indicate that mesenchyme specificity is highly conserved. For example, embryonic rabbit mammary gland mesenchyme has been shown to induce avian epithelium to differentiate into mammary gland-like structures.[46] More recently, embryonic mouse dental papilla mesenchyme has been shown to induce avian epithelium to differentiate into enamel-matrix producing cells — secretory ameloblasts.[86] Recent investigations in our laboratory have provided evidence for the isolation and partial characterization of enamel mRNAs.[79] Current methods of producing cDNAs enable more rigorous analysis of epithelial-mesenchymal interactions in mouse/chick chimeras using molecular probes. It is anticipated that the mechanism for direct cell-to-cell contact and the possible implications toward understanding differential gene expression during epithelial-mesenchymal interactions may soon be less obscure and more accessible to the methods of molecular biology.

ACKNOWLEDGMENTS

This research assessment was supported in part by research grants from the National Institute for Dental Research, U.S. Public Health Service (DE-02848, DE-03569, and DE-03513), and training grants DE-0094 and DE-00134. The authors wish to acknowledge the very important discussions which we had the pleasure of sharing with Drs. George Martin, Edward J. Kollar, and Jean-Victor Ruch related to selected topics in the review. The authors wish to acknowledge our appreciation to Bonnie Sudderth for preparation of the manuscript and to Pablo Bringas, Jr. for the illustrations and electron photomicrographs.

REFERENCES

1. **Spemann, H.,** *Embryonic Development and Induction,* Yale University Press, New Haven, Conn., 1938.
2. **Mangold, O.,** *Acta Genet. Med. Gemellol.,* 10, 1, 1961.
3. **Weiss, P.,** *Int. Rev. Cytol.,* 7, 391, 1958.
4. **Weiss, P.,** *Yale J. Biol. Med.,* 19, 235, 1947.
5. **Barth, L. G. and Barth, L. J.,** *Dev. Biol.,* 28, 18, 1972.
6. **Grobstein, C.,** *Nat. Cancer Inst. Monogr.,* 26, 279, 1967.
7. **Niu, M. C.,** in *Cellular Mechanisms in Differentiation and Growth,* Rudnick, D., Ed., Princeton University Press, New Haven, Conn., 1956, 155.
8. **Wolpert, L.,** *Curr. Top. Dev. Biol.,* 6, 183, 1971.
9. **Wolpert, L.,** in *Current Research Trends in Prenatal Craniofacial Development,* Pratt, R. and Christiansen, J. R., Ed., North-Holland/American Elsevier, New York, 1980, 89.
10. **Brachet, J.,** *Arch. Biol.,* 53, 207, 1942.
11. **Desphande, A. K., Niu, L. C., and Niu, M. C.,** in *The Role of RNA in Reproduction and Development,* Niu, M. C. and Segal, S. J., Eds., North-Holland/American Elsevier, New York, 1973, 229.
12. **Mishra, N. C., Niu, M. C., and Tatum, E. L.,** *Proc. Natl. Acad. Sci. U.S.A.,* 72, 642, 1975.
13. **Czihak, G. and Hörstadius, S.,** *Dev. Biol.,* 22, 15, 1970.
14. **Arnold, H. H., Innis, M. A., and Siddiqui, M. A. Q.,** *Biochemistry,* 17, 2050, 1978.
15. **Deshpande, A. K. and Siddiqui, M. A. Q.,** *Differentiation,* 10, 133, 1978.
16. **Amos, H.,** *Biochem. Biophys. Res. Commun.,* 5, 1, 1961.
17. **Bhargava, P. and Shanmugan, G.,** in *Progress in Nucleic Acid Research and Molecular Biology,* Davidson, J. N. and Cohn, W. E., Eds., Academic Press, New York, 1971, 103.
18. **Stebbing, N.,** *Cell Biol. Int. Rep.,* 3, 485, 1979.
19. **Slavkin, H. C. and Croissant, R.,** in *The Role of RNA in Reproduction and Development,* Niu, M. C. and Segal, S. J., Eds., North-Holland/American Elsevier, New York, 1973, 247.
20. **Kolodny, G. M.,** in *Cell Communication,* Cox, R. P., Ed., John Wiley & Sons, New York, 1974, 97.
21. **Kolodny, G. M.,** in *Extracellular Matrix Influences on Gene Expression,* Slavkin, H. C. and Greulich, R. C., Eds., Academic Press, New York, 1975, 75.
22. **Slavkin, H. C., Bringas, P., and Bavetta, L. A.,** *J. Cell. Physiol.,* 73, 179, 1969.
23. **Slavkin, H. C., Flores, P., Bringas, P., and Bavetta, L. A.,** *Dev. Biol.,* 23, 276, 1970.
24. **Kolodny, G. M., Culp, L. A., and Rosenthal, L. J.,** *Exp. Cell Res.,* 73, 65, 1972.
25. **Beierle, J. W.,** *Science,* 161, 798, 1968.
26. **Bennett, M., Mayhew, E., and Weiss, L.,** *J. Cell. Physiol.,* 74, 183, 1969.
27. **Montagneir, L.,** *C. R. Acad. Sci. Ser. D,* 267, 1968.
28. **Duesberg, P. H. and Colby, C.,** *Proc. Natl. Acad. Sci. U.S.A.,* 64, 396, 1969.
29. **Stern, R. and Friedman, R.,** *Nature (London),* 226, 612, 1970.

30. **Ro-Choi, T. S. and Busch, H.,** in *The Molecular Biology of Cancer,* Busch, H., Ed., Academic Press, New York, 1974, 241.
31. **Frederiksen, S., Pederson, T., Hellung-Larson, P., and Engberg, J.,** *Biophys. Acta,* 340, 64, 1974.
32. **Zeichner, M. and Breitkreutz, D.,** *Arch. Biochem. Biophys.,* 188, 410, 1978.
33. **Lee-Huang, S., Sierra, J., Naranjo, R., Filipowicz, W. and Ochoa, S.,** *Arch. Biochem. Biophys.,* 180, 276, 1977.
34. **Heywood, S. and Kennedy, D.,** *Biochemistry,* 15, 3314, 1976.
35. **Wessells, N. K.,** *Tissue Interactions and Development,* Benjamin/Cummings, Menlo Park, Calif., 1977.
36. **Moscona, A. A. and Hausman, R. E.,** in *Cell and Tissue Interactions,* Lash, J. W. and Burger, M. M., Eds., Raven Press, New York, 1977, 173.
37. **Saxen, L., Karkinen-Jaaskelainen, M., Lehtonen, E., Nordling, S., and Wartiovaara, J.,** in *The Cell Surface in Animal Embryogenesis and Development,* Poste, G. and Nicolson, G. L., Eds., North-Holland/American Elsevier, New York, 1976, 331.
38. **McMahon, D. and West, C.,** in *The Cell Surface in Animal Embryogenesis and Development,* Poste, G. and Nicolson, G. L., Eds., North-Holland/American Elsevier, New York, 1976, 449.
39. **Slavkin, H. C.,** *J. Biol. Buccale,* 6, 189, 1979.
40. **Slavkin, H. C.,** in *Developmental Aspects of Oral Biology,* Slavkin, H. C., and Bavetta, L. A., Eds., Academic Press, New York, 1972, 165.
41. **Pitts, J. D.,** in *International Cell Biology 1976—1977,* Brinkley, B. R. and Porter, K. R., Eds., Rockefeller Press, New York, 1977, 43.
42. **Subak-Sharpe, J. H., Burke, R. R., and Pitts, J. D.,** *J. Cell Sci.,* 4, 353, 1969.
43. **Gilula, N. B., Reeves, O. R., and Steinbach, A.,** *Nature (London),* 235, 262, 1972.
44. **Spemann, H. and Mangold, H.,** *Arch. Entwicklungsmeek. Orig.,* 100, 599, 1924.
45. **Grunz, H. and Staubach, J.,** *Differentiation,* 14, 59, 1979.
46. **Propper, A. Y. and Gomot, L.,** *Experientia,* 29, 1543, 1973.
47. **Thesleff, I., Barrach, H. J., Foidart, J. M., Vaheri, A., Pratt, R. M., and Martin, G. R.,** *Dev. Biol.,* 81, 182, 1981.
48. **Lesot, H., Osman, M., and Ruch, J. V.,** *Dev. Biol.,* 82, 371, 1981.
49. **Slavkin, H. C.,** *Oral Sci. Rev.,* 4, 1, 1974.
50. **Graver, H., Herold, R., Chung, T., Christner, P., Pappas, C., and Rosenbloom, J.,** *Dev. Biol.,* 63, 390, 1978.
51. **Slavkin, H. C., Weliky, B., Stellar, W., Slavkin, M. D., Zeichner-Ganz, M., Bringas, P., Hyatt-Fischer, H., Shimizu, M., and Fukae, M.,** *J. Biol. Buccale,* 6, 309, 1979.
52. **Guenther, H., Croissant, R. D., Shonfeld, S. E., and Slavkin, H. C.,** *Biochem. J.,* 163, 591, 1977.
53. **Yamada, M., Bringas, P., Grodin, M., MacDougall, M., Cummings, E., Grimmett, J., Weliky, B., and Slavkin, H. C.,** *J. Biol. Buccale,* 8, 127, 1980.
54. **Yamada, M., Bringas, P., Grodin, M., MacDougall, M., and Slavkin, H. C.,** *Calc. Tiss. Int.,* 31, 161, 1980.
55. **Schimke, R. T., Palmiter, R. D., Palacios, R., Rhoades, R. E., McKnight, S., Sullivan, D., and Summers, M.,** in *Genetic Mechanisms of Development,* Ruddle, F. H., Ed., Academic Press, New York, 1972, 225.
56. **Means, A. R. and O'Malley, B. W.,** in *Biochemistry of Cell Differentiation,* Vol. 9, Paul, J., Ed., University Park Press, Baltimore, 1974, 161.
57. **Towle, H. C., Tsai, M. J., Hirose, M., Tsai, S. Y., Schwartz, R. J., Parker, M. G., and O'Malley, B. W.,** in *The Molecular Biology of Hormone Action,* Papconstantinou, J., Ed., Academic Press, New York, 1976, 107.
58. **Rutter, W. J., Clark, W. R., Kemp, J. D., Bradshaw, W. S., Sanders, T. G., and Ball, W. D.,** in *Epithelial-Mesenchymal Interactions,* Fleischmajer, R. and Billingham, R. E., Eds., Williams & Wilkins, Baltimore, 1968, 114.
59. **Rosen, J. M.,** *Biochemistry,* 15, 5263, 1976.
60. **Terry, P. M., Banerjee, M. R., and Lui, R. M.,** *Proc. Natl. Acad. Sci. U.S.A.,* 74, 2441, 1977.
61. **Rutter, W. J., Pictet, R. L., Harding, J. D., Chivgwin, J. M., McDonald, R. J., and Brzybyla, A. E.,** in *Molecular Control of Proliferation and Differentiation,* Papaconstantinou, J., and Rutter, W. J., Eds., Academic Press, New York, 1978, 205.
62. **Gay, S. and Miller, E. J.,** *Collagen in the Physiology and Pathology of Connective Tissue,* Gustav Fischer Verlag, Stuttgart, 1978.
63. **Munksgaard, E. C., Rhodes, M., Mayne, R., and Butler, W. T.,** *Eur. J. Biochem.,* 82, 609, 1978.
64. **Lesot, H., Von der Mark, K., and Ruch, J. V.,** *C. R. Acad. Sci. Paris,* 286, 765, 1978.
65. **Engfeldt, B. and Hjerpe, A.,** *Calc. Tiss. Res.,* 10, 152, 1972.
66. **Linde, A.,** *Calc. Tiss. Res.,* 12, 281, 1973.
67. **Dimuzio, M. T., Veis, A.,** *Calc. Tiss. Res.,* 25, 169, 1978.
68. **Butler, W. T., Munksgaard, E. C., and Richardson, W. S.,** *J. Dent. Res.,* 58, 817, 1979.

69. **Jones, I. L. and Leaver, A. G.,** *Arch. Oral Biol.,* 19, 371, 1974.
70. **Fukae, M. and Shimizu, M.,** *Arch. Oral Biol.,* 19, 381, 1974.
71. **Termine, J. D., Torchia, D. A., and Conn, K. M.,** *J. Dent. Res.,* 58, 773, 1979.
72. **Termine, J. D., Belcourt, A. B., Christner, P. J., Conn, K. M., and Nylen, M. U.,** *J. Biol. Chem.,* 255, 9760, 1980.
73. **Slavkin, H. C., Trump, G. N., Schonfeld, E., Brownell, A. G., Sorgente, N., and Lee-Own, V.,** in *Cell Interactions in Differentiation,* Saxen, L. and Weiss, L., Eds., Academic Press, New York, 1978, 209.
74. **Koch, W. E.,** *J. Exp. Zool.,* 165, 155, 1967.
75. **Kollar, E. J.,** in *Developmental Aspects of Oral Biology,* Slavkin, H. C. and Bavetta, L. A., Eds., Academic Press, New York, 1972, 125.
76. **Lee-Own, V., Zeichner, M., Benveniste, K., Denny, P., Paglia, L., and Slavkin, H. C.,** *Biochem. Biophys. Res. Commun.,* 74, 849, 1977.
77. **Suga, S.,** *Arch. Oral. Biol.,* 15, 555, 1970.
78. **Shimizu, M., Tanabe, T., and Fukae, M.,** *J. Dent. Res.,* 58, 782, 1979.
79. **Zeichner-David, M., Weliky, B. G., and Slavkin, H. C.,** *Biochem. J.,* 185, 489, 1980.
80. **Hyatt, Fischer, H., Chrispens, J., O'Keefe, D., and Slavkin, H. C.,** *J. Dent. Res.,* 58(B), 1008, 1979.
81. **Slavkin, H. C., MacDougall, M., Bringas, P., Grodin, M., and Yamada, M.,** *J. Dent. Res.,* 59(A), 304, 1980.
82. **Hata, R. and Slavkin, H. C.,** *Proc. Natl. Acad. Sci. U.S.A.,* 75, 2790, 1978.
83. **Kollar, E. J. and Baird, G. R.,** *J. Embryol. Exp. Morphol.,* 24, 159, 1970.
84. **Waddington, C. H.,** *J. Exp. Biol.,* 11, 224, 1934.
85. **Garber, B., Kollar, E. J., and Moscona, A. A.,** *J. Exp. Zool.,* 168, 455, 1968.
86. **Kollar, E. J. and Fisher, C.,** *Science,* 207, 993, 1980.
87. **Romer, A. S. and Parsons, T. S.,** in *The Vertebrate Body,* W. B. Saunders, Philadelphia, 1978, 241.
88. **Karcher-Djuricic, V. and Ruch, J. V.,** *C. R. Seances Soc. Biol.,* 167, 915, 1973.
89. **Thesleff, I., Lehtonen, E., and Saxen, L.,** *Differentiation,* 10, 71, 1978.
90. **Slavkin, H. C., and Bringas, P.,** *Dev. Biol.,* 50, 428, 1976.
91. **Slavkin, H. C., Croissant, R. D., Bringas, P., Matosian, P., Wilson, P., Mino, W., and Guenther, H.,** *Fed. Proc.,* 35, 127, 1976.
92. **Croissant, R. D.,** *J. Dent. Res.,* 50, 1065, 1971.
93. **Kelley, R. O.,** *J. Exp. Zool.,* 172, 153, 1969.
94. **Tarin, D.,** *J. Anat.,* 111, 1, 1972.
95. **Grainger, R. M. and Wessells, N. K.,** *Proc. Natl. Acad. Sci. U.S.A.,* 71, 4747, 1974.
96. **Taderera, J. V.,** *Dev. Biol.,* 16, 489, 1976.
97. **Frenster, J. H.,** *Cancer Res.,* 31, 1128, 1971.
98. **Frenster, J. H.,** *Nat. New Biol.,* 236, 175, 1972.
99. **Frenster, J. H. and Herstein, P. R.,** *N. Eng. J. Med.,* 288, 1224, 1973.
100. **Lehmann, R., and Slavkin, H. C.,** *Dev. Biol.,* 49, 438, 1976.
101. **Brawerman, G.,** *Annu. Rev. Biochem.,* 46, 621, 1974.
102. **Pardue, M. L. and Gall, J. G.,** *Methods Cell Biol.,* 10, 1, 1975.
103. **Capco, D. G. and Jeffery, W. R.,** *Dev. Biol.,* 67, 137, 1978.
104. **Godard, C. and Jones, K. W.,** *Nucleic Acids Res.,* 6, 2849, 1979.

Chapter 7

CONTROL OF GENE EXPRESSION DURING EARLY MAMMALIAN EMBRYOGENESIS

Michael I. Sherman and Joel Schindler

TABLE OF CONTENTS

I. INTRODUCTION

Embryogenesis achieves the purpose of generating a complete organism from an egg. It is perhaps remarkable that, as any comparative embryology text will attest, there is such a diversity in the steps that embryos of different species undergo in reaching this objective. Embryogenesis in mammals is distinguished by the intimate interaction of the conceptus with the mother throughout the often lengthy period of development. This necessitates the formation of highly specialized accessory tissues which function in anchoring the conceptus to the mother and in controlling the uptake of nutrients from surrounding or proximal maternal fluids. It appears as though most of the earliest decisions concerning cell fate serve to establish separate embryonic and extraembryonic cell lineages (see Figure 1).[1-3] Therefore, efforts to understand the ways in which gene expression is regulated in early mammalian embryos are often concerned directly or indirectly with the origin of this dichotomy.

Studies on the regulation of gene expression in mammalian embryos have lagged behind those with nonmammalian embryos largely because of problems with tissue availability and culture procedures. Nevertheless, several experiments have now been carried out on early mammalian embryos with this subject in mind. In the following sections we shall examine the nature of gene products that appear in early mammalian (primarily mouse) embryogenesis, the levels at which their expression might be controlled, and the possible nature of the controlling elements. Finally, we shall consider whether the patterns of gene expression are the same in early mammalian and nonmammalian embryos despite obvious differences between the two with respect to duration of embryogenesis, environment, and cell types formed.

II. PATTERN OF TRANSCRIPTION DURING EARLY MAMMALIAN EMBRYOGENESIS

The appearance of developmental gene products (e.g., specific enzymes or antigens; see Section III) implies the prior synthesis of the appropriate mRNAs and the necessary translation apparatus. It is, therefore, constructive to review the information available on the production of various types of RNA during early mouse embryogenesis. Due to the paucity of available tissue, studies on RNA populations in early mammalian embryos have generally been limited to assessments of the time of initiation of synthesis of various types of RNA and determination of the relative proportions of each class of RNA at the various stages.

The net RNA content of the mouse embryo does not increase through the cleavage stages of development.[4] However, the synthesis of all of the commonly recognized classes of RNA begins shortly after the onset of embryogenesis (Table 1): ribosomal, 4S, and 5S RNA are all synthesized by the four-cell stage, perhaps even by the two-cell stage.[5-9] Most importantly in the context of the subjects to be discussed here, heterogeneous nuclear RNA (hn-RNA),[5,7,10,11] some of which presumably serves as precursor for mRNA, as well as poly-adenylated RNA (polyA$^+$ RNA)[12,13] per se, appear to be produced as early as the two-cell stage in mouse embryos.

DNA-dependent RNA polymerase activities have been detected in early mouse embryos.[14] RNA polymerase type II, the enzyme presumed to be involved in mRNA synthesis,[15] has been reported to be present at the two-cell stage.[16] Indeed, determinations of RNA polymerase activities under varying conditions of ammonium sulfate and α-amanitin treatment suggest that RNA polymerase I, II, and III are all active in the mouse embryo no later than the morula stage.[14,16,18]

Studies on RNA synthetic patterns in early rabbit embryos have also been carried out. Although there is evidence for early synthesis of most types of RNA in the embryos of this species, the production of rRNA appears to be delayed relative to the pattern observed in

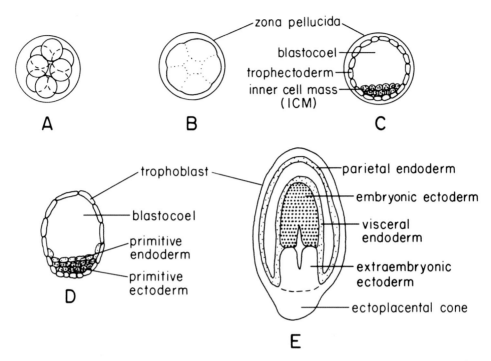

FIGURE 1. Stages in the early development of the mouse embryo. (A) Eight-cell embryo prior to compaction (3rd day of pregnancy; the day of observation of the sperm plug is considered the first day). No unequivocal differences in biochemistry or developmental potential have been found between blastomeres at this stage.[3] (B) Compact 8- to 16-cell embryo (late 3rd day of pregnancy). Blastomeres have become intimately apposed due to the formation of junctional complexes (see Ducibella[139]). (C) Preimplantation blastocyst (4th day of pregnancy). The embryo contains approximately 60 cells at this time.[38] An outer single layer of trophectoderm cells encloses an internal clump of cells (ICM) and a fluid-filled cavity, the blastocoel. (D) Implanting blastocyst (5th day of pregnancy). The embryo, containing about 100 cells,[38] has shed its acellular zona pellucida and has become adherent to the wall of the uterus (or to the substratum of the culture dish). At this stage of adhesion, by a convention of nomenclature adopted in this laboratory,[114] trophectoderm cells are now called trophoblast cells. The ICM has split into two layers: the primitive endoderm, which is destined to form cells which are extraembryonic, and the primitive ectoderm, which probably contains the only cells in the embryo at this stage which are fated to contribute to the embryo proper. (E) Early egg-cylinder stage (6th day of pregnancy). The embryo contains 200 cells or more.[123] The three trophectoderm derivatives (giant trophoblast cells, ectoplacental cone, and extraembryonic ectoderm) will all contribute to placental structures. The primitive endoderm layer splits into parietal endoderm (which forms a single-celled layer adherent to the trophoblast) and visceral endoderm (which forms a layer of the yolk sac). Again, only the embryonic ectoderm is destined to give rise to tissues in the embryo proper.[140,141] At a later stage, a mesoderm layer will appear in the egg cylinder, presumably derived from the embryonic ectoderm.[124] This layer probably gives rise to embryonic mesoderm and extraembryonic mesodermal elements (e.g., the mesodermal layer of the yolk sac; see Figure 1.7 of Rossant and Papaioannou[1]). (Adapted from Adamson, E. A. and Gardner, R. L., *Br. Med. Bull.*, 35, 113, 1979. With permission.)

mouse embryos (reviewed by Schultz and Church[19]). These observations suggest that even among mammalian species, differences can exist in patterns of transcription.

III. CHARACTERIZATION OF DEVELOPMENTAL GENE PRODUCTS DURING EARLY MAMMALIAN EMBRYOGENESIS

Gene expression in early embryos might be assessed in different ways. For example, genes could be categorized on the basis of the appearance of the proteins for which they code. Thus, embryonic proteins might be produced at all stages in all or most cells; these we shall refer to as **general** gene products. Other proteins, which we shall call **temporal**

Table 1
TIME OF SYNTHESIS OF DIFFERENT RNA
CLASSES DURING MOUSE EMBRYOGENESIS

Class of RNA	Earliest stage of detection	Ref.
4S	2- or 4-cell	5, 6, 9
5S	2- or 4-cell	5, 8, 9
rRNA	2- or 4-cell	5—9
hnRNA	2-cell	5, 6, 8, 10, 11
PolyA⁺ RNA	2-cell	12, 13

gene products, might be detectable only from a particular stage in embryogenesis. **Tissue-specific** proteins would be produced only by one or a few cell types in the embryo proper or extraembryonic structures.

Alternatively, in attempting to classify gene expression, one might take into account the time at which the mRNAs are produced, particularly relative to fertilization. Thus, constitutive genes would be transcribed and their mRNAs translated both prior to and following fertilization. Some proteins in the embryo might derive from the translation of transcripts during oogenesis but not during embryogenesis. In another type of "maternal" gene expression, genes might be transcribed during oogenesis but their mRNAs might not be translated until fertilization has taken place. Finally, in cases of true embryonic gene activation, transcription and translation would both occur only after fertilization.

All of the above classes of gene products probably occur in early mammalian embryos. Examples of many of them have already appeared in the literature.

A. General Gene Products

Evidence for the continuous expression of some genes prior to and during embryogenesis derives both from gel electrophoretic profiles of proteins labeled during incubation with radioactive amino acids (reviewed by Van Blerkom and Manes[20] and Sherman[21]) and from enzyme measurements (reviewed by Epstein[22]). It would be expected that enzymes essential for normal metabolic activity would fall into this general class. Structural proteins such as actin,[23] tubulin,[24] and ribosomal proteins[25] are also produced throughout oogenesis and embryogenesis, though not at constant rates.

B. Temporal Gene Products

Some proteins appear to be produced for the first time at specific periods during embryogenesis. It is tempting to assume that the time of appearance of a particular gene product is related to the embryo reaching a particular morphological stage, e.g., the morula stage, at which time compaction of blastomeres occurs (Figure 1B), or the blastocyst stage, characterized by the formation of a fluid filled cavity (Figure 1C). However, it is incautious to assume without further evidence that the appearance of such markers is stage- rather than time-dependent. The synthesis of a given protein might in fact occur a certain number of hours beyond fertilization regardless of the stage reached by the embryo; investigators generally do not test this possibility because they tend to discard abnormal or retarded embryos. This issue will be considered more fully in Section V.A. It should be understood, therefore, that when we refer to temporal gene products, we are not distinguishing between stage- and time-specific markers.

There are other limitations which must be recognized when searching for temporal gene products. The sensitivity of the assay is always an overriding factor. As mentioned above, the levels of synthesis of general proteins can vary dramatically at different stages of development. It is, therefore, conceivable that a general protein, present below detectable

Table 2
EXAMPLES OF DEVELOPMENTAL MARKERS OBSERVED
DURING EARLY MOUSE EMBRYOGENESIS

	Classification	
Type of marker	**Temporal**	**Tissue-specific**
Polypeptide	Fluorographic spots[a,20,26-28]	Fluorographic spots[a,44-49]
Surface, structural,	Embryonic surface antigen(s)[c,30-32]	PCC4 antigen(s)[51]
and secreted	Non-H-2 antigens[29]	α-Fetoprotein[53,54]
proteins[b]	Type III collagen[33]	
	Laminin[34]	
Enzymes	Lactate dehydrogenase 5[35-38]	Δ^5,3β-Hydroxysteroid
		dehydrogenase[55-57]
	Plasminogen activator[39-41]	Esterase G[59-61]

[a] These markers have been established by incubating embryos with radioactive amino acids and analyzing labeled polypeptides by two-dimensional polyacrylamide gel electrophoresis and fluorography.

[b] These markers have been established by immunological techniques.

[c] Although blastocysts were used as immunogen, reactivity of embryos with antiserum is observed as early as the two- to four-cell stage,[30,31] and radiolabeling studies indicate that the antigens involved are synthesized primarily between the two-cell and morula stages.[32]

levels by the assay used until a given stage is reached, will be considered incorrectly to be a temporal gene product. It is also possible that proenzymes present throughout early embryogenesis will only be activated at later stages or that new spots on gel electrophoretic profiles will reflect processing of a general polypeptide.

Table 2 lists a representative selection of presumptive temporal gene products observed during early mouse embryogenesis. The preimplantation temporal gene products, namely unidentified polypeptides observed on gel electrophoretic profiles,[20,26-28] "non-H-2" antigen(s),[29] embryo-specific surface antigens,[30-32] type III collagen,[33] and laminin, [34] are considered to be temporal gene products rather than tissue-specific ones since they are first detected at or before the eight-cell stage, when the blastomeres are presumably still pluripotent.[3] Lactate dehydrogenase 5 (LDH 5) appears at the time of implantation.[35-38] At this time, the embryo is at the late blastocyst stage and contains three distinct cell types (Figure 1D). It is likely that all cell types acquire LDH 5 activity at this stage, although Monk and Ansell[37] have claimed that the onset of activity in trophoblast cells depends upon contact with ICM-derived cells (see Section V.B.1).

Plasminogen activator, a serine protease, also appears first during periimplantation stages.[39-41] It is listed here as a temporal gene marker rather than a tissue-specific marker because activity is present in both trophoblast and primitive endoderem cells and lacking only in primitive ectoderm cells. On the basis of assays with tissue clumps, Bode and Dziadek[42] have claimed that most, if not all, cell types have plasminogen activator activity associated with them at the egg-cylinder stage (Figure 1E) and beyond; this is not in agreement with results obtained with cell cultures, in which activity is consistently restricted to parietal endoderm cells among ICM derivatives.[39,40,43] The discrepancy might be related to the method of assay: in the former case,[42] large numbers of cells were assayed together so that basal levels of enzyme activity in individual cells might together have given a positive response; in the latter studies[39,40,43] cells were analyzed individually.

Temporal gene products can be further subdivided into two classes: those which are produced continuously following their appearance, and those which are synthesized only for a specific interval during embryogenesis. Some temporal polypeptide markers fall into the first subclass whereas others are members of the second.[26-28] Plasminogen activator production by parietal endoderm cells is constituitive following its initial appearance, whereas it is produced by trophoblast cells only during their invasive phase.[39,40] The embryonic surface antigens described in Table 2 cease to be synthesized by the blastocyst stage,[32] but LDH 5 continues to be produced by many cell types, even in the adult.[37]

C. Tissue-Specific Gene Products

Once again caution must be exercised in assigning a protein the status of a tissue-specific gene product. Problems exist both with the sensitivity of the assays used and with the ability to separate cell types cleanly for analysis. In view of the latter difficulty, we shall consider a protein to be tissue-specific even if trace amounts are detected in other tissues.

As Table 2 illustrates, tissue-specific markers are detectable by all the techniques used to reveal temporal gene products. Van Blerkom et al.[44] first demonstrated that ICM cells and trophectoderm cells possess some different polypeptide spots. This observation has been expanded upon in a number of studies by Johnson and colleagues[45-49] which will be discussed in more detail in subsequent sections.

Immunological probes have been used extensively in efforts to find cell-specific surface antigens on pre- and periimplantation embryos (for reviews, see Sherman[21] and Wiley[50]). There has been controversy over the distribution and time of appearance of many of these antigens, and thus their status as developmental probes is unclear. One antigen, or group of antigens, present on PCC4 embryonal carcinoma cells appears also on ICM cells, but not on trophoblast cells or periimplantation embryos (Table 2).[51] Whether antiserum directed against PCC4 surface antigen(s) reacts with primitive endoderm cells alone or with primitive ectoderm cells as well has yet to be reported. A number of antisera directed against surface antigens of various cell types appear to have a generalized distribution on cells of preimplantation embryos but then become more restricted as development proceeds. Examples of antigens in this category include (1) antigens present on ectoplacental cone cells (antisera react initially with all cells at the eight-cell and preimplantation blastocyst stages but reactivity is restricted to trophectoderm-derived cells by the egg-cylinder stage);[52] (2) type IV collagen, which is associated with most blastomeres in early embryos[33,34] but eventually localizes in basement membrane-producing cell types, primarily of endodermal origin;[34] (3) laminin, a large glycoprotein which has a distribution similar to that of type IV collagen;[34] and (4) antigens present on mouse L-cell fibroblasts (antisera to these antigens react with blastomeres at the morula stage and with ICM and trophectoderm cells of blastocysts but only with mesodermal elements of later embryos).[50] The conversion of surface antigens from general gene products to tissue-specific gene products as embryogenesis proceeds suggests that repressive as well as activating influences are operative in developmental processes (see Section IV.A).

α-Fetoprotein is a useful marker for visceral endoderm cells since only these cells appear capable of synthesizing and secreting this protein.[53] Interestingly, Dziadek[54] has demonstrated that visceral endoderm cells surrounding the embryonic ectoderm layer at the egg-cylinder stage (see Figure 1E) produce α-fetoprotein whereas visceral endoderm cells adjacent to the extraembryonic ectoderm do not. However, as we shall discuss below (see Section V.B.2), this pattern can be reversed when reciprocal endoderm-ectoderm combinations are artificially made. Finally, tissue-specific enzyme activities have been reported. One example in Table 2 is $\Delta^5,3\beta$-hydroxysteroid dehydrogenase (3β-HSD) activity, which appears first following implantation in the mouse and is restricted to trophoblast cells during the first half of pregnancy.[55-57] This enzyme activity is detected prior to implantation in some other mammals in which implantation occurs substantially later than in rodents (for a review, see Bullock[58]). Another example is the so-called G-isozyme of nonspecific esterase activity (probably esterase-2) which appears to be largely restricted to visceral endoderm cells during the first half of pregnancy.[59-61] Specific enzyme markers have yet to be reported for embryonic ectoderm cells. Although these cells have been reported to have unusually large amounts of alkaline phosphatase activity,[62] so too do trophoblast cells.[63] However, since embryonal carcinoma cells, which resemble embryonic ectoderm cells in many respects (see Graham[64]), have a specific isozymic form of alkaline phosphatase,[65] it may be that this enzyme will be found to differ in embryonic ectoderm and trophoblast cells.

D. Maternal Gene Expression vs. Embryonic Gene Activation

The detection of particular antigens or enzymes in an early embryo does not necessarily indicate that the genes coding for them are active during embryogenesis. It is possible that the proteins in question were synthesized during oogenesis or that they were translated in the embryo from mRNAs already present in the unfertilized egg. LDH 1 appears to be a case of an enzyme synthesized prior to, but not for some time following, fertilization. The activity of this enzyme is relatively high in the fertilized egg and falls subsequently to levels almost undetectable by the late blastocyst stage,[66] at which time, as mentioned above, LDH 5 activity is observed.[34] Studies with antibody to LDH 1 have confirmed that although the enzyme is synthesized throughout oogenesis its production is sharply reduced following ovulation;[67] the subsequent decrease in enzyme activity correlates well with disappearance of the antigen.[68] It has been proposed that LDH 1 is only one of a group of enzymes whose synthesis falls off dramatically or ceases following ovulation or fertilization[22,67] although none of the other enzymes in this group has been studied thoroughly.

To date, there has been only one unequivocal demonstration of translation of previously inactive maternal mRNA by early mammalian embryos: Braude et al.[69] have found by in vitro translation studies that a group of polypeptides synthesized by two-cell embryos are encoded by mRNA present, but not used, prior to fertilization. Studies with inhibitors of mRNA synthesis suggest that this might be a generalized phenomenon in early mouse embryos as it is in nonmammalian embryos (see Sections IV.B and VI). Whereas the persistence of production of certain proteins in the presence of inhibitors of mRNA synthesis supports the view that they are translated from maternal messages, the interference with the synthesis of other proteins with the same antimetabolite treatments argues in favor of their being translated from embryonic mRNAs. Evidence supporting the existence of the latter class of gene products will be examined in more detail in Section IV.A.

E. Expression of Genes on the X Chromosome — A Special Case

Normally in female somatic cells only one of the two X chromosomes is actively transcribed.[70] It is commonly the case that inactivation is random, so that one X chromosome is active in half the cells whereas the homolog is active in the others. This general rule does not apply to early mouse embryos. One exception is that both X chromosomes appear to be active in preimplantation embryos; inactivation probably takes place in at least some cells by the late blastocyst stage (see Russell[71]). Monk[72] proposed that inactivation occurs unevenly and is related to the time at which the cells undergo differentiation, i.e., trophectoderm cells first, followed by primitive endoderm cells, and subsequently other cell types as they differentiate from the embryonic ectoderm; potential germ cells would retain both X chromosomes in an active state. Support for part of this hypothesis has come from the observation by Martin et al.[73] that genotypically female embryonal carcinoma cells have two active X chromosomes but only one remains active following differentiation. However, Monk and McLaren[74] have now obtained evidence that X inactivation does occur in female germ cells at some time prior to meiosis.

Takagi and colleagues[75-78] first proposed, on the basis of cytogenetic studies, that X-chromosome inactivation was nonrandom in extraembryonic tissues of rodents. Their experiments suggested that there was in fact preferential inactivation of the X chromosome inherited from the father. This claim has now been supported biochemically by Chapman and colleagues:[79-82] with the use of isozymic variants, it was determined that only maternal X-linked isozymic markers are produced by trophectoderm derivatives in mouse embryos. Furthermore, analysis of yolk sac tissue indicated that visceral endoderm cells also express only maternal X-chromosome markers whereas the mesoderm layer of the yolk sac shows an unbalanced mosaicism with the maternal X-chromosome being expressed more often than the paternally inherited homolog. In all of these studies, control analyses on preparations of the embryo proper showed no evidence of unbalanced X-chromosome inactivation.

It is, therefore, apparent again that in a thorough consideration of the control of gene expression during embryogenesis, it is necessary to account for inactivation as well as activation of certain genes.

IV. MECHANISMS FOR CONTROLLING THE EXPRESSION OF DEVELOPMENTAL GENE PRODUCTS

A. Transcriptional Controls

As indicated in Table 1 and Section II, mRNAs and their higher molecular weight precursors are synthesized from very early stages in mammalian embryogenesis. Information concerning the importance of these hnRNAs and mRNAs has been gained by blocking their synthesis with α-amanitin. In general, the antimetabolite interferes with mammalian embryonic development. However, it is notable that when α-amanitin is added at very early stages, its adverse effects are delayed: the embryos commonly undergo two or three cleavages following treatment before development is blocked.[83-85] As embryogenesis proceeds, the embryo becomes progressively more sensitive to α-amanitin.[48,86] The implication is that the contribution made by newly transcribed mRNA to embryonic development is greater at later stages than at earlier ones.

Braude and colleagues[45,47,48] have studied the effects of α-amanitin upon polypeptide synthetic profiles during the morula-to-blastocyst transition. They found that the majority of polypeptides are synthesized at apparently normal levels despite the presence of the antimetabolite (class I). On the other hand, α-amanitin blocks the normal increase in synthesis of some polypeptides (class II) and even prevents the expected decrease in synthetic levels of others (class III). The effect of α-amanitin upon class II polypeptides suggests that these species depend upon mRNA transcription for their continued synthesis. In more recent studies,[86] we have demonstrated that treatment with relatively low levels of α-amanitin during peri-implantation stages can almost completely block the normal appearance of 3β-HSD activity some 24 to 48 hr later. Although there is apparently a substantial delay between the time of synthesis and translation of 3β-HSD mRNA, the initial appearance of this enzyme activity nevertheless requires transcription of the embryonic genome.

The effect of α-amanitin on class III polypeptides is similar to that observed by Golbus et al.[83] with guanine deaminase activity: these investigators demonstrated that the normal drop in this enzyme activity as development proceeds can be blocked by the administration of α-amanitin. Once again, these results argue that the decline in production of these gene products is modulated at the level of transcription. The control need not be exerted by the repression of transcription of these genes; it is possible that the α-amanitin prevented the synthesis of mRNAs for proteases which would ultimately degrade class III polypeptides and guanine deaminase. Thus, α-amanitin studies strongly support the view that selective gene activation is operative in the early developmental program, but they do not provide irrefutable evidence for the suppression of transcription. Other experiments do, however, strengthen the view that gene repression is involved, e.g., the eventual restriction of originally common surface antigens to specific cell types as development proceeds and the inactivation of a previously active X chromosome (see Section III). Furthermore, there is evidence from recent studies with embryonal carcinoma cells, which possess the differentiative capacities of early embryonic cells (see Graham[64]), that differentiation is characterized not only by the appearance of new polypeptides but also by the cessation of production of several polypeptides normally produced in significant quantities by the stem cells.[87]

If one accepts the view that selective gene activation and repression are operative in the elaboration of the developmental program during early mammalian embryogenesis, one might ask how these processes are controlled. The simplest answer would be that inducers or repressors, either inherited from the oocyte, produced by the embryonic cells themselves, or acquired from the environment (see Section V.B), act directly by modulating transcription of the structural genes. However, the situation might not be so simple. It is known that the complexity of RNA in many cell types is much greater in the nucleus than in the cytoplasm (see Davidson and Britten[88]). In recent studies with sea urchin embryos, it was demonstrated that mRNAs which are found on polysomes only at restricted stages of embryogenesis or in selected cell types are, nevertheless, represented in the nuclear RNA fraction of most cells.[89,90] Accordingly, Davidson and Britten[88] have proposed a model in which structural genes are being constituitively transcribed; transcriptional controls would operate at the level of regulatory genes that code for RNAs capable of interacting with specific mRNA precursors to promote their processing and/or passage from nucleus to cytoplasm. The interaction would be via repetitive sequences present on the regulatory RNAs and their complements present on the nuclear mRNA precursors.

One attractive aspect of the Davidson-Britten model is that it can account for the simultaneous expression of batteries of gene products. Since it is proposed that regulatory RNAs contain sequences that can be recognized by a number of different nuclear mRNA precursors, the transcription of a regulatory gene could ultimately lead to the processing and transport to the cytoplasm at an appropriate developmental stage of several functionally related mRNAs that interact with the same regulatory RNA species. The kind of cascade effect which might be involved in the intrinsic programming of cells such as trophoblast (see Section V.B.1) could make use of such phenomena: following fertilization, synthesis of a regulatory RNA might be induced which would simultaneously promote translocation to the cytoplasm of a series of mRNAs. Their translational products might directly or indirectly activate new regulatory RNAs and/or suppress production of existing ones; this would lead to the appearance of a new population of gene products and the cascade could continue in this way.

Whether the regulatory model envisaged by Davidson and Britten is adequate to explain all transcriptional control is open to question. For example, the model does not readily account for control of ''superprevalent'' mRNA species which are present in much higher numbers than other mRNAs (see Davidson and Britten[88]). Another shortcoming of the model is the implication that suppression of gene activity occurs largely by passive means, depending upon the depletion of specific nuclear regulatory RNAs which are needed for activation and translocation of the various mRNAs. Caplan and Ordahl[91] have made a strong case for the **active** regulation of gene repression in developing systems; they conclude that there should be two mechanisms for suppressing gene expression, one irreversibly, the other reversibly. As we have pointed out in Section III, the cessation of production of certain gene products in some or all cells appears to play an important role in developmental patterns during early mammalian embryogenesis. The fact that X-chromosome inactivation involves physical changes in chromatin structure and results in late replication of the inactivated DNA[78] speaks in favor of a repressive process which operates at the level of the chromosome rather than on nuclear RNA transcripts. On the other hand, the Caplan-Ordahl model is inconsistent with the observed complexity of nuclear RNAs which presumably led Davidson and Britten to propose their model. It would appear that neither model alone adequately accounts for transcriptional regulation during early embryogenesis, although elements of each might be operative.

Finally, recent studies on the immunoglobulin gene have provided evidence that gene expression might be regulated by physical rearrangement of the DNA.[92] There is little evidence at this time to support the view that this is a general phenomenon. Nevertheless, we should not eliminate *a priori* the possibility that gene expression in early mammalian embryos is controlled to some degree by the rearrangement of parts of the genome.

B. Post-Transcriptional Controls

As mentioned in the previous section, the synthesis of the majority of polypeptides in mouse morulae and blastocysts appears to be resistant to α-amanitin treatment.[30,45,47] In view of the mode of action of α-amanitin, the logical conclusion from these observations is that the synthesis of many proteins in preimplantation mouse embryos does not depend upon continuing mRNA transcription. Other studies have substantiated the views that preformed mRNAs play an important role in gene expression during early mammalian embryogenesis and that post-transcriptional controls are therefore in effect. An analysis of the distribution of radiolabeled RNA in newly ovulated eggs indicates a substantial carryover from the oocyte to the zygote.[93] Although the bulk of this labeled material apparently belongs to the tRNA and rRNA classes, 10 to 15% is associated with the hnRNA fraction. It was concluded from subsequent studies that a significant amount of maternal polyA$^+$ RNA persists even to the blastocyst stage.[94] Thus, maternal mRNAs are presumably available for translation through-out the preimplantation period, and perhaps beyond it. The most direct demonstration of both the utilization of preformed maternal mRNAs and the post-transcriptional control of gene expression by early mammalian embryos was provided by Braude et al.[69] As mentioned above, these investigators demonstrated by in vitro translation studies that the mRNAs for a group of proteins first synthesized at the two-cell stage are present (but not expressed) in the unfertilized egg.

Whereas the above studies provide a strong case for the storage and subsequent utilization of maternal mRNAs, experiments performed in this laboratory[86] suggest that even mammalian embryonic mRNAs are subject to post-transcriptional regulation. We investigated the effect of sublethal levels of α-amanitin treatment during periimplantation stages on the simultaneous and subsequent appearance of several developmental markers. Under conditions in which the synthesis of polyA$^+$ RNA was inhibited by approximately 70%, protein synthesis was also depressed, but only after a substantial delay (24 to 48 hr) and to a lesser extent (Table 3). Similarly, the activities of two enzymes, plasminogen activator and β-glucuronidase, were only minimally affected either during or after termination of the α-amanitin treatments (see Reference 86 and Table 3). Implantation-related events (hatching from the zona pel-lucida, attachment to the substratum, and trophoblast outgrowth) were undisturbed. Even in instances where the expression of markers was affected, namely 3β-HSD production and the bilayering of the ICM (presumably indicative of the appearance of primitive endoderm cells and their production of a basement membrane), the most effective time of treatment was 24 to 48 hr in advance of their scheduled appearance (Table 3). Because it was necessary to use sublethal levels of α-amanitin in these studies, the inhibition of polyA$^+$ RNA synthesis was incomplete.[86] Technically, this prevents us from eliminating certain explanations for these results. However, the interpretation that we consider most likely is that the expression of all the developmental markers we studied was regulated at a posttranscriptional level. Since the data support the view that the expression of at least some of the markers requires embryonic transcription, it therefore appears that embryonic as well as maternal mRNAs are subject to these controls. One might, therefore, envision a scheme wherein the tran-scription of developmental genes (or their regulatory RNAs; see previous section) and the subsequent translation of their mRNAs are routinely separated temporally by many hours.[86]

If the expression of developmental genes is generally subject to post-transcriptional reg-ulation, at what level(s) do the controls act? In order for a structural gene transcript to be translated, it must be processed: this usually involves the shortening of the transcript by removal of either terminal or intervening sequences, the addition of a methylated m^{7}GpppX

Table 3
EFFECTS OF α-AMANITIN TREATMENT DURING MOUSE
POSTBLASTOCYST DEVELOPMENT

	% Inhibition on equivalent gestation day[a]	
Property studied	6	8
[35]S-Methionine incorporation into protein	10	38
Plasminogen activator activity	13	14
Increase in β-glucuronidase activity	0	32
Δ^5,3β-Hydroxysteroid dehydrogenase activity	ND[b]	77
Two-layered ICMs	ND[b]	87

Note: Blastocysts were removed from uteri on the 4th day of pregnancy and cultured overnight. Some of the blastocysts were subsequently exposed to α-amanitin (1 μg/mℓ) for 24 hr from equivalent gestation day 5 to 6. At the end of the period of exposure to α-amanitin, the rate of polyA-containing RNA synthesis was found to have been reduced by 69% in comparison to embryos not exposed to the antimetabolite (Schindler and Sherman, 1981). This table illustrates the effects of the α-amanitin treatment upon various developmental parameters as measured on the 6th and 8th equivalent gestation days. Data are presented more fully by Schindler and Sherman.[86]

[a] Equivalent gestation day refers to the age of the embryos had they been maintained *in utero*.
[b] ND, not determined. Δ^5,3β-Hydroxysteroid dehydrogenase activity was undetectable on equivalent gestation day 6 in these experiments, and two-layered ICMs are not visible by phase contrast microscopy prior to the seventh equivalent gestation day.

cap at the 5′, end and the introduction of a polyA tract at the 3′ end.[95-97] The mRNA must also be transported from nucleus to cytoplasm where it can ultimately enter the polysomal RNA pool. Processing and translocation of mRNAs from nucleus to cytoplasm can be accomplished in a matter of minutes;[98] although Tobin[99] has concluded that there is a variation in the time required for processing and translocation of mRNAs in some cells, the estimated differences in rate are only two- to threefold. However, we have little information concerning processing and translocation of mRNAs in the early mammalian embryo. The proposal[100] that maternal mRNAs are activated in mouse embryos by capping after fertilization is unlikely to be correct based upon results from a careful study by Schultz et al.[101] Neither does there appear to be a burst of polyadenylation of preformed mRNAs following fertilization of the mouse egg.[85]

It seems, therefore, most reasonable to assume that post-transcriptional control over mRNA expression in early mammalian embryos is exerted primarily in the cytoplasm, after the mRNAs have been processed. There have been many reports based upon studies with nonmammalian embryos that some mRNAs are "masked", i.e., they are sequestered from the polysomal RNA pool by incorporation into protein-containing structures called "informasomes". mRNAs appear to be fully processed upon release from informasomes since they are active in in vitro translation systems following purification by salt and/or phenol extraction (e.g., see Rosenthal et al.[103] and Jackle[104]). There is as yet no evidence that informasomes or similar structures are involved in the post-transcriptional control of gene expression in early mammalian embryos.

V. FACTORS INFLUENCING THE DEVELOPMENTAL PROGRAM OF CELLS IN THE EARLY MAMMALIAN EMBRYO

A. General vs. Developmental Gene Expression

Before considering the nature of epigenetic cues which might play a role in establishing the developmental program in early mammalian embryos, we shall consider two questions.

The first is whether there is an absolute correlation between molecular and morphological development; the second is whether the types of control exerted over the expression of general and temporal (or tissue-specific) genes are the same or different.

In Section III.B we attempted to draw a distinction between time- and stage-specific products. Although the appearance of temporal gene products is most often associated with the embryos reaching a given developmental stage, evidence is accumulating that at least some markers appear according to a time schedule rather than in response to morphological or organizational considerations. Two striking examples of this are (1) unfertilized oocytes which after prolonged culture seem to synthesize polypeptides whose appearance is normally associated with postimplantation embryos[105] and (2) nondividing blastomeres dissociated at the 16-cell stage (from embryos which had been prevented from undergoing compaction) which acquire trophoblast-specific properties after appropriate periods of culture.[106] The objective of citing these examples is not to convey the idea that cellular organization is unimportant in developmental processes, but rather to suggest that time-dependent processes, including cascade effects (see next section), could also be involved.

It has been known for some time that 5-bromodeoxyuridine treatment of several cell types often suppresses the appearance of differentiative gene products whereas general gene products are unaffected.[107] This raises the possibility that the transcription of general and specialized genes is differentially regulated in these cells. There is, on the other hand, no such indication of a dichotomy in gene regulation in early mouse embryonic cells similarly treated: the incorporation of 5-bromodeoxyuridine into the DNA of trophoblast cells fails to prevent the normal appearance of developmental markers, and for ICM cells even low doses of the antimetabolite appear to be lethal in a relatively nonspecific manner.[108]

Following viral infection there is usually a dramatic shift in the nature of translation products formed as viral polypeptides appear in place of host cell polypeptides (see Koch et al.[109]). Studies of translation in various systems suggest that such a shift might result from differing efficiencies of mRNAs for translation and particularly for initiation of translation.[109] In order to test the possibility that mRNAs for general, temporal, and tissue-specific genes have different translational efficiencies, we studied gene expression in mouse blastocysts subjected to a variety of suboptimal culture conditions which would be expected to interfere with protein synthesis and perhaps other processes as well. Although the various gene products under study showed different metabolic thresholds for expression, there was no clear indication that specialized gene products had a higher threshold than general gene products.[38]

Whereas neither of these experiments eliminates the possibility of differential controls over general and specialized gene products in early mammalian embryogenesis, they provide no evidence that profound differences exist.

B. Intrinsic vs. Extrinsic Factors

Regulatory signals, whether they are operative at or beyond the level of the genome in the early embryo, might originate outside an affected cell or be inherent to the cell itself. For example, cell-cell interaction is an environmental influence which is often cited (but rarely rigorously proven) as being involved in determining the developmental program of a cell. At the molecular level, hormones, metabolites, and even ions are known to affect cellular programs. It would be natural to assume that such regulatory signals would be readily available to, and operative in, mammalian embryos since the conceptus develops in the genital tract, in intimate contact with maternal cells and/or components of maternal fluids. On the other hand, as we have discussed above, gene expression in the oocyte undoubtedly plays a role in the program subsequently elaborated by the early mammalian embryo. It is conceivable, therefore, that ovulated oocytes, by virtue of previous transcriptional and translational activity, need only to receive an appropriate stimulus (normally fertilization) in order to undergo an interconnected cascade of events that we associate with early embryonic development. Such a chain of events might be able to continue reliably without external influence for a considerable period of time.

We consider it likely that both intrinsic and extrinsic controls are operative during the first half of pregnancy in the mammalian embryo. However as will be discussed below, external signals might be much more important in the control of gene expression in some cell types than in others.

1. The Trophectoderm-to-Trophoblast Conversion

Several experiments have suggested that the trophectoderm-to-trophoblast conversion and the subsequent differentiative processes in trophoblast cells do not require extrinsic signals. Trophectoderm cells isolated from maternal and ICM cells,[108,110] or even from each other,[106] appear to undergo a normal program of differentiation to trophoblast. In fact, in blastomere disaggregation experiments it can be shown that the trophoblast program is dominant since in some cases **all** the isolated blastomeres from a given embryo will give rise to trophec-todermal vesicles (structures containing only an outer layer of cells i.e., without an ICM) which will in turn produce only trophoblast cells.[111] The trophectoderm-to-trophoblast conversion will occur in serum-free media,[38,112] reducing further the likelihood that external signals are involved.

On the basis of the experiments described above, it has been proposed (see Sherman[3]) that the potential to form trophoblast is inherent in all blastomeres from very early stages of embryogenesis and that the realization of that potential does not depend upon external cues. It should be stressed, however, that even though environmental factors do not seem to promote trophoblast differentiation, they can in some cases delay, block, or even override the trophoblast developmental program. The trophectoderm-to-trophoblast conversion, as monitored by morphological, functional, and biochemical criteria, can be prevented *in utero* by ovariectomizing the mother on the third day of pregnancy.[38,113] Although this has led to the speculation by some that trophectoderm cells must receive specific maternal signals (hormonal or otherwise) in order to undergo the transition to trophoblast, it is more likely that trophoblast differentiation is prevented in ovariectomized mothers by suboptimal conditions in the uterus which suppress overall blastocyst metabolism, not just differentiative events (see Sherman and Wudl[114]); in fact we can simulate the ovariectomy effect on blastocysts by culturing them in the absence of amino acids and serum.[38]

The ICM seems to exert a more specific regulatory effect upon trophoblast differentiation. Gardner[115] showed clearly by microsurgery and microinjection experiments that the trophectoderm cells overlying the ICM at the blastocyst stage remain in a proliferative mode, whereas those distal from the ICM enter a phase of polyploidization, causing them ultimately to become giant trophoblast cells. The giant cell population possesses 3β-HSD activity, whereas the cells derived from the proliferative population (the ectoplacental cone and so-called secondary giant trophoblast cells) do not appear to acquire this enzyme activity.[55,116] Thus, cells not in contact with the ICM at the blastocyst stage quickly lose their proliferative capabilities and undergo differentiation, whereas the progeny of at least some cells originally in contact with the ICM continue to divide and presumably remain relatively undifferentiated for a longer period of time.

Finally, studies on LDH 5 expression in vitro suggest that environmental factors can influence trophoblast programming in subtle ways. As mentioned above, peri-implantation blastocysts characteristically show a shift from LDH 1 to LDH 5 activity (Table 2). However, Monk and Ansell[37] have claimed that unlike intact blastocysts, trophectodermal vesicles fail to produce LDH 5 during implantation-related steps in vitro. Furthermore, Monk and Petzholdt[117] reported that blastocysts prevented from attaching to a substratum in vitro do not undergo the shift from LDH 1 to LDH 5; the same is observed of blastocysts prevented from implanting *in utero* by ovariectomy of the mother.[38] If, however, the blastocysts are allowed to aggregate or if their blastocoels collapse, then LDH 5 activity is observed.[38,117,118] Taken together, these results raise the possibility that trophoblast cells produce LDH 5 in response to some environmental stimulus.

2. Determination and Differentiation of the Inner Cell Mass

It was stated above that the trophoblast program is overridden in blastomeres which occupy an internal position at the blastocyst stage. ICMs can be isolated from the trophectoderm layer by a process called immunosurgery.[119] If immunosurgery is carried out upon early blastocysts, the resultant ICMs will regenerate a layer of trophectoderm, presumably because some or all cells of the ICM have not yet lost the potential to form trophoblast; on the other hand, if immunosurgery is carried out 6 to 12 hr later, at the expanded blastocyst stage, the outer layer of the ICM will develop as endoderm.[45,120] The implication from these experiments is that the change in fate results from some imprinting event(s) which occur(s) at a precise period in embryogenesis, between the early and expanded blastocyst stages.

After the determinitive change has occurred, ICMs isolated by immunosurgery become completely surrounded by a layer of endoderm-like cells.[119] It is only these outer cells which resemble endoderm morphologically and biochemically.[39,40,121] Similarly, in unperturbed blastocysts, primitive endoderm cells form along the exposed surface of the ICM but not at the ICM-trophoblast interface (Figure 1D). One logical interpretation of these results is that blastomeres of the ICM require a certain period of enclosure by trophectoderm cells in order to lose their propensity for forming trophoblast, but they must assume an external position in order to undergo overt differentiation.

Dziadek[122] has suggested a role of cellular interactions in the development of visceral endoderm cells. She has used α-fetoprotein as an indicator of the differentiation of these cells (Table 2). She has found that the cells of the primitive endoderm layer formed by ICMs after immunosurgery do not develop readily into α-fetoprotein-producing cells, consistent with previous evidence[39] that these cells develop mainly into parietal endoderm. On the other hand, if the initial endoderm layer around the ICMs is removed by a further round of immunosurgery,[39] under certain conditions a second layer of endoderm-like cells will appear.[122,123] The vast majority of these cells react with anti-α-fetoprotein serum.[122] Dziadek[122] has raised the possibility that these visceral endoderm-like cells differentiated in response to contact with overlying parietal endoderm cells. It is not clear whether potential visceral endoderm cells must also be externalized before they can differentiate overtly: some internal cells were reactive with anti-α-fetoprotein serum in these studies, although this might have been due to absorption of secreted α-fetoprotein.[122]

Finally, in another study Dziadek[54] demonstrated convincingly that interaction of extraembryonic ectoderm cells with visceral endoderm has a suppressive effect upon the production of α-fetoprotein by the latter cells. Dziadek and Adamson[53] had reported earlier that on the 8th day of pregnancy, visceral endoderm cells in contact with embryonic ectoderm tissue (to be called VE-E cells) produce α-fetoprotein whereas cells in contact with extraembryonic ectoderm (to be called VE-X cells) do not. To determine the cause of the dichotomy, Dziadek[54] separated VE-E and VE-X layers from their underlying embryonic or extraembryonic ectodermal tissues, respectively. She found that all isolated visceral endoderm layers subsequently produced α-fetoprotein (VE-E layers immediately; VE-X layers within 12 hr of culture). VE-E and VE-X layers also produced α-fetoprotein when cultured in association with embryonic ectoderm tissue. Conversely, neither VE-E nor VE-X cells produced α-fetoprotein when cultured in association with extraembryonic ectoderm tissue. Thus, extraembryonic cells appear to exert the same kind of repressive influence upon at least part of the visceral endoderm differentiation program as do ICM cells upon potential trophoblast cells.

C. Speculations

We have virtually no concrete information involving the nature of factors influencing the developmental program of cells in the early mammalian embryo. Any proposals must, therefore, be speculative. The experiments described do, however, suggest the participation of both intrinsic and extrinsic factors in early differentiative events.

As far as trophoblast cells are concerned, it appears as though elaboration of the program of differentiation does not depend upon external cues, although trophoblast development can be impeded by environmental factors. On the other hand, the differentiation of cells in the ICM does appear to require environmental stimuli. Based on interpretations of experiments described in the previous section one can propose a progressive sequence of differentiation events which depends upon cellular interaction: i.e., progenitor parietal endoderm cells differentiate in response to interaction with trophoblast cells; subsequently, precursor visceral endoderm cells differentiate as a result of contact with parietal endoderm. It is conceivable that mesoderm cells will ultimately be shown to differentiate from progenitor cells in response to contact with visceral endoderm (see Snow[124]).

One feature that appears to be consistent in early embryogenesis is that cells differentiate overtly only when they have at least one surface free of contact with other cells. Why this should be the case is not at all clear; it seems a paradox that cells should receive the impetus for differentiation through interaction with other developmentally more advanced cells but proceed to differentiate only when that contact is severed.

We have very few examples of cases in which cells exert negative influences upon the development of other cell types in the mammalian embryo. In the instances cited (suppression of differentiation of trophoblast cells in contact with ICM cells and suppression of α-fetoprotein production by visceral endoderm cells in contact with extraembryonic ectoderm cells), the affected cells are acted upon by other cells which are probably at an earlier developmental stage. Whether there is a tendency for more advanced cells to be retarded in their differentiative program by contact with less advanced ones remains to be determined.

What is the nature of the regulatory factors? Models to explain developmental fate based upon gradients of determinants throughout the embryo and even within each blastomere are attractive because they are consistent both with normal development as well as with alterations to the developmental pattern resulting from manipulation of blastomeres (see Johnson et al.,[2] Johnson,[125] and Sherman[3]). However, there are no firm data concerning the nature of determinants, or even to prove their existence. If determinants exist, they might be located on the cell membrane (see Handyside[126]) or they might be cytoplasmic. If determinants are on the cell membrane, then they would presumably influence synthetic processes indirectly (e.g., by alteration of cyclic nucleotide levels or ion flux). Cytoplasmic effectors could modulate translation or ultimately penetrate the nucleus and influence gene expression directly. Studies on intercellular communication in early mouse embryos are consistent with information transfer either via membrane interactions or by the direct passage of cytoplasmic factors from one cell to another (e.g., see Lo and Gilula[127,128]).

VI. OVERVIEW

Although many genes are expressed constituitively during mammalian oogenesis and embryogenesis, other genes are expressed only at certain times and/or in particular cell types. Therefore, in order to understand at the most elementary molecular level how the developmental program operates during early mammalian embryogenesis, we must account for both activation of selected genes as well as repression of others when their expression is no longer required. We are also faced with having to answer specific kinds of questions. How are maternal mRNAs, and presumably some embryonic mRNAs as well, stabilized and what cues lead ultimately to their being utilized or degraded? Are control processes the same for general genes as for temporal or tissue-specific genes? Why do some genes remain active once they have been turned on, whereas others are turned off again shortly after being activated? What governs the pattern of expression of some genes from nontissue-specific early in embryogenesis to tissue-specific at later stages? Is the repression of individual genes achieved in the same way as genes inactivated in tandem (e.g., those on the X chromosome)? Models have been put forward which are concerned with translational controls (e.g., Spirin[102]), transcriptional controls (e.g., Davidson and Britten[88]), and the selective repression of genes (e.g., Caplan and Ordahl[91]) during development. In some cases these diverse models are complementary and in other cases they are contradictory; however, none of these models attacks the problem of differentiation at a level that is adequately global to satisfactorily explain the complexities of gene regulation during early mammalian embryogenesis.

It was stated at the beginning of this review that mammalian embryos can be distinguished from their nonmammalian counterparts by the environment in which they develop, the cell types formed, and the span of time required for the process to reach completion. It is now appropriate to consider whether these factors suggest that developmental processes in the two types of embryos differ in any significant way. We have briefly described experiments which indicate that the maternal environment can influence embryonic development. The state of the uterus is important for the realization of implantation. Embryos which are grossly abnormal at later stages can presumably be recognized as such by the mother, resulting in the termination of pregnancy by spontaneous abortion. The importance of the uterine environment in controlling implantation is underscored by the fact that the process can be delayed for lengthy periods when the uterine milieu is altered artificially, e.g., by ovariectomy in some mammals, or normally, as during lactation or hibernation in others (for a review, see Aitken[129]). Procedures leading to abortion might be triggered by changes in the maternal milieu if the conceptus fails to deliver temporal assurances (hormonal signals?) that it is developing normally. These lines of evidence and others support the view that there is crosstalk between mother and conceptus in mammals (see Reference 130 for further consideration of this subject). However, whereas embryonic signals might induce specific gene functions in tissues of the maternal genital tract, we see little evidence that maternal signals play a necessary role in modulating **specific** gene expression in early mammalian embryogenesis (see Section V.B.1). Perhaps the strongest support for this statement derives from our ability to culture early mammalian embryos successfully through the stages under consideration in this review, often with media lacking complex macromolecular factors (e.g., see References 112 and 131).

Mammalian conceptuses contain accessory cell types which are highly specialized in supporting development of the embryo proper. There appears to be no indication that growth and differentiation of these tissues are regulated by processes other than those operative in the embryo proper or in embryos of lower animals. In fact, one might compare the supporting role of mammalian extraembryonic tissues to accessory tissues which are formed during insect larval stages and then discarded as metamorphosis proceeds (even the polyploidization of trophoblast cells has a counterpart in polyteny which occurs in larval tissues of several insects[132]).

Embryogenesis in some mammals occurs over a time interval at least an order of magnitude greater than that of most nonmammalian species. The rates of growth and development of nonmammalian embryos differ most notably at the beginning of embryogenesis. It is because of this difference that it becomes so difficult to draw temporal comparisons between mammalian and nonmammalian embryos: Can regulatory mechanisms in the mouse embryo be the same as, e.g., the embryo of *Spisula*, the surf clam, when the former embryo develops to 200 cells in 5 days,[38] whereas the latter embryo requires 5 *hr*[103] to reach the same cell number? However, when processes are considered **in terms of stage rather than age**, differences between the two types of early embryos are minimized. For example, we have indicated that polypeptide synthetic profiles change markedly in mouse embryos after the first cleavage has occurred;[26-28,69] in *Spisula*, similar experiments indicate that at the two-cell stage the synthesis of some proteins is curtailed while new species occur.[103] Braude et al.[69] have demonstrated further that post-transcriptional controls are responsible for at least some of the observed shift in mouse embryos; the same is true for *Spisula* embryos.[103] Finally, there is evidence that embryonic mRNA transcripts appear in mouse embryos by the four-cell stage (see Section II and Sherman[21]) and this too has been reported to occur in *Spisula* (Tansey and Ruderman, unpublished data cited by Rosenthal et al.[103]). It should not be construed from these correlations that all embryos show identical developmental patterns. In the sea urchin, for example, the dramatic shift in polypeptide synthetic profiles does not occur prior to the gastrula stage.[133] On the other hand, there is evidence that subtle differences can be detected in polypeptide synthetic profiles from the different cell types (micromeres, mesomeres, and macromeres) in the sea urchin embryo at the 16-cell stage,[134] which again, albeit probably due to coincidence, coincides with the stage at which cell-specific polypeptide markers (ICM vs. trophectoderm) can be detected in mouse embryos.[45,46] Notwithstanding the relatively late stage at which the polypeptide synthetic profile shifts in sea urchin embryos, there is evidence for at least some transcriptional activity during cleavage.[134-137] In fact, as summarized in Table 5.4 in Davidson,[138] mRNAs synthesized by the embryonic genome can be detected on polysomes early in embryogenesis in several nonmammalian species.

One gets the impression from the thorough descriptions by Davidson[138] and from the above examples that regardless of species, the intracellular environment in the egg is changed by ovulation and/or fertilization in such a way that there follow not only increases in translational and transcriptional rates, but also shifts in preferences for existing mRNAs to be translated and genes to be transcribed. As development proceeds and maternal mRNAs break down or become inactivated, the relative importance of embryonic mRNAs increases progressively. Although there might be slight differences in the rates at which these shifts occur in terms of cell number or stage reached, these differences among mammalian and nonmammalian embryos are undoubtedly much more subtle than they would be if assessed in terms of time from fertilization. It is conceivable, therefore, that the control mechanisms operative in early mammalian embryogenesis are nothing more than a slow-motion recapitulation of those regulating embryogenesis in nonmammalian embryos. It remains to be determined whether the lengthy cell cycle time in early mammalian embryos is the cause or the result of the protracted unfolding of their developmental program. If the premise is valid that regulatory patterns are the same in all embryonic species, the relatively slow rate of development in mammalian embryos might ultimately prove to be a great advantage in studies aimed at elucidating the nature of the various control mechanisms and the ways in which they might be integrated and coordinated to ensure that embryogenesis will be virtually error-proof.

REFERENCES

1. **Rossant, J. and Papaioannou, V. E.,** The biology of embryogenesis, in *Concepts in Mammalian Embryogenesis,* Sherman, M. I., Ed., MIT Press, Cambridge, 1976, 1.
2. **Johnson, M. H., Pratt, H. P. M., and Handyside, A. H.,** The generation and recognition of positional information in the preimplantation mouse embryo, in *Cellular and Molecular Aspects of Implantation,* Glasser, S. R. and Bullock, D. W., Eds., Plenum, New York, 1981, 55.
3. **Sherman, M. I.,** Control of cell fate during early mouse embryogenesis, in *P & S Biomedical Sciences Symposium on Bioregulators of Reproduction,* Jagiello, G. and Vogel, H. J., Eds., Academic Press, New York, 1981, 559.
4. **Olds, P. J., Stern, S., and Biggers, J. D.,** Chemical estimates of the RNA and DNA contents of the early mouse embryo, *J. Exp. Zool.,* 186, 39, 1973.
5. **Woodland, H. R. and Graham, C. F.,** RNA synthesis during early development in the mouse, *Nature (London),* 221, 327, 1969.

6. **Church, R. B.,** Differential gene activity, in *Congenital Malformations,* Fraser, F. C., McKusick, V. A., and Robinson, R., Excerpta Medica, Amsterdam, 1970, 19.

7. **Monesi, V. and Molinaro, M.,** Macromolecular synthesis and effect of metabolic inhibitors during preimplantation development in the mouse, *Adv. Biosci.,* 6, 101, 1970.

8. **Knowland, J. and Graham, C.,** RNA synthesis at the two-cell stage of mouse development, *J. Embryol. Exp. Morphol.,* 27, 167, 1972.

9. **Clegg, K. B. and Pikó, L.,** Patterns of RNA synthesis in early mouse embryos, *J. Cell Biol.,* 67(Suppl.), 72a, 1975.

10. **Ellem, K. A. O. and Gwatkin, R. B. L.,** Patterns of nucleic acid synthesis in the early mouse embryo, *Dev. Biol.,* 18, 311, 1968.

11. **Pikó, L.,** Synthesis of macromolecules in early mouse embryos cultured *in vitro:* RNA, DNA and a polysaccharide component, *Dev. Biol.,* 21, 257, 1970.

12. **Warner, C. M.,** RNA polymerase activity in preimplantation mammalian embryos, in *Development in Mammals,* Vol. 1, Johnson, M. H., Ed., North-Holland, Amsterdam, 1977, 99.

13. **Levey, I. L., Stull, G. B., and Brinster, R. L.,** Poly(A) and synthesis of polyadenylated RNA in the preimplantation mouse embryo, *Dev. Biol.,* 64, 140, 1978.

14. **Versteegh, L. R., Hearn, T. F., and Warner, C. M.,** Variations in the amounts of RNA polymerase forms I, II and III during preimplantation development in the mouse, *Dev. Biol.,* 46, 430, 1975.

15. **Roeder, R. G. and Rutter, W. J.,** Multiple ribonucleic acid polymerases and ribonucleic acid synthesis during sea urchin development, *Biochemistry,* 9, 2543, 1970.

16. **Moore, P. G. M.,** The RNA polymerase activity of the preimplantation mouse embryo, *J. Embryol. Exp. Morphol.,* 34, 291, 1975.

17. **Warner, C. M. and Versteegh, L. R.,** *In vivo* and *in vitro* effect of α-amanitin on preimplantation mouse embryo RNA polymerase, *Nature (London),* 248, 678, 1974.

18. **Siracusa, G. and Vivarelli, E.,** Low-salt and high-salt RNA polymerase activity during preimplantation development in the mouse, *J. Reprod. Fert.,* 43, 567, 1975.

19. **Schultz, G. A. and Church, R. B.,** Transcriptional patterns in early mammalian development, in *The Biochemistry of Animal Development,* Vol. 3, Weber, R., Ed., Academic Press, New York, 1975, 47.

20. **Van Blerkom, J. and Manes, C.,** The molecular biology of the preimplantation embryo, in *Concepts in Mammalian Embryogenesis,* Sherman, M. I., Ed., MIT Press, Cambridge, 1977, 37.

21. **Sherman, M. I.,** Developmental biochemistry of preimplantation mammalian embryos, *Annu. Rev. Biochem.,* 48, 443, 1979.

22. **Epstein, C. J.,** Gene expression and macromolecular synthesis during preimplantation embryonic development, *Biol. Reprod.,* 12, 82, 1975.

23. **Abreu, S. L. and Brinster, R. L.,** Synthesis of tubulin and actin during the preimplantation development of the mouse, *Exp. Cell Res.,* 114, 135, 1978.

24. **Schultz, R. M., Letourneau, G. E., and Wassarman, P. M.,** Program of early development in the mammal: changes in patterns and absolute rates of tubulin and total protein synthesis during oogenesis and early embryogenesis in the mouse, *Dev. Biol.,* 68, 341, 1979.

25. **LaMarca, M. J. and Wassarman, P. M.,** Program of early development in the mammal: changes in absolute rates of synthesis of ribosomal proteins during oogenesis and early embryogenesis in the mouse, *Dev. Biol.,* 73, 103, 1979.

26. **Van Blerkom, J. and Brockway, G. O.,** Qualitative patterns of protein synthesis in the preimplantation mouse embryo. I. Normal pregnancy, *Dev. Biol.,* 44, 157, 1975.

27. **Levinson, J., Goodfellow, P., Vadeboncoeur, M., and McDevitt, H.,** Identification of stage-specific polypeptides synthesized during murine preimplantation development, *Proc. Natl. Acad. Sci. U.S.A.,* 75, 3332, 1978.

28. **Cullen, B., Emigholz, K., and Monahan, J.,** The transient appearance of specific proteins in one-cell mouse embryos, *Dev. Biol.,* 76, 215, 1980.

29. **Muggleton-Harris, A. L. and Johnson, M. H.,** The nature and distribution of serologically detectable alloantigens on the preimplantation mouse embryo, *J. Embryol. Exp. Morphol.,* 35, 59, 1976.

30. **Wiley, L. M. and Calarco, P. G.,** The effects of anti-embryo sera and their localization on the cell surface during mouse preimplantation development, *Dev. Biol.,* 47, 407, 1975.

31. **Johnson, L. V. and Calarco, P. G.,** Immunological characterization of embryonic cell surface antigens recognized by antiblastocyst serum, *Dev. Biol.,* 79, 208, 1980.

32. **Johnson, L. V. and Calarco, P. G.,** Stage-specific embryonic antigens detected by an antiserum against mouse blastocysts, *Dev. Biol.,* 79, 224, 1980.

33. **Sherman, M. K., Gay, R., Gay, S., and Miller, E. J.,** Association of collagen with preimplantation and peri-implantation mouse embryos, *Dev. Biol.,* 74, 470, 1980.

34. **Leivo, I., Vaheri, A., Timpl, R., and Wartiovaara, J.,** Appearance and distribution of collagens in the early mouse embryo, *Dev. Biol.,* 76, 100, 1980.

35. **Auerbach, S. and Brinster, R. L.**, Lactate dehydrogenase isozymes in the early mouse embryo, *Exp. Cell Res.*, 46, 89, 1967.
36. **Auerbach, S. and Brinster, R. L.**, Lactate dehydrogenase isozymes in mouse blastocyst cultures, *Exp. Cell Res.*, 53, 313, 1968.
37. **Monk, M. and Ansell, J.**, Patterns of lactate dehydrogenase isozymes in mouse embryos over the implantation period *in vivo* and *in vitro*, *J. Embryol. Exp. Morphol.*, 36, 653, 1976.
38. **Sellens, M. H. and Sherman, M. I.**, Effects of culture conditions on the developmental programme of mouse blastocysts, *J. Embryol. Exp. Morphol.*, 56, 1, 1980.
39. **Strickland, S., Reich, E., and Sherman, M. I.**, Plasminogen activator in early embryogenesis: enzyme production by trophoblast and parietal endoderm, *Cell,* 9, 231, 1976.
40. **Sherman, M. I., Strickland, S., and Reich, E.**, Differentiation of early mouse embryonic and teratocarcinoma cells *in vitro:* plasminogen activator production, *Cancer Res.*, 36, 4208, 1976.
41. **Sherman, M. I.**, Studies on the temporal correlation between secretion of plasminogen activator and stages of early mouse embryogenesis, *Oncodev. Biol. Med.*, 1, 7, 1980.
42. **Bode, V. C. and Dziadek, M. A.**, Plasminogen activator secretion during mouse embryogenesis, *Dev. Biol.*, 73, 272, 1979.
43. **Jetten, A. M., Jetten, M. E. R., and Sherman, M. I.**, Analyses of cell surface and secreted proteins of primary cultures of mouse extraembryonic membranes, *Dev. Biol.*, 70, 89, 1979.
44. **Van Blerkom, J., Barton, S. C., and Johnson, M. J.**, Molecular differentiation in the preimplantation mouse embryo, *Nature (London),* 259, 319, 1976.
45. **Johnson, M. H., Handyside, A. H., and Braude, P. R.**, Control mechanisms in early mammalian development, in *Development in Mammals,* Vol. 2, Johnson, M. H., Ed., North-Holland, Amsterdam, 1977, 67.
46. **Handyside, A. H. and Johnson, M. H.**, Temporal and spatial patterns of the synthesis of tissue-specific polypeptides in the preimplantation mouse embryo, *J. Embryol. Exp. Morphol.*, 44, 191, 1978.
47. **Braude, P. R.**, Control of protein synthesis during blastocyst formation in the mouse, *Dev. Biol.*, 68, 440, 1979.
48. **Braude, P. R.**, Time-dependent effects of α-amanitin on blastocyst formation in the mouse, *J. Embryol. Exp. Morphol.*, 52, 193, 1979.
49. **Johnson, M. H.**, Molecular differentiation of inside cells and inner cell masses isolated from the preimplantation mouse embryo, *J. Embryol. Exp. Morphol.*, 53, 335, 1979.
50. **Wiley, L. M.**, Early embryonic cell surface antigens as developmental probes, *Curr. Top. Dev. Biol.*, 13, 167, 1979.
51. **Gachelin, G., Kemler, R., Kelly, F., and Jacob, F.**, PCC4, a new cell surface antigen common to multipotential embryonal carcinoma cells, spermatozoa, and mouse early embryos, *Dev. Biol.*, 57, 199, 1977.
52. **Searle, R. F. and Jenkinson, E. J.**, Localization of trophoblast-defined surface antigens during early mouse embryogenesis, *J. Embryol. Exp. Morphol.*, 43, 147, 1978.
53. **Dziadek, M. and Adamson, E. D.**, Localization and synthesis of alphafetoprotein in post-implantation mouse embryos, *J. Embryol. Exp. Morphol.*, 43, 289, 1978.
54. **Dziadek, M.**, Modulation of alphafetoprotein synthesis in the early postimplantation embryo, *J. Embryol. Exp. Morphol.*, 46, 135, 1978.
55. **Chew, N. J. and Sherman, M. I.**, Biochemistry of differentiation of mouse trophoblast: $\Delta^5,3\beta$-Hydroxysteroid dehydrogenase, *Biol. Reprod.*, 12, 351, 1975.
56. **Sherman, M. I. and Atienza, S. B.**, Production and metabolism of progesterone and androstenedione by cultured mouse blastocysts, *Biol. Reprod.*, 16, 190, 1977.
57. **Sherman, M. I., Atienza, S. B., Salomon, D. S., and Wudl, L. R.**, Progesterone formation and metabolism by blastocysts and trophoblast cells *in vitro*, in *Development in Mammals,* Vol. 2, Johnson, M. H., Ed., North-Holland, Amsterdam, 1977, 209.
58. **Bullock, D.**, Steroids from the pre-implantation blastocyst, in *Development in Mammals,* Vol. 2, Johnson, M. H., Ed., North-Holland, Amsterdam, 1977, 199.
59. **Sherman, M. I.**, Biochemistry of differentiation of mouse trophoblast: esterase, *Exp. Cell Res.*, 75, 449, 1972.
60. **Sherman, M. I.**, Esterase isozymes during mouse embryonic development *in vivo* and *in vitro*, in *Isozymes,* Vol. 3, Markert, C. L., Ed., Academic Press, New York, 1975, 83.
61. **Sherman, M. I. and Atienza-Samols, S. B.**, Enzyme analysis of mouse extra-embryonic tissues, *J. Embryol. Exp. Morphol.*, 52, 127, 1979.
62. **Solter, D., Damjanov, I., and Škreb, N.**, Distribution of hydrolytic enzymes in early rat and mouse embryos: a reappraisal, *Z. Anat. Entwicklungsgesch.*, 132, 291, 1970.
63. **Sherman, M. I.**, The biochemistry of differentiation of mouse trophoblast: alkaline phosphatase, *Dev. Biol.*, 27, 337, 1972.

64. **Graham, C. F.,** Teratocarcinoma cells and normal mouse embryogenesis, in *Concepts in Mammalian Embryogenesis,* Sherman, M. I., Ed., MIT Press, Cambridge, 1977, 315.

65. **Wada, H. G., Vandenberg, S. R., Sussman, H. H., Grove, W. E., and Herman, M. M.,** Characterization of two different alkaline phosphatases in mouse teratoma: partial purification, electrophoretic, and histochemical studies, *Cell,* 9, 37, 1976.

66. **Epstein, C. J., Wegienka, E. A., and Smith, C. W.,** Biochemical development of preimplantation mouse embryos: *in vivo* activities of fructose 1,6 diphosphate aldolase, glucose-6-phosphate dehydrogenase, malate dehydrogenase and lactate dehydrogenase, *Biochem. Genet.,* 3, 271, 1969.

67. **Mangia, F., Erickson, R. P., and Epstein, C. J.,** Synthesis of LDH-1 during mammalian oogenesis and early development, *Dev. Biol.,* 54, 146, 1976.

68. **Spielmann, H., Erickson, R. P., and Epstein, C. J.,** Immunochemical studies of lactatedehydrogenase and glucose-6-phosphate dehydrogenase in preimplantation mouse embryos, *J. Reprod. Fert.,* 40, 367, 1974.

69. **Braude, P., Pelham, H., Flach, G., and Lobatto, R.,** Post-transcriptional control in the early mouse embryo, *Nature (London),* 282, 102, 1979.

70. **Lyon, M. F.,** X-chromosome inactivation and developmental patterns in mammals, *Biol. Rev.,* 47, 1, 1972.

71. **Russell, L. B., Ed.,** *Genetic Mosaics and Chimeras in Mammals,* Plenum Press, New York, 1978.

72. **Monk, M.,** Biochemical studies on X-chromosome activity in preimplantation mouse embryos, in *Genetic Mosaics and Chimeras in Mammals,* Russell, L. B., Ed., Plenum Press, New York, 1978, 239.

73. **Martin, G. R., Epstein, C. J., Travis, B., Tucker, G., Yatziv, S., Martin, D. W., Jr., Clift, S., and Cohen, S.,** X-chromosome inactivation during differentiation of female teratocarcinoma stem cell *in vitro,* *Nature (London),* 271, 329, 1978.

74. **Monk, M. and McLaren, A.,** X-chromosome activity in foetal germ cells of the mouse, *J. Embryol. Exp. Morphol.,* 63, 75, 1981.

75. **Takagi, N. and Sasaki, M.,** Preferential inactivation of the paternally derived X-chromosome in the extraembryonic membranes of the mouse, *Nature (London),* 256, 640, 1975.

76. **Wake, N., Takagi, M., and Sasaki, M.,** Non-random inactivation of X-chromosome in the rat yolk sac, *Nature (London),* 262, 580, 1976.

77. **Takagi, N., Wake, N., and Sasaki, M.,** Cytologic evidence for preferential inactivation of the paternally derived X-chromosome in XX mouse blastocysts, *Cytogenet. Cell Genet.,* 20, 240, 1978.

78. **Takagi, N.,** Preferential inactivation of the paternally derived X-chromosome in mice, in *Genetic Mosaics and Chimeras in Mammals,* Russell, L. B., Ed., Plenum Press, New York, 1978, 341.

79. **West, J. D., Frels, W. I., Chapman, V. M., and Papaioannou, V. E.,** Preferential expression of the maternally derived X-chromosome in the mouse yolk sac, *Cell,* 12, 873, 1977.

80. **West, J. D., Papaioannou, V. E., Frels, W. I., and Chapman, V. M.,** Preferential expression of the maternally derived X-chromosome in extraembryonic tissues of the mouse, in *Genetic Mosaics and Chimeras in Mammals,* Russell, L. B., Ed., Plenum Press, New York, 1978, 361.

81. **Frels, W. I. and Chapman, V. M.,** Expression of the maternally derived X-chromosome in the mural trophoblast of the mouse, *J. Embryol. Exp. Morphol.,* 56, 179, 1980.

82. **Frels, W. I., Rossant, J., and Chapman, V. M.,** Maternal X-chromosome expression in mouse chorionic ectoderm, *Dev. Genet.,* 1, 123, 1979.

83. **Golbus, M., Calarco, P. G., and Epstein, C. J.,** The effects of inhibitors of RNA synthesis (α-amanitin and actinomycin D) on preimplantation mouse embryogenesis, *J. Exp. Zool.,* 186, 207, 1973.

84. **Manes, C.,** The participation of the embryonic genome during early cleavage in the rabbit, *Dev. Biol.,* 32, 453, 1973.

85. **Levey, I. L., Troike, D. E., and Brinster, R. L.,** Effects of α-amanitin on the development of mouse ova in culture, *J. Reprod. Fert.,* 50, 147, 1977.

86. **Schindler, J. and Sherman, M. I.,** Effects of α-amanitin on programming of mouse blastocyst development, *Dev. Biol.,* 84, 332, 1981.

87. **Schindler, J.,** unpublished data, 1980.

88. **Davidson, E. H. and Britten, R. J.,** Regulation of gene expression: possible role of repetitive sequences, *Science,* 204, 1052, 1979.

89. **Wold, B. J., Klein, W. H., Hough-Evans, B. R., Britten, R. J., and Davidson, E. H.,** Sea urchin embryo mRNA sequences expressed in the nuclear RNA of adult tissues, *Cell,* 14, 941, 1978.

90. **Lev, Z., Thomas, T. L., Lee, A. S., Angerer, R. C., Britten, R. J., and Davidson, E. H.,** Developmental expression of two cloned sequences coding for rare sea urchin embryo messages, *Dev. Biol.,* 76, 322, 1980.

91. **Caplan, A. I. and Ordahl, C. P.,** Irreversible gene repression model for control of development, *Science,* 201, 120, 1978.

92. **Tonegawa, S., Maxam, A., Tisard, R., Bernard, O., and Gilbert, W.,** Sequence of a mouse germ-line gene for a variable region of an immunoglobulin light chain, *Proc. Natl. Acad. Sci. U.S.A.,* 75, 1485, 1978.
93. **Bachvarova, R.,** Incorporation of tritiated adenosine into mouse ovum RNA, *Dev. Biol.,* 40, 52, 1974.
94. **Bachvarova, R. and DeLeon, V.,** Polyadenylated RNA of mouse ova and loss of maternal RNA in early development, *Dev. Biol.,* 74, 1, 1980.
95. **Darnell, J. E., Jelinek, W. R., and Molloy, G. R.,** Biogenesis of mRNA: genetic regulation in mammalian cells, *Science,* 181, 1215, 1973.
96. **Shatkin, A. J.,** Capping of eukaryotic mRNAs, *Cell,* 9, 645, 1976.
97. **Crick, F.,** Split genes and RNA splicing, *Science,* 204, 264, 1979.
98. **Palmiter, R. D., Moore, P. B., Mulvihill, E. R., and Emtage, S.,** A significant lag in the induction of ovalbumin messenger RNA by steroid hormones: a receptor translocation hypothesis, *Cell,* 8, 557, 1976.
99. **Tobin, A. J.,** Evaluating the contribution of posttranscriptional processing to differential gene expression, *Dev. Biol.,* 68, 47, 1979.
100. **Young, R. J.,** Appearance of 7-methylguanosine-5'-phosphate in RNA of mouse 1-cell embryos three hours after fertilization, *Biochem. Biophys. Res. Commun.,* 76, 32, 1977.
101. **Schultz, G. A., Clough, J. R., and Johnson, M. H.,** Presence of cap structures in messenger RNA of mouse eggs, *J. Embryol. Exp. Morphol.,* 56, 139, 1980.
102. **Spirin, A. S.,** On "masked" forms of messenger RNA in early embryogenesis and in other differentiating systems, *Curr. Top. Dev. Biol.,* 1, 1, 1966.
103. **Rosenthal, E. T., Hunt, T., and Ruderman, J. V.,** Selective translation of mRNA controls the pattern of protein synthesis during early development of the surf clam, Spisula solidissima, *Cell,* 20, 487, 1980.
104. **Jäckle, H.,** In vitro translatability of RNP-particles from insect eggs and embryos (*Smittia spec.,* Chironomidae, Diptera), *J. Exp. Zool.,* 212, 177, 1980.
105. **Van Blerkom, J.,** Patterns of molecular differentiation during oogenesis and embryogenesis in mammals, in *Cellular and Molecular Aspects of Implantation,* Glasser, S. R. and Bullock, D. W., Eds., Plenum Press, New York, 1981, 155.
106. **Sherman, M. I. and Atienza-Samols, S. B.,** Differentiation of mouse trophoblast does not require cell-cell interaction, *Exp. Cell Res.,* 123, 73, 1979.
107. **Rutter, W. J., Pictet, R. L., and Morris, P. W.,** Toward developmental mechanisms of developmental processes, *Annu. Rev. Biochem.,* 42, 601, 1973.
108. **Sherman, M. I. and Atienza, S. B.,** Effects of bromodeoxyuridine, cytosine arabinoside and Colcemid upon *in vitro* development of mouse blastocysts, *J. Embryol. Exp. Morphol.,* 34, 467, 1975.
109. **Koch, G., Oppermann, H., Bilello, P., Koch, F., and Nuss, D.,** Control of peptide chain initiation in uninfected and virus infected cells by membrane mediated events, in *Modern Trends in Human Leukemia,* Vol. 2, Neth, R., Gallo, R. C., Mannweiler, K., and Moloney, W. C., Eds., J. F. Lehmans, Munich, 1976, 541.
110. **Sherman, M. I.,** The role of cell-cell interaction during early mouse embryogenesis, in *The Early Development of Mammals,* Balls, M. and Wild, A. E., Eds., Cambridge University Press, London, 1975, 145.
111. **Wudl, L. R. and Sherman, M. I.,** *In vitro* studies of mouse embryos bearing mutations in the T complex: *t⁶, J. Embryol. Exp. Morphol.,* 48, 127, 1978.
112. **Rizzino, A. and Sherman, M. I.,** Development and differentiation of mouse blastocysts in serum-free medium, *Exp. Cell Res.,* 121, 221, 1979.
113. **Bergström, S.,** *Surface Ultrastructure of Mouse Blastocysts Before and at Implantation,* Ph.D thesis, University of Uppsala, Uppsala Offset Center, 1971.
114. **Sherman, M. I. and Wudl, L. R.,** The implanting mouse blastocyst, in *The Cell Surface in Animal Embryogenesis and Development,* Poste, G. and Nicolson, G., Eds., North-Holland, Amsterdam, 1976, 81.
115. **Gardner, R. L.,** An investigation of inner cell mass and trophoblast tissue following their isolation from the mouse blastocyst, *J. Embryol. Exp. Morphol.,* 28, 279, 1972.
116. **Salomon, D. W. and Sherman, M. I.,** The biosynthesis of progesterone by cultured mouse midgestation trophoblast cells, *Dev. Biol.,* 47, 394, 1975.
117. **Monk, M. and Petzholdt, I.,** Control of inner cell mass development in cultured mouse blastocysts, *Nature (London),* 265, 338, 1977.
118. **Spielmann, H., Eibs, H.-G., Jacob-Müller, U., and Bischoff, R.,** Expression of lactate dehydrogenase isozyme 5 (LDH-5) in cultured mouse blastocysts in the absence of implantation and outgrowth, *Biochem. Genet.,* 16, 191, 1978.
119. **Solter, D. and Knowles, B.,** Immunosurgery of mouse blastocyst, *Proc. Natl. Acad. Sci. U.S.A.,* 72, 5099, 1975.
120. **Handyside, A. H.,** Time of commitment of inside cells isolated from pre-implantation mouse embryos, *J. Embryol. Exp. Morphol.,* 45, 37, 1978.

121. **Wiley, L. M., Spindle, A. I., and Pedersen, R. A.,** Morphology of isolated inner cell masses developing *in vitro, Dev. Biol.,* 63, 1, 1978.

122. **Dziadek, M.,** Cell differentiation in isolated inner cell masses of mouse blastocysts *in vitro:* onset of specific gene expression, *J. Embryol. Exp. Morphol.,* 53, 367, 1979.

123. **Pedersen, R. A., Spindle, A. I., and Wiley, L. M.,** Regeneration of endoderm by ectoderm isolated from mouse blastocysts, *Nature (London),* 270, 435, 1977.

124. **Snow, M. H. L.,** Embryo growth during the immediate postimplantation period, *Ciba Foundation Symp.,* 40, 53, 1976.

125. **Johnson, M. H.,** Intrinsic and extrinsic factors in preimplantation development, *J. Reprod. Fert.,* 55, 255, 1979.

126. **Handyside, A. H.,** Distribution of antibody and lectin binding sites on dissociated mouse blastomeres: evidence for polarisation at compaction, *J. Embryol. Exp. Morphol.,* 60, 99, 1980.

127. **Lo, C. W. and Gilula, N. B.,** Gap junctional communication in the preimplantation mouse embryo, *Cell,* 18, 399, 1979.

128. **Lo, C. W. and Gilula, N. B.,** Gap junctional communication in the post-implantation mouse embryo, *Cell,* 18, 411, 1979.

129. **Aitken, R. J.,** Embryonic diapause, in *Development in Mammals,* Vol. 1, Johnson, M. H., Ed., North-Holland, Amsterdam, 1977, 307.

130. **Whelan, J., Ed.,** *Ciba Foundation Symp.,* 64, 1979.

131. **Whitten, W. K.,** Nutrient requirements for the culture of preimplantation embryos *in vitro, Adv. Biosci.,* 6, 129, 1971.

132. **Pavan, C. and Brito da Cunha, A.,** Gene amplification and phylogeny of animals, *Genetics,* 61(Suppl.), 1, 1969.

133. **Brandhorst, B. P.,** Two-dimensional gel patterns of protein synthesis before and after fertilization of sea urchin eggs, *Dev. Biol.,* 52, 310, 1976.

134. **Senger, D. R. and Gross, P. R.,** Macromolecule synthesis and determination in sea urchin blastomeres at the sixteen-cell stage, *Dev. Biol.,* 65, 404, 1978.

135. **Senger, D. R., Arceci, R. J., and Gross, P. R.,** Histones of sea urchin embryos: transients in transcription, translation, and the composition of chromatin, *Dev. Biol.,* 65, 28, 1978.

136. **Rodgers, W. H. and Gross, P. R.,** Inhomogeneous distribution of egg RNA sequences in the early embryo, *Cell,* 14, 279, 1978.

137. **Ernst, S. G., Hough-Evans, B. R., Britten, R. J., and Davidson, E. H.,** Limited complexity of the RNA in micromeres of sixteen-cell sea urchin embryos, *Dev. Biol.,* 79, 119, 1980.

138. **Davidson, E. H.,** *Gene Activity in Early Development,* Academic Press, New York, 1976.

139. **Ducibella, T.,** Surface changes of the developing trophoblast cell, in *Development in Mammals,* Vol. 1, Johnson, M. H., Ed., North-Holland, Amsterdam, 1977, 5.

140. **Diwan, S. B. and Stevens, L. C.,** Development of teratomas from the ectoderm of mouse egg cylinders, *J. Natl. Cancer Inst.,* 57, 937, 1976.

141. **Škreb, N., Švajger, A., and Levak-Švajger,** Developmental potentialities of the germ layers in mammals, *Ciba Foundation Symp.,* 40, 27, 1976.

142. **Adamson, E. A. and Gardner, R. L.,** Control of early development, *Br. Med. Bull.,* 35, 113, 1979.

Chapter 8

ON POSSIBLE ROLE OF RNAs IN THE PROCESSES OF CELL DIFFERENTIATION

Tamito Noto

TABLE OF CONTENTS

I. INTRODUCTION

The emphasis on embryological research has been changing in recent years from the morphological aspects of differentiation to the genetic aspects of differentiation. The analysis of protein, RNA, or DNA synthesis in differentiating cells provided useful information on the nature and control of gene expression. Many speculations were made on the control mechanisms operating during the terminal differentiation or the differentiation of early embryogenesis, in which the proteins, hormones, RNAs, or DNAs were the candidates for the control substances. Although it might be somewhat premature at present to discuss all of these speculations, an attempt is made in the second half of this chapter to present some of these hypotheses, including our own on "free gene complexes", with emphasis on the RNA-mediated control of gene expression during differentiation. In the first part of this paper, a brief review on various systems utilized for analysis of the mechanism(s) controlling gene transcription will be presented.

II. RECENT ADVANCES IN VARIOUS GENE EXPRESSION SYSTEMS

A. Crystallin Genes

The investigations on crystallins from chick, rat, and calf can be classified into two groups. One is concerned with the differential patterns of crystallin synthesis during the terminal differentiation of lens cells,[1-5] and the other is concerned with the accumulation of crystallin RNAs.[6-10] The results of these investigations can be summarized as follows: (1) the α-, β-, and δ- (or γ-) crystallins show differential synthesis depending on normal developmental stages of lens cells; and (2) the accumulation of crystallin mRNAs coding for these proteins can be analyzed.

Okada and co-workers[4] studied by immunofluorescence and electrophoresis observations the process of differentiation or dedifferentiation of adult rat lens epithelial cells into the lentoid body specific in γ-crystallin in cell culture. It is noteworthy that in long-term cell cultures, a new type of crystallin, βμ-crystallin, was detected in the dedifferentiating cell line. Okada's laboratory[5] also demonstrated that the transdifferentiation was associated with the preferential synthesis of δ-crystallin. Vermorken et al.[1] performed a more quantitative analysis on the calf crystallin subunits showing that α A-chains and several β-crystallin polypeptides increased during epithelial fiber-like lens cell transition. Mcavoy[3] analyzed, by an indirect immunofluorescence method, the coordinated expression of rat crystallin genes with the proliferation or elongation factors of the lens cells. α-Crystallin was synthesized paralleling the proliferation of epithelial cells and β- or α-crystallins with the elongation factors of the retina. Courtois et al.[2] demonstrated that in vitro morphological transformation of bovine lens epithelial cells into actively proliferating tumors did not affect the synthesis of α-crystallin. Beebe and Piatigorsky[8] observed that in the serum-stimulated explantation of chick lens epithelial cells the δ-crystallin synthesis was not only a function of δ-crystallin mRNA accumulation (see also References 6 and 7), but was also controlled by either post-transcriptional or a translational regulatory mechanism. Thomson et al.[9] emphasized a similar aspect of chick crystallin gene expression based on observation of differential turnover of crystallin mRNS, but suggested that the regulation of individual crystallin polypeptide synthesis could occur prior to translation. Finally, Jackson et al.,[10] using hybridization with a cDNA probe, estimated the mRNA sequence complexity. In the chick lens mRNAs, 5800 to 7200 sequences of average size were found arranged in three abundance classes, the most abundant of which represented at least 4 sequences coding for the 3 crystallin proteins.

B. Fibroin Gene

The fibroin gene has also been analyzed in depth. In a review on *Bombyx mori* silk fibroin gene, Suzuki and Ohshima[11] discussed the gene, its structural features with its flanking sequences, and their potential role in the regulation of fibroin gene transcription. Structural analysis of fibroin-producing (active) and nonproducing (silent) genes[12] focusing especially on the intervening sequence and the sequence around the 5′-end demonstrated, unfortunately, no significant difference between the two. Manning and Gage's results[13] on physical mapping of the silk fibroin gene were similar to those of Suzuki's.

C. Globin Genes

This system has been analyzed exclusively. Only a few selected reports will be discussed here. Battaglia and Melli[14] made the observation that the translational products of isolated globin mRNA consist of four different species. They found that the mRNA corresponding to the four globin subunits varied in proportion according to the development stage of the red blood cells, suggesting a mechanism of differential gene control operating at the translational level. Microinjection of rabbit globin mRNA into one- to two-cell *Xenopus laevis* embryos resulted in globin synthesis distributed to the animal hemisphere cells of the developed gastrula in a greater proportion than in the other zones.[15] As for the rabbit globin gene, it was suggested[16] that the flanking DNA sequences apparently located at the 3′-end of the rabbit β-globin gene might link the other globin genes. This linking might be essential for transcription, translation, and processing of mRNAs to produce active β-globin mRNA. Brotherton et al.[17] demonstrated differential patterns of globin chain synthesis among various types of hemoglobin during normal mouse fetal development. The three embryonic hemoglobins of primitive nucleated erythrocytes derived from yolk sac blood islands are EI (x_2y_2), EII (α_2y_2), and EIII (α_2z_2), and the adult hemoglobin of the nonnucleated definitive erythrocyte originated from the fetal liver is ($\alpha_2\beta_2$). The transitional accumulation of hemoglobin, from the embryonic to adult type, is interesting from the viewpoint of differential expression of globin genes. They observed that in 13-day-old fetuses Hb EI contained only y chains, but Hb EII contained both α and y chains, and Hb III only α chains. On day 11 the ratio (0.29) between α/β globin was markedly lower than on day 13 (0.93), indicating a differential synthesis of globin chains during differentiation. Leder et al.[18] found that the globin mRNA sequences were interrupted by an approximately 550-base long intervening sequence and suggested that the clipping-out of the RNA after transcription, the processing reactions, might play a role during differentiation of β-globin gene. In spite of the profound analysis of globin genes, mechanisms of the globin gene control operating during the embryonic to adult stages are poorly understood and an assumption is made that humoral factors might play important roles in differentiation of hemoglobin.

D. Globulin Genes

This system proved to be one of the most attractive ones for developmental biologists. Hozumi and Tonegawa[19] found that the V_k and C_k chains of the early mouse globulin genes were located away from each other and were joined to form a contiguous polynucleotide stretch during differentiation of lymphocytes. They observed differences in hybridization patterns between the genomes of embryonic cells and of the plasmacytoma. The embryo DNA showed two components that hybridized with C- and V-gene sequences, respectively, while the pattern of the tumor DNA showed a single component smaller than either of the embryonic components hybridizable with both V- and C-gene sequences. Furthermore, they suggested various models for V-C gene joining, the most interesting of which is the excision-insertion model. They speculated in this model that if a promoter site is created by inserting the excised portion of the multiple V genes, this might result in activation of V gene. Thus, the gene is created by excision and by joining. Whether or not this mechanism of gene alteration during differentiation can be found in other genes remains to be seen. Tonegawa et al.[20] hypothesized that (1) most higher cell genes consist of informational DNA (the introns) interspersed with silent sequences (the exons); (2) if the splicing mechanism is general and independent of the gene structure, a gene which corresponds to one transcription unit can also correspond to many polypeptide chains of related or differing functions; and finally (3) if specific new splicing patterns can be turned on by special gene products, new splicing enzymes can control gene expression during differentiation.

E. Ovalbumin Gene

Garel and Axel[23] proposed the possibility of structural changes in nucleoprotein within the ovalbumin gene. By DNase-I digestion of oviduct nuclei and annealing to ovalbumin cDNA active ovalbumin genes were revealed which were sensitive to nuclease attack. Bellard et al.[24] reported on a selective accessibility of actively transcribed ovalbumin genes to DNase I digestion proposing that the modified subunit of transcribing chromatin corresponded to an extended form of chromatin in the absence of any particulate nucleosomal structure (see also Reference 25). Schütz et al.[26] demonstrated the accumulation of ovalbumin mRNAs during steroid hormone stimulation (see also Reference 25). O'Malley et al.,[27] who seemed to have maintained that the nonhistone proteins of chromatin play a central role in the control of differential gene expression, suggested the possibility that RNA polymerase could initiate randomly on the ovalbumin DNA sequences. Mapping analysis of the ovalbumin DNA by Breathnach et al.[28] and Weinstock et al.[29] indicated that the ovalbumin gene in chicken was split showing two interruptions in the sequences coding for ovalbumin mRNA,[28] or that the linear DNA sequence coding for ovalbumin was interrupted by at least two intragenic DNA spacers absent from the corresponding RNA.[29] No difference of DNA sequences was found between the ovalbumin producer and nonproducer somatic cells. As in the previously described fibroin or globin genes, the spacer sequences of the ovalbumin gene also indicate that these unique sequences are fixed within the germ line and do not result from genomic rearrangement.

F. Enamel Genes

Amelogenesis, especially ameloblast differentiation, is undoubtedly a promising system for investigation on differentiation, but at present the work on enamel gene expression is still at preliminary stages.[30] Previous work on extracellular matrix vesicles containing RNA potentially involved in transfer of genetic information during epithelial-mesenchymal interactions[31] presents an attractive hypothesis. Fukae and Shimizu[32] reported that at four different stages of an early developing bovine incisor, ten or more electrophoretically separable protein components were obtained and that the proportion of slow-moving components decreased during the course of mineralization while there was a marked increase in the proportion of the fast and intermediate components. This writer and co-workers[33] also observed, in their experiments on the pattern analysis of young enamel proteins isolated from neonatal mouse incisors, that eight bands were recognizable, both on acetate cellulose strip and polyacrylamide gel electrophoresis. In wheat germ cell-free translation systems using embryonic rabbit molar tooth organ mRNAs, there were seven polypeptides which had properties similar to enamel protein and some extracellular matrix proteins.[34]

G. Myosin and Actin Genes

Paterson and Bishop,[35] based on analysis of the sequence complexity, abundance class distribution, and coding capacity of the myosin and actin mRNAs from primary culture of 12-day embryonic chick breast muscle, suggested that the differential synthesis of muscle-specific proteins is controlled at the transcriptional level, not at the translational level. They showed that the differentiated myofibrils from myoblasts demonstrated a new abundance class consisting of six or seven different mRNA sequences, each present in 15,000 copies per cell and its accumulation (a tenfold increase probably due to the accumulation of a large number of copies of a small number of mRNA sequences) appeared to be correlated with the increased synthesis of muscle-specific proteins. John et al.[36] demonstrated by cDNA annealing to an excess of HMC mRNA from 14-day chick embryonic leg skeletal muscle that replicating mononucleate myogenic cells contained no detectable MHC mRNA, while in the mononucleate cells or multinucleate myotubes a relatively large amount of MHC mRNA was found to accumulate. Strohman et al.[37] also made a similar observation at the translational level of MHC mRNAs in cell cultures of 12-day chick embryonic breast muscle.

H. Histone Genes

Three reviews seem to focus on recent advances on the histone gene expression during differentiation. Newrock et al.[38] examined differential histone synthesis during the early development of sea urchin, *S. purpuratus*, using immunofluorescence techniques. The three histone-like proteins (probably HI, H2A, and H2B histones) are differentially synthesized and incorporated into chromatin during early development by a switching on and off mechanism, i.e., through an interplay between maternal mRNA information stored in the cytoplasm of the unfertilized egg and new information transcribed from the genome of the embryo. Based on the differential synthesis of histone subtypes, the authors proposed a model for initial gene expression. The initial key substance can be the maternal stored mRNAs for CS proteins which are translated immediately after fertilization and affect the earliest cleavage stage. The chromatin remoldings will elicit the next cellular generations of morula chromatin to transcribe new mRNAs for β, γ, or δ histone subtypes. Weinberg et al.[39] demonstrated, by sequence analysis, a heterogeneity in histone genes. On the other hand, Kressmann et al.[40] presented the viewpoint that the detection of gene expression at protein level is not a good indicator of the fidelity of in vitro transcription. Mertz and Gurdon[41] injected DNAs, including virus or bacteriophage DNAs, into *Xenopus* oocytes and found that the first step in transcription would be association of the injected DNA with histones to form transcriptionally active nucleoprotein complex.

I. Ribosomal Genes

The mechanisms of transcription of the isolated or cloned ribosomal DNAs have been vigorously under investigation. Higashinakagawa et al.[42] described a procedure to isolate amplified nucleoli from immature oocytes of *Xenopus laevis* free of any DNA component other than rDNA, which also contained intact ribosomal RNA precursor and an active endogenous RNA polymerase. Scalenghe et al.[43] performed an incubation of isolated germinal vesicles of *Xenopus* oocytes containing nucleoli with an exogenously added template of *Xenopus* DNA obtained from erythrocytes, and observed a stimulation of endogenous RNA transcription. Brown and Gurdon[44] reported that purified 5S DNA from *Xenopus* erythrocytes was transcribed in vitro with high fidelity into full-length 5S RNA of the correct sequence. They also suggested that some of the genes isolated from a cell type (the nucleated erythrocyte) which were nonfunctional in vivo were transcribed in the oocyte. A regulatory model proposed by Gurdon and Brown[45] suggested that the injected 5S DNA was converted into a minichromosome-like structure (DNA-protein complex), its transcriptional properties being the same as those of the oocyte's native 5S genes on chromosome. The proteins contained in the minichromosome might be similar to the regulatory substances which affect *Xenopus* ribosomal RNA synthesis during oogenesis[46] or to the DNA-binding proteins (DB-1),[47] an acidic protein of unfertilized *Drosophila* eggs which bind to sequences associated with rDNA. Brown and Gurdon[48] also demonstrated the presence of multigene families of repeating units of cloned *Xenopus* 5S DNA. The circular recombinant rDNA configurations[49] were demonstrated by direct visualization of *Xenopus* chromatin transcription by electron microscopy. The results showed that in the injected oocytes transcriptionally active rDNA could be identified by the presence of lengthened ribonucleoprotein fibrils and the closely packed polymerases.

J. Mitochondrial Genes

The distribution of mitochondrial 4S RNA (mostly organelle-specific tRNAs) gene sites was demonstrated[50] in *Xenopus* ovary by electron microscopy. Using in vitro ferritin-4S RNA hybridization on *Xenopus* mitochondrial DNA the number of *Xenopus* 4S RNA sites was shown to be 21 in total, 15 on the H strand, and 6 on the L strand, 1 of which was located in the gap between 2 rRNA genes and another 1 flanked each outside end of the rRNA genes. Ramirez and Dawid[51] demonstrated by hybridization annealings and electromicroscopically the functional map of mitochondrial rRNA genes which correlated with the physical map of 4S RNA genes.[50] The replication origin of mitDNA whose replicating direction proceeds opposite to rRNA transcription is 700 to 950 base-pair distal from the end of the small rRNA gene. The mitDNA replication origin shows a high evolutionary conservation in the two *Xenopus* species. Bonner et al.[52] reviewed the general aspect of the mitochondrial genes in *Drosophila melanogaster*.

III. RECENT STUDIES ON THE RNA-MEDIATED CONTROL OF GENE EXPRESSION DURING DIFFERENTIATION

A. Biological Functions of RNAs

Sufficient evidence exists on biological activity and properties of RNAs with potential functions in regulation of differentiation.[53-56] More recent evidence came from Siddiqui's laboratory.[57-60] They demonstrated a cardiac transformation in the explanted chick postnodal piece by a poly(A)-containing RNA, 7S-RNA of the 16-day-old chick embryonic heart tissue, which was clearly different from actin or myosin mRNA. The transformation was demonstrated by electron microscopic and biochemical observations, the appearance of myofibrils and glycogen granules, increased synthesis of actin- and myosin-like proteins, and elaboration of acetylcholinesterase activity. At present, they seem to assume that the RNA might participate at the translational level or by its resulting products in regulating gene expression during early cardiac differentiation. Mclean et al.,[61] who demonstrated a similar cardiac transformation in the explanted chick postnodal piece exposed to an RNA [poly(A)-containing RNA]-enriched extract obtained from adult chicken heart, suggested that the exogenous heart RNA extract might lead to *de novo* synthesis of new mRNA specific for membrane proteins involved in the electrogenesis of action potentials. Finally, Tung and Niu[62,63] demonstrated that mRNA from carp's mature ovarian eggs, when injected into fertilized eggs from different strains of goldfish, caused transformation of the tail from double to single. They also demonstrated biological functions of DNA. DNA extracted from visceral organs of urodele was injected into fertilized eggs of goldfish. Although in significantly low frequency, the young fish receiving the DNA developed one balancer similar to that of urodilian species.

B. Possible Mechanisms of RNA Functions

Davidson and Britten,[64,65] in a comprehensive review, provided an in-depth analysis of the possible mechanism of regulation of gene expression. Key points of their hypothesis are (1) a given mRNA precursor with its intervening or flanking sequences contains special short spacer sequences; (2) the repetitive DNA sequences of animal genomes synthesize a "sensor-controlled" nuclear RNA transcript which contains sequences complementary to the short spacer sequences of the mRNA precursor (i.e., integrating regulatory transcription unit); and (3) an interaction occurs between the mRNA precursor and the complementary transcript resulting in the formation of spacer RNA-complementary RNA duplexes. Since repetitive DNA sequences could produce various integrating regulatory transcription units with different combinations of complementary RNA sequences, it would be possible that the formation of RNA-RNA duplexes should control the production of pre-mRNA at the post-transcriptional level during differentiation. Frenster and Herstein[66] hypothesized that certain species of nuclear RNA (derepressor RNA which has complementary sequence to mRNA) should hybridize selectively to a complementary DNA strand thereby activating mRNA synthesis on another DNA strand. They suggested a control of gene transcription by operator RNA-derepressor RNA duplex under the condition when excessive operator RNA is accumulated. Tseng et al.[67] reported a unique type of small RNA, the initiator RNA originally proposed by Reichard et al.[68] isolated from a human lymphoblastoid cell line consisting of approximately nine ribonucleotide residues which might function as a primer in discontinuous synthesis. Goldstein[69] demonstrated the presence of another type of small RNA, small nuclear RNAs (snRNA), 65 to 200 nucleotides long, during mitosis of *Amoeba proteus*. The snRNAs were shown to be released to the cytoplasm during cell division and then to return virtually intact to the postdivision daughter nuclei. Both middle anaphase and late prophase chromosomes were found to have a large amount of snRNAs associated suggesting a particular affinity for snRNAs to the incompletely condensed (or decondensed) chromosomes. The snRNAs were considered to open up selected loci to transcriptive activity by specific base-pairing mechanism.

C. A General Hypothesis on Gene Differentiation

I would like to close this review speculating on various theoretical considerations for regulation of differentiation.[70] New evidences for the hypotheses described here will be added in the following section. First of all, the shifting chromatin architectures illustrated by Mott and Callan[71] in the Newt lampbrush chromosomes can be pictured to provide us a fine morphological background for the architectures of functioning chromatin during developmental events of a given somatic cell. An interphase chromatin architecture consists of a continuous fibril made up of unit fibers of the genome attached to the inner surface of the nuclear membrane. Lateral loops from chromomere sites arranged on the fibril extend toward a center of nuclear space. These lateral loops might also form subchromomeres (see Figure 1). In an ultimate interphase, the loops develop fully into a three-dimensional network framing a nucleolar architecture. These interphase architectures were actually observed in the embryonic chick cardiac chromatin and other chromatin preparations using SDS-squashing and glyoxal-glutaraldehyde fixation techniques.[72-74]

In order to simplify our discussion, let us focus on one interphase chromomere site along which extends laterally and forms "primitive loops" consisting of polycistron-like gene sites (see the looping model shown in Figure 1). These primitive loops, which were proposed by the author as a model, can be set free into cellular environments and form "free gene complexes". These complexes might play important roles not only in cellular metabolism such as a circular ribosomal DNA structure,[75,76] but also in gene activation during developmental processes. A distinction between a given primitive loop and its deriving free gene complex can be made such that the gene sequence structures of a given primitive loop are permanent or conserved during generations to generations, while the free gene complex contains only informative sequences,[77] which are temporarily functionable during a few successive generations. At an initial phase of free gene complex formation these could be the DNA-RNA-protein complexes formed which would perform transcriptional control activities within the nuclear environment. We can see a good example of free gene complexes. One could also speculate that mitochondria might act as a free gene complex whose organizers were lost during evolution. The hypothetical free gene complexes can be classified into three types: surviving, invading, and adapting free gene complexes. The surviving free gene complexes would be transferred at high concentrations from the mother cell to the daughter cells with the same RNA sequences as those in the mother generation (surviving RNAs). The invading free gene complexes would similarly have the "invading RNAs", migrate from adjacent heterogeneous cells to the mother cell, and then transfer to daughter cells.

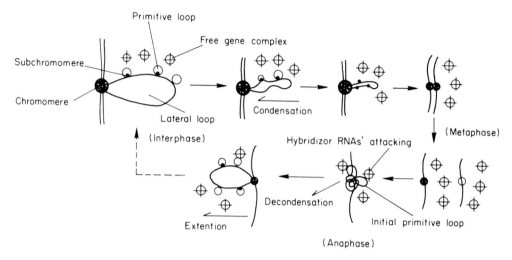

FIGURE 1.   Diagrammatic representation of chromatin morphology and behavior of free gene complexes ($\oplus$) in the lateral loop model of developing nuclei. In the metaphase of mother cell generation, various types of free gene complexes derived from a primitive loop carrying different genetic information are released into cellular environments. Accumulation of these complexes is maximum at the metaphase when the condensation of the lateral loop is completed (see the upper part of the figure). In the interphase of a daughter cell generation, the free gene complexes introduced into daughter nucleus will activate the formation of initial primitive loops and decondense the chromatin architectures of chromomere or subchromomeres resulting in extension of the lateral loop of the daughter cell (see the lower part of the figure).

The invading RNA sequences could be totally different from those in the target mother cell. The "adapting RNAs" in their free gene complexes could combine with the very proteins that were translated by the RNAs. The preadapting RNAs might lose their sequences or add new sequences, resulting in transformations of the preadapting RNAs into the adapting RNAs (see B and C in Figure 2). These adapting RNAs as well as the invading RNAs could play roles in activation of new gene sites.

An earlier finding by Neyfakh et al.[78] suggests that most of the newly synthesized RNAs in a mother cell should be found in daughter nucleus. The observations on RNA migration during cell division suggest that the parental RNAs could probably play a role by interacting with the preactivating daughter gene sites at a certain critical phase of mitosis (see diagram in Figure 1).

I propose a general working hypothesis: the primary gene activations could be caused by specific DNA-RNA hybridization in vivo at the "initial primitive loop" of decondensing parts of anaphase chromomere. This primitive loop that has the same gene sequences as in the loop of interphase can start to extend from a decondensing anaphase chromomere and develop into a lateral loop of interphase as shown in Figure 1. Probably at anaphase DNA double strands of the initial primitive loops are closed as a rule, but in the following phase the closed strands should be unwound partially, being quite probable that in such a critical moment of anaphase chromatin, short pieces of RNA derived and translocated from the mother free gene complexes which are close to the coding sites of the initial daughter primitive loops. This might open DNA double strands of the loops along the various regions selecting exactly the same sequence positions as their mother DNA codings (see diagram in Figure 2). I would like to call such RNAs "hybridizer RNAs", which may in general survive, invade, and adapt RNAs. Frenster's derepressor RNA, activator RNA of Davidson and Britten, and Reichard-Tseng's initiator RNA, could all be included in this category. All

An initial daughter primitive loop

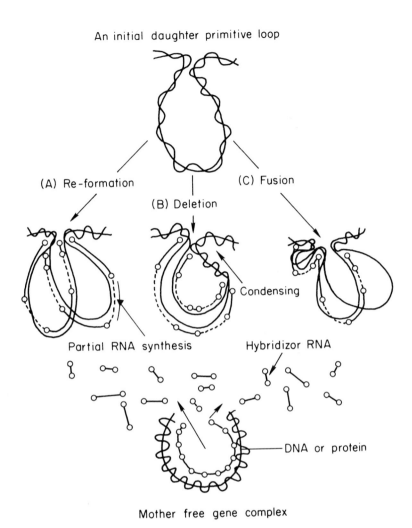

(A) Re-formation   (C) Fusion

(B) Deletion

Condensing

Partial RNA synthesis   Hybridizor RNA

DNA or protein

Mother free gene complex

FIGURE 2. Diagram showing the mechanisms of gene activation by hybridizor RNAs. Small RNA molecules can hybridize to the initial primitive loops (see Figure 1). Different groups of hybridizor RNA, surviving RNAs (A), adapting RNAs (B and C), which are released respectively from different free gene complexes (in this figure only one free gene complex is shown for convenience), can activate the initial primitive loops by hybridizing with open DNA loops. Each hybridizor RNA can recognize the coding region of DNA molecules active in the mother cell. As shown in (A), if surviving RNAs hybridize to the initial primitive loop, the daughter primitive loop would have produced the same RNA molecules as those in the primitive loops of the mother cell generation. Elongation of RNA molecules is also possible, as shown in B or C.

of the RNAs should possess the ability to hybridize in vivo. The maternal RNAs stored in oocytes might be the initial hybridizer RNAs which could regulate early development immediately after fertilization.

As a result of initial hybridization, partial discontinuous RNA synthesis will be initiated along the DNA strands. Thus, after the completion of RNA synthesis, the RNAs will be released into nucleus or cytoplasm as daughter free gene complexes and the activated gene sites will continue to synthesize the same types of RNAs until the end of interphase. If the hybridizer RNAs behave and function like Reichard-Tseng's initiator RNAs, the above-mentioned process of RNA synthesis might be trans-switched on DNA synthesis resulting in a gradual development of a fully extending lateral loop with primitive loops shown in Figure 1. In each of such extending processes of looping architectures, the hybridizer RNAs could switch stepwise paralleling the RNA synthesis or trans-switch the DNA synthesis, which would bring functional or structural gene amplifications. Thus, it would be possible that the hybridizer RNAs in free gene complexes could have the ability to give specific information from mother cells to daughter gene machinery and to evoke a gene regulation at the transcriptional level.

The mechanism of gene activation utilizing hybridizer RNAs is illustrated in Figure 2. In the mechanism (A), the surviving RNAs can activate exactly the same type of RNA synthesis as that in the mother cells. In the deletion mechanism (B), the initial primitive loop has a complete gene sequence inherited to the daughter generation while the invading or adapting RNAs can activate only partial coding information. Noninformative parts might be forced to be condensed into a part of chromomere. In a fusion mechanism (C), the possibility of total (or partial, not shown) combination of two gene sequences is shown demonstrating new gene activation by the invading or adapting RNAs.

REFERENCES

1. **Vermorken, A. J. M., Helderink, J. M. H. C., de Ven, W. J. M. V., and Bloemendal, H.,** Lends differentiation: crystallin synthesis in isolated epithelia from calf lenses, *J. Cell Biol.,* 76, 175, 1978.
2. **Courtois, Y., Simoneau, L., Tassin, J., Laurent, M. V., and Malaise, E.,** Spontaneous transformation of bovine lens epithelial cells, *Differentiation,* 10, 23, 1978.
3. **Mcavoy, J. W.,** Cell division, cell elongation and distribution of α-, β- and γ-crystallins in the rat lens, *J. Embryol. Exp. Morphol.,* 44, 149, 1978.
4. **Hamada, Y., Watanabe, K., Aoyama, H., and Okada, T. S.,** Differentiation and dedifferentiation of rat lens epithelial cells in short- and long-term cultures, *Dev. Growth Diff.,* 21(3), 205, 1979.
5. **Araki M., Yanagida, M., and Okada, T. S.,** Crystallin synthesis in lens differentiation in cultures of neural retinal cells of chick embryos, *Dev. Biol.,* 69, 170, 1979.
6. **Shinohara, T. and Piatigorsky, J.,** Quantitation of δ-crystallin messenger RNA during lens induction in chick embryos, *Proc. Natl. Acad. Sci. U.S.A.,* 73(8), 2808, 1976.
7. **Milstone, L. M., Zelenka, P., and Piatigorsky, J.,** δ-Crystallin mRNA in chick lens cells: mRNA accumulates during differential stimulation of δ-crystallin synthesis in cultured cells, *Dev. Biol.,* 48, 197, 1976.
8. **Beebe, D. C. and Piatigorsky, J.,** The control of δ-crystallin gene expression during lens cells development: dissociation of cell elongation, cell division, δ-crystallin synthesis, and δ-crystallin mRNA accumulation, *Dev. Biol.,* 59, 174, 1977.
9. **Thomson, I., Wilkinson, C. E., Jackson, J. F., de Pomerai, D. I., Clayton, R. M., Truman, D. E. S., and Williamson, R.,** Isolation and cell-free translation of chick lens crystallin mRNA during normal development and transdifferentiation of neural retina, *Dev. Biol.,* 65, 372, 1978.
10. **Jackson, J. F., Clayton, R. M., Williamson, R., Thomson, I., Truman, D. E. S., and dePomerai, D. I.,** Sequence complexity and tissue distribution of chick lens crystallin mRNAs, *Dev. Biol.,* 65, 383, 1978.
11. **Suzuki, Y. and Ohshima Y.,** Isolation and characterization of the silk fibroin gene with its flanking sequences, *Cold Spring Harbor Symp. Quant. Biol.,* 42, 947, 1978.
12. **Tsujimoto, Y. and Suzuki, Y.,** Structural analysis of the fibroin gene at the 5' end and its surrounding regions, *Cell,* 16, 425, 1979.
13. **Manning, R. F. and Gage, L. P.,** Physical map of the *Bombyx mori* DNA containing the gene for silk fibroin, *J. Biol. Chem.,* 253(6), 2044, 1978.
14. **Battaglia, P. and Melli, M.,** Isolation of globin messenger RNA of *Xenopus laevis, Dev. Biol.,* 60, 337, 1977.
15. **Froehlich, J. P., Browder, L. W., and Schultz, G. A.,** Translation and distribution of rabbit globin mRNA in separated cell types of *Xenopus laevis* gastrulae, *Dev. Biol.,* 56, 356, 1977.
16. **Flavell, R. A., Jeffreys, A. J., and Grosveld, G. C.,** Physical mapping of repetitive DNA sequences neighboring the rabbit β-globin gene, *Cold Spring Harbor Symp. Quant. Biol.,* 42, 1003, 1978.
17. **Brotherton, T. W., Chui, D. H. K., Gauldie, J., and Patterson, M.,** Hemoglobin ontogeny during normal mouse fetal development, *Proc. Natl. Acad. Sci. U.S.A.,* 76(6), 2853, 1979.

18. **Leder, P., Tilghman, S. M., Tiemeier, D. C., Polsky, F. I., Seidman, J. G., Edgell, M. H., Enquist, L. W., Leder, A., and Norman, B.,** The cloning of mouse globin and surrounding gene sequences in bacteriophage λ, *Cold Spring Harbor Symp. Quant. Biol.,* 42, 915, 1978.

19. **Hozumi, N. and Tonegawa, S.,** Evidence for somatic rearrangement of immunoglobulin genes coding for variable and constant regions, *Proc. Natl. Acad. Sci. U.S.A.,* 73(10), 3628, 1976.

20. **Tonegawa, S., Maxam, A., Tizard, R., Bernard, O., and Gilbert, W.,** Sequence of a mouse germ-line gene for a variable region of an immunoglobulin light chain, *Proc. Natl. Acad. Sci. U.S.A.,* 75(3), 1485, 1978.

21. **Weigert, M., Gatmaitan, L., Loh, E., Schilling, J., and Hood, L.,** Rearrangement of genetic information may produce immunoglobulin diversity, *Nature (London),* 276, 785, 1978.

22. **Stein, G. S., Spelsberg, T. C., and Kleinsmith, L. J.,** Nonhistone chromosomal proteins and gene regulation, *Science,* 183, 817, 1974.

23. **Garel, A. and Axel, R.,** The structure of the transcriptionally active ovalbumin genes in chromatin, *Cold Spring Harbor Symp. Quant. Biol.,* 42, 701, 1978.

24. **Bellard, M., Gannon, F., and Chambon, P.,** Nucleosome structure III: the structure and transcriptional activity of the chromatin containing the ovalbumin and globin genes in chick oviduct nuclei, *Cold Spring Harbor Symp. Quant. Biol.,* 42, 779, 1978.

25. **Palmiter, R. D., Mulvihill, E. R., McKnight, G. S., and Senear, A. W.,** Regulation of gene expression in the chick oviduct by steroid hormones, *Cold Spring Harbor Symp. Quant. Biol.,* 42, 639, 1978.

26. **Schütz, G., Nguyen-Huu, M. C., Giesexke, K., Hynes, N. E., Groner, B., Wurtz, T., and Sppel, A. E.,** Hormonal control of egg white protein messenger RNA synthesis in the chicken oviduct, *Cold Spring Harbor Symp. Quant. Biol.,* 42, 617, 1978.

27. **O'Malley, B. W., Tsai, M. J., Tsai, S. Y., and Towle, H. C.,** Regulation of gene expression in chick oviduct, *Cold Spring Harbor Symp. Quant. Biol.,* 42, 605, 1978.

28. **Breathnach, R., Mandel, J. L., and Chambon, P.,** Ovalbumin gene is split in chicken DNA, *Nature (London),* 270, 314, 1977.

29. **Weinstock, R., Sweet, R., Weiss, M., Cedar, H., and Axel, R.,** Intragenic DNA spacers interrupt the ovalbumin gene, *Proc. Natl. Acad. Sci. U.S.A.,* 75(3), 1299, 1978.

30. **Slavkin, H. C.,** Amelogenesis *in vitro, J. Dent. Res.,* 58(B), 735, 1979.

31. **Slavkin, H. C. and Crossant, R.,** Intercellular communication during odontogenic epithelial-mesenchymal interactions: isolation of extracellular matrix vesicles containing RNA, in *The Role of RNA in Reproduction and Development,* Niu, M. C. and Segal, S. J., Eds., North-Holland, Amsterdam, 1973, 247.

32. **Fukae, M. and Shimizu, M.,** Studies on the proteins of developing bovine enamel, *Arch. Oral Biol.,* 19, 381, 1974.

33. **Noto, T., Barrichello, T. L. S., and Cury, J. A.,** *In vitro* incorporation of ^{3}H-uridine into nucleus or chromosome of ameloblasts and analysis of young enamel proteins in the isolated neonatal mouse incisor, *Acta Med. Univ. Kagoshima,* 20, 107, 1978.

34. **Lee-Own, V., Zeichner, M., Benveniste, K., Denny, P., Paglia, L., and Slavkin, H. C.,** Cell-free translation of messenger RNAs of embryonic tooth organs: synthesis of the major extracellular matrix proteins, *Biochem. Biophys. Res. Commun.,* 74(3), 849, 1977.

35. **Paterson, B. M. and Bishop, J. O.,** Changes in the mRNA population of chick myoblasts during myogenesis *in vitro, Cell,* 12, 751, 1977.

36. **John, H. A., Patrinou-georgoulas, M., and Jones, K. W.,** Detection of myosin heavy chain mRNA during myogenesis in tissue culture by *in vitro* and *in situ* hybridization, *Cell,* 12, 501, 1977.

37. **Strohman, R. C., Moss, P. S., Micou-Eastwood, J., Spector, D., Przybyla, A., and Paterson, B.,** Messenger RNA for myosin polypeptides: isolation from single myogenic cell cultures, *Cell,* 10, 265, 1977.

38. **Newrock, K. M., Alfageme, C. R., Nardi, R. V., and Cohen, L. H.,** Histone changes during chromatin remodeling in embryogenesis, *Cold Spring Harbor Symp. Quant. Biol.,* 42, 421, 1978.

39. **Weinberg, E. S., Overton, G. C., Hendricks, M. B., Newrock, K. M., and Cohen, L. H.,** Histone gene heterogeneity in the sea urchin *Strongylocentrotus purpuratus, Cold Spring Harbor Symp. Quant. Biol.,* 42, 1093, 1978.

40. **Kressmann, A., Clarkson, S. G., Telford, J. L., and Birnstiel, M. L.,** Transcription of *Xenopus* tDNA$_1$^{met} and sea urchin histone DNA injected into the *Xenopus* oocyte nucleus, *Cold Spring Harbor Symp. Quant. Biol.,* 42, 1077, 1978.

41. **Mertz, J. E. and Gurdon, J. B.,** Purified DNAs are transcribed after microinjection in *Xenopus* oocytes, *Proc. Natl. Acad. Sci. U.S.A.,* 74, 1502, 1977.

42. **Higashinakagawa, T., Wahn, H., and Reeder, R. H.,** Isolation of ribosomal chromatin, *Dev. Biol.,* 55, 375, 1977.

43. **Scalenghe, F., Buscaglia, M., Steinheil, C., and Crippa, M.,** Large scale isolation of nuclei and nucleoli from vitellogenic oocytes of *Xenopus laevis, Chromosoma,* 66, 299, 1978.

44. **Brown, D. D. and Gurdon, J. B.,** High-fidelity transcription of 55 DNA injected into *Xenopus* oocytes, *Proc. Natl. Acad. Sci. U.S.A.,* 74(5), 2064, 1977.

45. **Gordon, J. B. and Brown, D. D.,** The transcription of 5s DNA injected into *Xenopus* oocytes, *Dev. Biol.,* 67, 346, 1978.

46. **Shiokawa, K., Kawahara, A., Misumi, Y., Yasuda, Y., and Yamana, K.,** Inhibitor of ribosomal RNA synthesis in *Xenopus laevis* embryos, *Dev. Biol.,* 57, 210, 1977.

47. **Weideli, H., Schedl, P., Artavanis-Tsakonas, S., Steward, R., Yuan, R., and Gehring, W. J.,** Purification of a protein from unfertilized eggs of *Drosophila* with specific affinity for a defined DNA sequence and the cloning of this DNA sequence in bacterial plasmids, *Cold Spring Harbor Symp. Quant. Biol.,* 42, 693, 1978.

48. **Brown, D. D. and Gurdon, J. B.,** Cloned single repeating units of 5S DNA direct accurate transcription of 5S RNA when injected into *Xenopus* oocytes, *Proc. Natl. Acad. Sci. U.S.A.,* 75, 2849, 1978.

49. **Trendelenburg M. F. and Gurdon, J. B.,** Transcription of cloned *Xenopus* ribosomal genes visualized after injection into oocyte nuclei, *Nature (London),* 276, 292, 1978.

50. **Ohi, S., Ramirez, J. L., Upholt, W. B., and Dawid, I. B.,** Mapping of mitochondrial 4S RNA genes in *Xenopus laevis* by electron microscopy, *J. Mol. Biol.,* 121, 299, 1978.

51. **Ramirez, J. L. and Dawid, I. B.,** Mapping of mitochondrial DNA in *Xenopus laevis* and *Xenopus borealis, J. Mol. Biol.,* 119, 133, 1978.

52. **Bonner, J. J., Berninger, M., and Pardue, M. L.,** Transcription of polytene chromosomes and of the mitochondrial genome in *Drosophila melanogaster, Cold Spring Harbor Symp. Quant. Biol.,* 42, 803, 1978.

53. **Butros, J.,** Action of heart and liver RNA on the differentiation of segments of chick blastoderms, *J. Embryol. Exp. Morphol.,* 13(1), 119, 1965.

54. **Niu, M. C. and Mulherkar, L.,** The role of exogeneous heart-RNA in development of the chick embryo cultivated *in vitro, J. Embryol. Exp. Morphol.,* 24, 33, 1970.

55. **Niu, M. C. and Segal, S. J., Eds.,** *The Role of RNA in Reproduction and Development,* North-Holland, Amsterdam, 1973.

56. **Noto, T. and Shiotsuki, M. L. N.,** The effects of embryonic heart RNA and actinomycin D both on the isolated tissue-pieces of the chick blastoderm and on the isolated primitive tubular heart of the early chick embryo, *Dev. Growth Diff.,* 15(4), 329, 1973.

57. **Siddiqui, M. A. Q., Deshpande, A. K., Arnold, H. H., Jakowlew, S. B., and Crawford, P. A.,** RNA-mediated manipulation of early chick embryonic gene functions *in vitro, Brookhaven Symp. Biol.,* 29, 62, 1977.

58. **Deshpande, A. K. and Siddiqui, M. A. Q.,** A reexamination of heart muscle differentiation in the postnodal piece of chick blastoderm mediated by exogeneous RNA, *Dev. Biol.,* 58, 230, 1977.

59. **Deshpande, A. K., Jakowlew, S. B., Arnold, H. H., Crawford, P. A., and Siddiqui, M. A. Q.,** A novel RNA affecting embryonic gene functions in early chick blastoderm, *J. Biol. Chem.,* 252(18), 6521, 1977.

60. **Deshpande, A. K. and Siddiqui, M. A. Q.,** Acetylcholinesterase differentiation during myogenesis in early chick embryonic cells caused by an inducer RNA, *Differentiation,* 10, 133, 1978.

61. **Mclean, M. J., Renaud, J. F., and Sperelakis, N.,** Cardiac-like action potentials recorded from spontaneously contracting structures induced in postnodal pieces of chick blastoderm exposed to an RNA-enriched fraction from adult heart, *Differentiation,* 11, 13, 1978.

62. **Tung, T. C. and Niu, M. C.,** The effect of carp egg-mRNA on the transformation of goldfish tail, *Sci. Sin.,* 20(1), 59, 1977.

63. **Tung, T. C. and Niu, M. C.,** Organ formation caused by nucleic acid from different class, *Sci. Sin.,* 20(1), 56, 1977.

64. **Davidson, E. H. and Britten, R. J.,** Note on the control of gene expression during development, *J. Theor. Biol.,* 32, 123, 1971.

65. **Davidson, E. H. and Britten, R. J.,** Regulation of gene expression: possible role of repetitive sequences, *Science,* 204, 1052, 1979.

66. **Frenster, J. H. and Herstein, P. R.,** *The Role of RNA in Reproduction and Development,* Niu, M. C. and Segal, S. J., Eds., North-Holland, Amsterdam, 1973, 330.

67. **Tseng, B. Y., Erickson, J. M., and Gloulian, M.,** Initiator RNA of nascent DNA from animal cells *J. Mol. Biol.,* 129, 531, 1979.

68. **Reichard, P., Eliasson, R., and Söderman, G.,** Initiator RNA in discontinuous polyoma DNA synthesis, *Proc. Natl. Acad. Sci. U.S.A.,* 71(12), 4901, 1974.

69. **Goldstein, L.,** Role for small nuclear RNAs in "programming" chromosomal information, *Nature (London),* 261, 519, 1976.

70. **Noto, T.,** Theoretical consideration on gene differentiation (II): free gene complexes and hybridizer RNAs, *Acta Med. Univ. Kagoshima,* 20, 111, 1978.

71. **Mott, M. R. and Callan, H. G.,** An electron-microscope study of the lampbrush chromosomes of the Newt *Tritrus cristatus, J. Cell Sci.,* 17, 241, 1975.

72. **Noto, T. and Merzel, J.,** Chromosomal structures of ameloblasts in the neonatal mouse incisor, *Acta Med. Univ. Kagoshima,* 21, 25, 1979.

73. **Noto, T.,** RNA incorporation into embryonic chromatin structures of the chick, *Acta Med. Univ. Kagoshima,* 21, 31, 1979.

74. **Noto, T.,** Light microscopic visualization of overall chromatin architectures in cardiac cells of the chick embryo, *Acta Med. Univ. Kagoshima,* 22, 135, 1980.

75. **Trendelenburg, M. F., Scheer, U., Zentgraf, H., and Franke, W. W.,** Heterogeneity of spacer length in circles of amplified ribosomal DNA of two insect species, *Dytiscus marginalis* and *Acheta domesticus, J. Mol. Biol.,* 108, 453, 1976.

76. **Scheer, U., Trendelenburg, M. F., and Franke, W. W.,** Regulation of transcription of genes of ribosomal RNA during Amphibian oogenesis, *J. Cell Biol.,* 69, 465, 1976.

77. **Georgiev, G. P.,** On the structural organization of operon and the regulation of RNA synthesis in animal cells, *J. Theor. Biol.,* 25, 473, 1969.

78. **Neyfakh, A. A. and Kostomarova, A. A.,** Migration of newly synthesized RNA during mitosis. I. Embryonic cells of the loach (*Misgurnus fossilis* L.), *Exp. Cell Res.,* 65, 340, 1971.

Chapter 9

CONTROL OF GENE ACTIVITY IN EARLY CARDIAC MUSCLE DEVELOPMENT

M.A.Q. Siddiqui

TABLE OF CONTENTS

I. INTRODUCTION

One of the central problems in cell biology is the mechanism whereby an individual organism develops and maintains the diversity of cell types characteristic of an adult state out of the original single cell, the fertilized ovum. The process of embryonic cell development and differentiation was analyzed in early days in terms of spatial pattern formation, and models were proposed based on gradients of influence and cell interactions.[1-5] According to current concepts, changes in embryonic development are caused by differential gene activity which is controlled, in some way, by elements in cytoplasm of the fertilized egg.[6-8] It is assumed that cytoplasmic factors, activators, and repressors of gene activity are synthesized in the oocytes and subsequently segregate into different cells. The existence of such gene-controlling elements has been established[9] and their contribution to postfertilization development is the subject of several preceding chapters in this volume. Little is known, however, of the molecular mechanism(s) that controls the gene activity.

II. EXAMPLES OF POTENTIAL GENE-CONTROLLING RNA

A. Repetitive DNA Transcripts

One of the stages at which gene control can be exercised is transcription of DNA into RNA. It is generally assumed that cell-specific population of mRNAs results from the transcription of a set of structural genes specific for the cell type. Current data, however, suggest that variations between nuclear RNA transcripts from cell to cell are minimum. In nuclei, at least five times more unique DNA transcripts are present and a greater complexity exists for nuclear RNA than is found in cytoplasm.[10] In addition, nuclei also contain transcripts of repetitive DNA, which is present in a tissue-specific pattern.[10-14] The structural genes in eukaryotic genome constitute only a minor fraction of total DNA and a large percentage of the repetitive DNA, present in a few hundred copies per haploid genome, is found interspersed among single copy sequences, a feature presumably shared by organisms from most branches of evolutionary tree with some exceptions.[15-17] Whether this feature of gene organization has a direct relevance to gene control is far from clear. But the presence of large excess of repetitive DNA and the disparity between the RNA population of nuclei and cytoplasm certainly allow speculations on the potential role of repetitive DNA sequence in gene regulation. Models have been proposed that envisage gene regulation by specific interactions at repetitive sequences in the genomic DNA.[6,8,18,19] More recently, it was proposed that regulatory interactions occur between single copy gene transcripts and the complementary regions present in the repetitive gene transcripts.[20] The functional expression of a particular set of genes is thus regulated by the transcription of regulatory segment of repetitive DNA into a "control RNA". This model is based on the premise that nuclear RNA transcripts of single copy DNA are common between tissues, but RNA consisting of nonmessage sequences is not. Although there is no clear evidence yet available in support of the main theme of the model, recent reports have described the presence of distinct low molecular weight RNAs in human, mice, and chicken cells that appear to have been transcribed from the Alu family of interspersed repetitive DNA.[21-24] An abundant 7S RNA was isolated and purified from uninfected HeLa cells which can form strong hybrids with the dominant family of repetitive human genome.[23] Chinese hamster ovary cells contain a homogenous 4.5S RNA, which appears to be hydrogen bonded to poly(A) hnRNA, and shares homology to the Alu family DNA.[24] The sequences of these RNAs are apparently highly conserved through evolution. They are also found near the origin of replication of several onconaviruses.[25] Although the cellular function of these RNAs is not yet known, the specific association of these molecules with poly(A)-containing cytoplasmic and nuclear RNAs suggests that they may have a role in regulation of mRNA synthesis and/or its function.

B. Small Nuclear RNAs

Another class of low-molecular weight RNA is small nuclear RNA (SnRNA)[26-29] whose function remained obscure until recently. The SnRNAs exist as complexes of RNA and protein and fall into discreet classes according to the proteins they bind.[30] Two recent developments have generated new interest in the function of SnRNA. First, based on the observation that the nucleotide sequence at the 5'-end of Ul RNA of SnRNA family shares homology with the consenses sequences of a number of splice junctions in precursor mRNAs, it was proposed that SnRNAs are utilized in intermolecular RNA interactions to provide the precise recognition sites necessary for the accurate processing of mRNA.[28,30] Second, antibodies produced by patients with autoimmune disease, systemic lupus erythematosus, have been shown to recognize specific SnRNPs.[31] The addition of these antibodies to adenoviral RNA inhibits RNA splicing.[32] Data concerning the organization and expression of SnRNA are scant, but recently human genomic fragments containing sequences complementary to Ul, U2, U3 RNAs, and chicken U1 DNA fragments were isolated, cloned, and characterized.[29,33] In at least two of human SnRNA genes, the repetitive Alu family DNA segments were found located upstream in the same orientation as the SnRNA gene. In chicken U1 gene, no intervening sequences are present, and neither are the consensus TATAA sequence and the putative polyadenylation signal, AATTAAA, present. In chicken, the U1 RNA is present in relatively high concentration which remains constant under different states of hormonal stimulation. Although the sequence of chicken U1 RNA differs significantly from that of other SnRNAs, there is almost 90% homology between the 5'-end of U1 RNA and the exon-intron junctions of the ovalbumin and ovomucoid genes in chicken. These data suggest strongly that U1 RNA is involved in processing of chicken ovalbumin and ovomucoid mRNAs. Whether this is actually the case in vivo is yet to be ascertained.

C. Viroid RNA

Viroid RNA, the smallest known agents of infectious disease, are known to exist only in higher plants.[34] Despite their small size (1×10^5 daltons), viroid RNA replicate autonomously in host plants. Viroid RNA are single-stranded molecules, free of modified bases and capped termini, and exist both as linear and covalently linked circular molecules. They exhibit extensive intramolecular complementarity and in their native state are found as highly base-paired rod-like structures. The reasons for the structural transition and its relation to the functional role of the RNA are not known, but the double-stranded structure of potato spindle tuber viroid RNA has been implicated to play a role in viroid replication.[35] The origin of viroids is also obscure. One might speculate that viroids are the products of spliced out intervening sequences in precursor mRNAs. Although introns are generally regarded as nonsense sequences and are rapidly degraded, it is possible that intramolecular base pairing in viroid RNA affords stabilization to the molecule which can escape degradation. It is, nevertheless, a mystery as to how viroid produce specific disease symptoms in host plants. The nuclear location of viroids and their ability to replicate autonomously would suggest that viroid might interfere in some way with the regulatory processes of the host cells.

D. Immune RNA

Substantial evidence exists on the role that RNA plays in the transfer of immunity.[36] Early observations were confirmed by a number of investigators who were able to show that nonimmune lymphoid cells can be converted to specific immunologic activity by treatment with extracts rich in RNA prepared from macrophage or other lymphoid cell previously exposed to antigen. Rabbit peritoneal exudate (PE) cells when incubated with bacteriophage yield RNA capable of inducing the formation of phage-specific antibodies in rabbit spleen cell cultures.[37] The immunogenic activity of the PE cells homogenate resides in two RNA fractions, one of which is capable of inducing the formation of 19S immunoglobulins

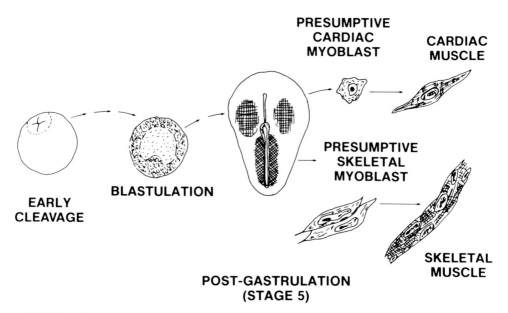

FIGURE 1. Chick embryonic muscle development. The presumptive cardiac muscle cells are localized on lateral sides (shaded oval areas) adjacent to Hensen's node in stage 5 chick embryo.

containing the allotypic markers of the RNA donor rabbit.[38] The involvement of RNA in other cellular control functions has been postulated[39-46] (see References 47 to 52 for review).

In spite of the intriguing examples mentioned above, there is no clear evidence that would shed light on the mechanism(s) by which RNA would serve a physiological function in regulatory processes during development. It is clear that the complex series of early morphogenetic and developmental patterns in an embryo reflect the changes in transcriptional activity of the genome. But it is not clear how the classical embryological concepts on determination of morphogenesis can be interpreted in molecular terms. The basic understanding of the mechanism(s) that control development would require a careful analysis of RNA transcription throughout the early stages of embryonic development. Several laboratories have studied regulation of gene transcription in embryonic cells utilizing specific DNA probes. Many of these studies, however, concern cells which were partly differentiated, so that the genome was already active for transcription of specific proteins. Therefore, the question as to how and when the initial activation of specific population of genes occurs in early embryo remains unclear. Embryonic systems which permit analysis of gene transcription in early stages of development and identification of potential regulatory molecules would be highly suitable for elucidating the underlying mechanism.

III. CHICK HEART DEVELOPMENT: A MODEL SYSTEM

In the postgastrulation stage, stage 5,[53] of the chick embryo, the precardiogenic cells are known to be localized in bilateral regions adjacent to Hensen's node (Figure 1).[54,55] This apparently homogeneous and undifferentiated population of cells is already programmed to develop into cardiac myoblasts, since the explants when grown in culture in a variety of media or as chorioallantoic transplants differentiate into well-defined cardiac muscle tissue. Although cells in culture isolated from differentiated muscle tissues have been used for the study of gene expression during muscle differentiation, the availability of primitive form of cells in the precardiogenic region of stage 5 chick blastoderm provides a unique advantage as a model system to probe into the mechanism of induction of cardiac muscle gene transcription in embryonic development.

It is believed that the precardiac mesoderm of the chick begins to invaginate during stage 4 of development, moving anteriorly along the growing primitive streak. Based on [³H] thymidine-labeled grafts obtained from various stages, it was observed[54] that the precardiogenic region extends at this stage halfway between anterior and posterior ends of the streak. When tissue fragments isolated by transverse cuts 0.4 to 0.6 mm posterior to Hensen's node are grown separately in vitro, the posterior fragment, the postnodal piece (PNP), does not develop into any histologically identifiable structure.[56-62] In contrast, the anterior portion (AP) containing Hensen's node develops readily into axial structures. The PNP when combined with the AP, however, develops into embryonic tissues, including a well-defined heart tubular structure, suggesting that the anterior portion carries the putative information needed for the precardiac cells located in the PNP to transform into cardiac muscle tissue. Later, it was shown that the PNP cells can also be induced to develop into cardiac muscle-like tissue when the explants are cultured in the presence of RNA isolated from differentiated chick embryonic heart.[63-65] The difficulty in understanding the mechanism of this novel mode of differentiation was the lack of information on the nature and properties of the inducing agent and the biochemical criteria for establishing unequivocally the morphogenetic transition. We have reexamined the phenomenon critically, and have recently reported[66-69] that a distinct low molecular weight RNA (7S CEH-RNA) appears to play a crucial role in embryonic heart muscle formation, since addition of the RNA to PNP cells does indeed promote a specific mode of changes that is similar to embryonic myogenic process. Neither synthetic polynucleotides nor a variety of RNA isolated under identical conditions from several sources could replace the RNA from chick heart tissue (see Section V).

In order to relate the process of embryonic heart cell differentiation to changes in specific gene activity in both the RNA-induced system in vitro, as well as in vivo, our experimental approach was to monitor the syntheses of mRNAs for several muscle-specific marker proteins in parallel with that of 7S CEH-RNA as a function of early embryonic differentiation (Figure 2).

IV. BIOGENESIS OF MYOSIN mRNAs IN CARDIAC MUSCLE TISSUE

The synthesis of cardiac muscle proteins and the appearance of myofibrillar structures are early events during chick embryonic development.[70] Muscle differentiation is characterized by the production of large quantities of myofibrillar proteins, myosin light (MLC) and heavy chain (MHC) polypeptides, actin, tropomyosin, troponin, etc. There is sufficient evidence that the dramatic changes that occur in muscle cell differentiation and in cultured cells of myogenic origin are transcriptionally and/or translationally controlled.[71-78] It is believed that muscle-specific mRNAs are expressed coordinately at the time of myoblast fusion when the protein synthesis is activated,[78-80] but others[81-83] have postulated that skeletal muscle myosin mRNA is cytoplasmically stored in an inactive form and its activation is controlled posttranscriptionally. It was also reported that more than one (MHC) polypeptide is present in muscle cells, implying that there are multiple structural genes for MHC.[83-85] Relatively little information is, however, available on the molecular events that occur during cardiogenic muscle cell differentiation, particularly during stages when precardiogenic cells transform into cardiac myoblasts and then differentiate into well-defined cardiac muscle cells. We, therefore, initiated studies on cardiac muscle development since it offers a unique advantage in the availability of a population of early embryonic cells which are predetermined and can be grown to differentiate successfully in the laboratory.

A. Isolation and Purification of Myosin mRNAs

Myosin light chain (MLC_1 and MLC_2) mRNAs were isolated from the MLC synthesizing polysomes of the 16-day-old chick embryonic heart tissue by immunoadsorption of total polysomes to MLC-specific antibodies prepared in our laboratory.[86,87] The RNA from

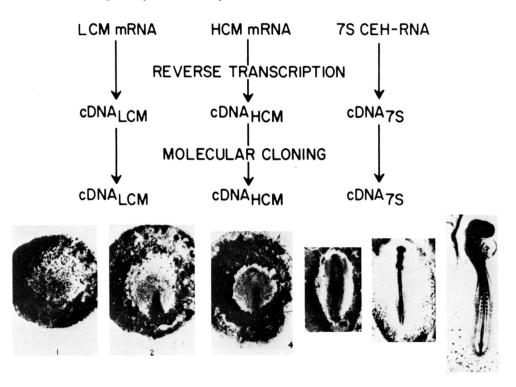

FIGURE 2.   Scheme for quantitation of muscle-specific RNA synthesis during early stages of chick embryonic development (see text): MLC mRNA, myosin light-chain messenger RNA; MHC mRNA, myosin heavy-chain messenger RNA; cDNA, complementary DNA. The early stages of chick embryonic development are shown on the bottom.

immunoadsorbed polysomes was fractionated by centrifugation on isokinetic sucrose gradients and the RNA fraction sedimenting between 9S and 14S was purified by a second sucrose gradient centrifugation and oligo (dT) cellulose chromatography. The purified fraction directed the translation of polypeptides that were indistinguishable from the authentic myosin light chains, MLC_1 and MLC_2, when examined by two-dimensional separation using isoelectrofocusing and polyacrylamide gel electrophoresis (Figure 3). The purified mRNA migrated as two distinct bands on polyacrylamide-formamide and on agarose-urea gels. The electrophoretic mobilities of these two mRNA species correspond to 360,000 and 320,000 daltons accounting for approximately 1,090 and 980 nucleosides per chain. The molecular weight of the translation products of these two RNA were 24,000 and 18,000, respectively, identical to those of MLC subunits of 16-day-old chick embryonic heart. This would suggest that the mRNAs contain 370 and 430 nucleotides more than expected to code for these polypeptides, respectively.

In order to achieve the maximum purification of MHC mRNA, it was necessary to fractionate the total poly(A) RNA by two successive formamide-containing sucrose gradient centrifugations.[88] Fractionation in nondenaturing conditions resulted in distribution of MHC mRNA activity throughout the gradient. The fractions sedimenting faster than 28S rRNA and exhibiting MHC activity were pooled from several gradients and recentrifuged under identical conditions. The second gradient resolved the MHC mRNA from MLC mRNA and the mRNAs of intermediate sizes (Figure 4). No detectable level of contamination by other mRNAs was observed upon gel electrophoretic separation under denaturing conditions. A significant portion of MHC translation activity, however, remained in the unbound fraction. Since the activity could not be depleted by repeating chromatography on affinity columns, it was assumed that MHC mRNA in the unbound fraction lacks poly(A) or contains a very short poly(A) tail.

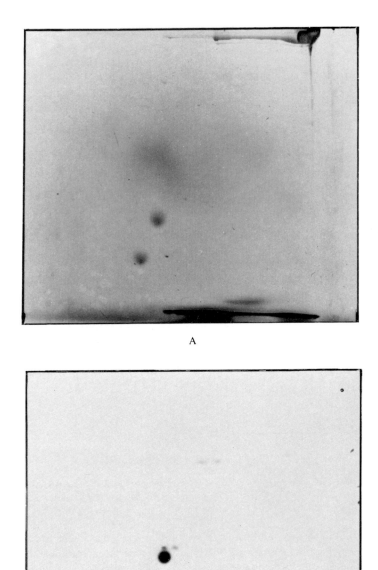

A

B

FIGURE 3.   Two-dimensional separation of labeled translation products and purified myosin subunits. An aliquot of the rabbit reticulocyte lysate translation mixture with the purified MLC mRNA and purified myosin preparation was subjected to the two-dimensional separation by isoelectrofocusing in the first dimension and polyacrylamide gel electrophoresis in the second dimension.[87] The labeled products were identified by fluorography and the marker MLC subunits by staining with Coomassie Blue. (A) Marker-purified myosin subunits; (B) [35S]methionine-labeled translation products of MLC mRNA. Migration on the first dimension was from right to left and on the second dimension from top to bottom. HCM, heavy-chain myosin; MLC₁ and MLC₂, light-chain myosin subunits 1 and 2.

An interesting feature of MHC mRNA biogenesis during skeletal muscle development is that the majority of cytoplasmic MHC mRNA appears to lack a poly(A) terminus or contain

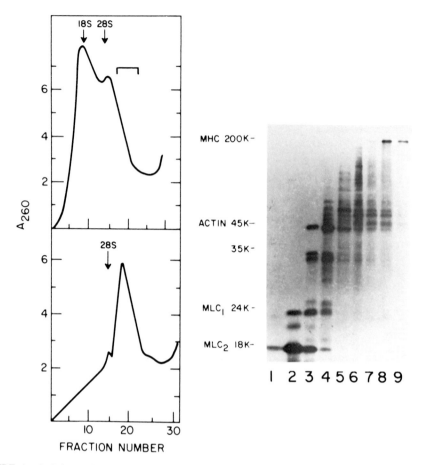

FIGURE 4. Isolation and purification of MHC mRNA. Total poly(A)-containing RNA was fractionated by successive sucrose gradient centrifugations,[88] and the fractions were analyzed by translation assay.[87] The translation products were examined by polyacrylamide gel electrophoresis.

shortened poly(A) tails, the significance of which is not clear.[89,90] This appeared to be true in the case of chick cardiac MHC mRNA, since a significant amount of MHC translation activity was found in RNA fractions that failed to bind to oligo(dT)-cellulose or to poly(U) Sephadex. Although repeating the cycles of chromatography depleted the levels of other mRNA activity, MHC mRNA levels in the unbound fraction remained unchanged. This apparently was not due to secondary structure of the RNA, since heat denaturation or chromatography at elevated temperatures did little to improve its binding to affinity columns. The size of the poly(A) tract in MHC mRNA that bound to oligo(dT)-cellulose was about 160 to 180 nucleotides.[88] Additional spots with faster mobility were also found during the poly(A) size estimation which suggested that the oligo(dT)-cellulose bound MHC mRNA fraction might be of more than one class with respect to the poly(A) size. Previous reports[90] have shown that three discrete classes of MHC mRNA with mean lengths of 13, 30, and 230 adenosine residues in the poly(A) tail are present in skeletal muscle cells.

B. cDNA Synthesis and Quantitation of Myosin mRNAs

cDNAs were synthesized using AMV reverse transcriptase and mRNAs as templates.[86,88] A high proportion of cDNA for MLC mRNA was of the average length of 1050 nucleosides. The complexity of MLC mRNA, based on the hybridization kinetics of globin and MLC back hybridizations, was 2225 nucleosides, which is in excellent agreement with the combined sizes of the two MLC mRNAs (MLC$_1$ = 1090 nucleosides and MLC$_2$ = 980 nucleosides) based on electrophoretic separation on denaturing gels. The reassociation kinetics of MLC cDNA with various RNA preparations also allowed quantitation of MLC-specific mRNA sequences in these RNAs. In 16-day-old chick embryonic heart muscle cell, MLC mRNA comprises about 2% of total poly(A) RNA and 0.02% of polysomal RNA.

To synthesize a cDNA of maximum size for MHC which can serve as an authentic probe for MHC mRNA, we utilized several parameters which reportedly influence the size distribution of cDNA made from large size template mRNAs. When we examined the cDNA synthesized under optimized conditions, a major portion of cDNA was of the size of 1000 to 2000 nucleotides, although there was some synthesis of cDNA reaching up to 4000 nucleotides. Upon back hybridization of cDNA to MHC mRNA, 60 to 70% of the radioactivity in cDNA became S_1 resistant and the reassociation approached slowly to this saturation level. Under identical conditions, globin cDNA reassociated with globin mRNA to the almost 100% level and showed a sharp transition. When we performed hybridization of MHC cDNA with oligo(dT)-cellulose unbound RNA fractions to a relatively low Rot value (9.5×10^{-2}), we found that the cross hybridization was unexpectedly high (53%).[88] Ribosomal RNA, which is about the same size as MHC mRNA and is likely to be present in MHC mRNA even after several passages through oligo(dT)-cellulose, is generally assumed to be a poor template for reverse transcription. It is, nevertheless, utilized in the reaction and its template activity would, in fact, be increased relative to MHC mRNA when the latter itself serves as a poor template. This would obviously affect the specificity of MHC cDNA as a probe and would lead to misinterpretation of hybridization data. Upon fractionation of cDNA on an alkaline sucrose gradient, which was done for the purpose of size selection and subsequent cloning, we found a substantial difference in the kinetics of back hybridization between the fast- and slow-moving cDNA fractions. The slow-moving cDNA, which represented only a small fraction of the total cDNA, hybridized with kinetics similar to unfractionated cDNA, whereas the majority of cDNA (fast moving and intermediate fractions) displayed pseudo first-order kinetics with a sharp transition occurring within the 100-fold range of Rot values characteristic of pure mRNA. The slow-moving cDNA fraction may represent transcripts for 28S rRNA. Small size cDNA transcripts are indeed expected when 28S rRNA is utilized as template. Since our objective was to use large size cDNA as a probe and the hybridization kinetics of the fractionated cDNA were indicative of faithful cDNA copies and homogeneity in mRNA (see below), we used the fast-moving fraction combined with the intermediate fraction for further work.

Under the conditions of hybridization, MHC cDNA reassociated with MHC mRNA with an observed $Rot_{1/2}$ of 2.52×10^{-2}, whereas the $Rot_{1/2}$ for globin mRNA:DNA hybridization done in parallel was 5.63×10^{-3} (Figure 5). The theoretically expected value for MHC mRNA:DNA reassociation is 2.60×10^{-2} based on the estimated molecular weight of 1.9×10^6 daltons for MHC mRNA. The observed $Rot_{1/2}$ for MHC mRNA was very close to the expected value and gave a complexity of 5370 nucleosides for the mRNA which is in good agreement with the calculated complexity of 5800 nucleosides and suggests that MHC mRNA is a single RNA sequence. If we correct for the difference in the average length of globin cDNA (600 bases) and MHC cDNA (1,500 bases), assuming that Rot is inversely proportional to the square root of cDNA length,[91] the calculated complexity becomes 6900 nucleosides. In either case, the complexity measurement supports the conclusion that the cardiac MHC mRNA in chick embryonic tissue is a single mRNA species.

C. Identification of Recombinant DNA Plasmids Containing MHC and MLC Gene Sequences

The cDNAs synthesized using purified MLC and MHC mRNAs were converted to double-stranded cDNAs which were then cloned in *E. coli* strain X1776 or HB101 using the Hind III or PstI cleavage sites in the plasmid pBR322.[87,92] Of the few transformants that contained

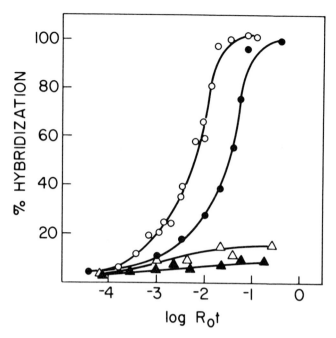

FIGURE 5. Hybridization kinetics of fractionated MHC and globin cDNAs with MHC and globin mRNA. RNA-excess back-hybridizations were carried out using MHC and globin cDNA with the respective mRNA templates in the presence of 50% formamide as described earlier.[88] ○, globin cDNA; mRNA; ●, MHC cDNA:mHC mRNA; △, MHC rDNA:Reovirus RNA; and ▲ MHC cDNA:Yeast RNA.

large size DNA inserts specific for MLC, clone pML10 insert was 950 ± 50 base pains in length.[87] There are 340 extra nucleosides in the mRNA than required to code for MLC_2 polypeptide. On the assumption that most of these noncoding nucleosides are located in the 3′ terminus, more than half of the coding region of MLC_2 mRNA would be represented in the clone DNA insert. Indeed, the base sequence analysis presently in progress in our laboratory indicated that coding sequence for all but 15 N-terminal amino acids is represented in pML10. The cDNA generated from the purified MHC mRNA contained DNA transcripts that hybridized effectively with λGdl DNA containing the chick 28S rRNA gene, suggesting that the mRNA, in spite of the apparent functional purity, contained trace amounts of 28S rRNA. The poor efficiency of MHC mRNA as template for reverse transcription combined with the possible contamination of the enzyme with nucleases resulted in the synthesis of ribosomal cDNA. As such, the transformants obtained by cloning of ds-cDNA were screened with both [^{32}P]cDNA prepared using purified MHC mRNA and nick-translated [^{32}P]λGdl DNA in order to eliminate 28s rDNA-containing clones. Colonies, which hybridized to MHC cDNA and not to λGdl DNA, were selected as candidates for MHC clones.

The fidelity of MHC gene sequences in the recombinant plasmid, pMHC8, was demonstrated by the ability of the plasmid DNAs to arrest the translation of MHC mRNA in total poly(A) RNA as before.[92] The plasmid DNA pMHC8 was further characterized by restriction map analysis and heteroduplex formation. A partial restriction map for clone pMHC8 is shown in Figure 6. The size of the insert in pMHC8 DNA is about 850 bp, which represents approximately 12% of the size of chick heart MHC mRNA. Since about 5800 nucleosides are required to code for MHC (220,000 daltons), there are 800 to 1000 extra bases present in MHC mRNA. Some of these extra bases may represent the 3′-noncoding region of the MHC mRNA and, as such, the majority of DNA in pMHC8 would represent the noncoding sequence. The restriction maps of other MHC clones were identical with differences only in chain lengths. The pML10 DNA insert, on the other hand, represents the majority of the coding sequence, and the complete 3′-noncoding region. Heteroduplexes

pMHC 8

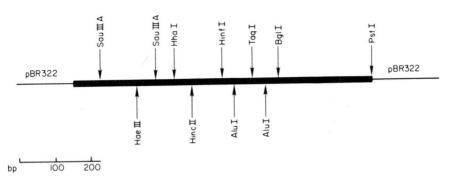

FIGURE 6.   Restriction map of plasmid pMHC8. The cDNA shown by the heavy bar was inserted in pBR322 at Pst I site, 3608. The DNA insert did not contain sites for E.CoR1, Bam HI, Hind III, SalG I, Bgl II, Pvu I, Pvu II, Ava I. Note that one of the two Pst I sites was not regenerated.

of pMHC8 with pBR322 DNA were made and the representative electron micrographs are shown in Figure 7. All heteroduplexes observed contained single-stranded deletion loops of approximately 0.8 kb for pMHC8 and 0.9 kb for pML10 DNAs.

D. Myosin Gene Transcription in Chick Heart Cells

In order to ascertain whether mHC mRNA sequences can be identified in a mixed population of chick heart RNA, nick-translated [32P] pMHC DNA was hybridized to RNAs which were fractionated on formaldehyde-agarose gel and transferred to nitrocellulose paper. When hybridized to chick heart muscle poly(A) RNA and oligo(dT) cellulose unbound RNA, the latter contains significant amounts of MHC mRNA because of its poor binding capacity to affinity columns;[88] hybridization occurred only to RNA with mobility of about 29 to 30S corresponding to the mobility of purified MHC mRNA. In order to test the tissue specificity of plasmid pMHC8, equal amounts of RNAs from chick skeletal muscle, adult rat heart muscle, mouse liver, E. coli, yeast, MOPC fibroblast cells, rabbit globin mRNA, and RNA from differentiated chick heart cells in culture were fractionated as above and hybridized to nick-translated [32P]pMHC DNA. Figure 8 shows that pMHC DNA hybridized effectively to RNA from chick cardiac muscle tissue but very poorly to mRNA from chick skeletal muscle cells. Hybridization also occurred to the RNA isolated from the chick heart RNA from the differentiated myoblast (lane 11). Both hybridizations occurred in the 29 to 30S region on the gel corresponding to the mobility of authentic MHC mRNA. The faster-moving bands present only in RNA extracted from the cultured cells may be due to partial degradation of the MHC mRNA during cell collection and RNA extraction procedures. The DNA, surprisingly, did not hybridize to rat heart muscle RNA. The RNA sequence represented in clone pMHC8 is, therefore, highly tissue specific and affords a reliable detection of cardiac MHC sequence during chick embryonic development.

Cultured chick embryonic heart cells which undergo a distinct morphogenetic differentiation could, in principle, provide a useful experimental system to study the molecular events associated with cardiac muscle differentiation. To examine the appearance of MHC mRNA in these cells, total RNA was isolated from cells in culture at 16 hr, 48 hr, 4 days, and 7 days of development. Upon hybridization with [32P]pMHC8 DNA probe as before, no MHC mRNA appeared up to 48 hr of incubation. At about 48 hr the single myoblast cells begin to undergo morphological differentiation followed by acquisition of rhythmic pulsations in 3- to 4-day-old culture. A distinct band with mobility similar to purified MHC mRNA appeared to be of maximum intensity in 4-day-old cultured cells which corresponds

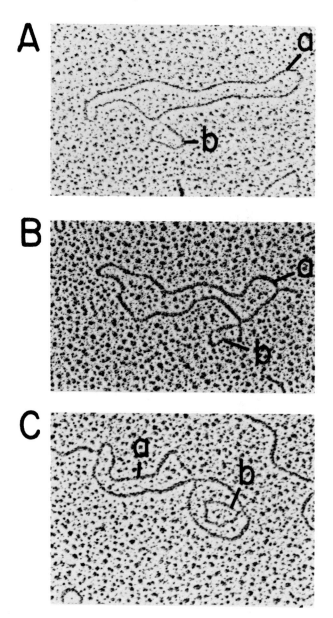

FIGURE 7. Electron micrographs of the heteroduplexes between pMHC8 and pBR322. Heteroduplexes were formed between pBR322 and pMHC8 (A and B) and pMHC4 (C) DNAs. a, duplex plasmid DNA; b, deletion loop.

to the stage where differentiated cardiac myotubular structures are visible. The high molecular weight hybridization band was still present, although to a significantly lower extent, in the cultured cells as old as 7 days. When MHC-specific RNA sequences in developing chick embryonic cells in culture were detected by *in situ* hybridization using [³H]pMHC DNA, only 4-day-old cells were heavily labeled (Figure 9D). Prior treatment of cells with pancreatic RNase caused significant loss of hybridizable mRNA (Figure 9A) noticeable by the lack of silver grains. The specificity of hybridization was demonstrated by using labeled pBR322 DNA (Figure 9A and C).

Recent observations on appearance of mRNAs for the cardiac muscle structural proteins suggest that these proteins are synthesized coordinately by the burst of corresponding mRNAs

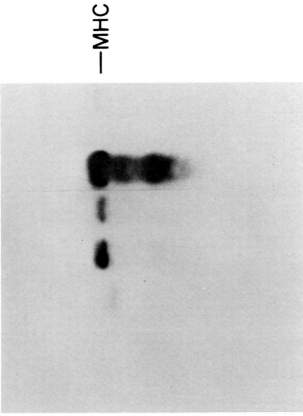

FIGURE 8. Hybridization of pMHC8 DNA to various RNAs. Lanes 1 to 4 represent hybridization of pMHC8 DNA to 2.5 μg of total RNA from 16-hr, 40-hr, 4-day, and 6-day-old chick heart cells in culture, respectively. Lanes 5 to 8 contain globin mRNA (1.0 μg), MOPC-41 fibroblast RNA (5 μg), mouse liver RNA (5.0 μg), and *E. coli* RNA (10 μg), respectively. Lanes 9 and 10 represent chick heart poly (A) RNA (2 and 1 μg, respectively); lane 11 contains RNA from 7-day-old chick heart cells in culture (10 μg); and lanes 12 and 13 represent 2.5 μg each of RNAs from adult rat heart and chick skeletal muscle tissues, respectively.

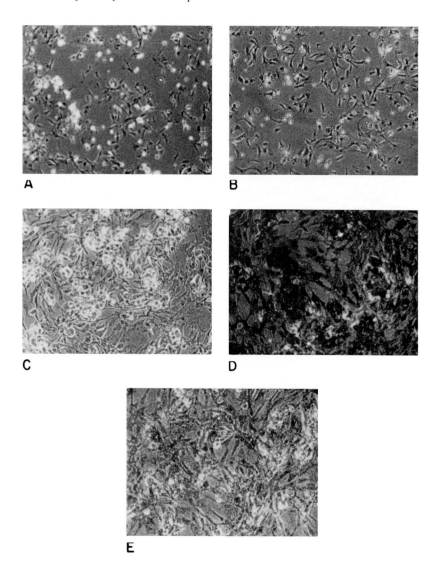

FIGURE 9.   Hybridization *in situ* of pMHC8 and pBR322 [³H]DNA. Chick embryonic heart cells (12-day-old) were grown in culture in Lab-Tek® (Miles Laboratories, Inc.) slide chambers, immobilized, and processed for hybridization *in situ*.[88] Hybridization was done in 100 µℓ of buffer containing 50% deionized formamide in 10 mm Tris-HCl (pH 7.5), 200 mm NaCl, 5 mm MgCl₂, at 37°C for 40 hr using 100,000 cts/min of ³H-labeled, nick-translated pMHC8 DNA. One set of slide chambers was treated with pancreatic RNAase (10 µg/mℓ) before hybridization. The unhybridized nucleic acids were removed by treatment with nuclease and the slides were exposed to Kodak® NTB-2 emulsion and developed after 2 to 3 weeks. (A and C) Hybridized with control pBR322 [³H]DNA; (B and D) hybridized with pMHC8[³H]DNA; and (E) hybridized with pMHC8 [³H]DNA after treatment with pancreatic RNAase.

at the time of myoblast differentiation.[75,79] Our results on *in situ* hybridization suggested that MHC mRNA is transcribed after 48 hr of incubation paralleling morphogenetic differentiation of myoblasts into myocytes. In contrast to MHC, however, the chick muscle glyceraldehyde-3-phosphate dehydrogenase (GAPDH) mRNA transcripts were apparently present in cells as early as 5-hr-old myoblasts, suggesting that expression of GAPDH genes precedes that of myosin.[93] GAPDH, a noncontractile protein, and specific recombinant plasmids (pGAP30 and pGAP36) were recently constructed in our laboratory.[93]

V. IDENTIFICATION AND CHARACTERIZATION OF INDUCER 7S RNA

Earlier workers had suggested[63-65] that the RNA capable of inducing specific changes in chick blastodermal cell located in PNP might have mRNA-like properties. We reexamined the phenomenon and found that the transition is indeed dependent upon the addition of an RNA fraction isolated from embryonic heart tissue to culture medium supporting the PNP explants.[66] The inducing molecule is a low molecular weight of about 7S in size with distinct properties. 7S RNA has been purified and characterized for its structural properties (see below).[94] In order to understand fully the mechanism of induction of embryonic heart muscle gene transcription in chick embryo and the role the putative inducer RNA plays in the process, we asked the questions listed below:

1. What is the nature of RNA-induced changes in the PNP cells?
2. Is the transition in PNP dependent upon a specific RNA molecule?
3. Do all PNP cells respond to inducer RNA?
4. Is the RNA taken up by the PNP cells?
5. What are the structural and functional properties of inducer RNA?
6. Is the inducer RNA a physiological entity?

In the following, we shall attempt to answer these questions, at least in part, and summarize experiments currently in progress.

A. What Is the Nature of RNA-Induced Changes in PNP?

In previous studies, identification of RNA-induced changes was based mainly on gross morphological criteria and on acquisition of spontaneous beating. In addition to these changes, we redefined the PNP differentiation by the following criteria.

1. Appearance of myofibrils and glycogen granules — It is well established that the appearance of myofibrils and glycogen particles are two identifying characteristics for differentiation of embryonic muscle cells in vivo, in cell suspensions, and in vitro.[95,96] Ultrastructural examination of RNA-treated and control PNPs showed normal cell components, e.g., endoplasmic reticulum, nuclei, mitochondria, Golgi complex, etc., but the RNA-treated PNPs which exhibited spontaneous pulsations contained highly differentiated myocytes similar in ultrastructural complements to embryonic myocardial cells. Clearly evident (Figure 10) were the orderly arranged myofibrils with thick and thin filaments and Z bands. Histochemical examination of chick embryonic heart tissue and the beating PNPs revealed rich deposition of glycogen particles, whereas the control PNP were relatively free of glycogen.[66,67]

2. Synthesis of myosin and actin polypeptides — When the PNPs were labeled with [^{3}H] alanine, there was a 2.5- to 3.0-fold increase in label incorporated into myosin polypeptides. Fractionation of partially purified muscle proteins from induced and uninduced PNPs on SDS/polyacrylamide gel showed the presence of myosin and actin-like polypeptides.[66,67]

3. Acetylcholinesterase activity — The myogenic process involved, in addition to increased synthesis of muscle proteins, elaboration of specialized membrane components and the enzyme acetylcholinesterase. Cells of myogenic origin differentiated in culture in absence of neuronal elements produce acetylcholinesterase similar to muscle tissue *in situ*.[97] When the PNP was cultured in presence of the 7S RNA and the beating PNPs were pooled and processed for acetylcholinesterase activity, there was a 3.4-fold increase in the rate of hydrolysis of acetylcholine compared to the nonbeating PNP, and a 6-fold increase over that of the unincubated PNPs.[66] The increase was always found in the same batches of explants that also exhibited specific morphological and biochemical changes, whereas the PNPs grown in absence of inducer RNA or with carrier RNA distinctly lacked these properties. Histochemical examination for acetylcholinesterase provides a direct visualization of the locale

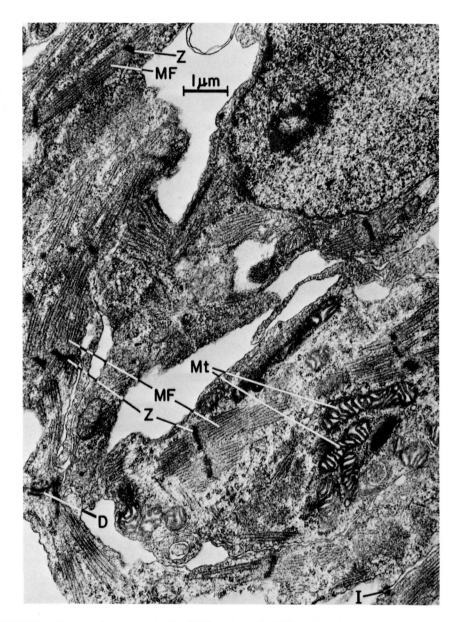

FIGURE 10.   Electron micrographs of cells of PNP explants. The PNPs cultivated in the presence of CEH-RNA were examined for ultrastructural components.[66] MF, myofibrils; D, desmosomes; Z, Z bands; and Mt, mitochondria.

of the enzyme during various stages of development. The explants when stained according to Karnovsky and Roots[98] after incubation with the inducer RNA and the carrier RNA separately exhibited differences between the intensity and distribution of the enzyme positive granular material.

B. Is the Transition in PNP Dependent upon a Specific RNA Molecule?

As shown in Table 1, none of the explants cultivated in absence of inducer RNA (or in presence of carrier RNA) exhibited the characteristics described above. In contrast, the total RNA from the embryonic heart or any of the RNA fractions obtained during purification containing the active RNA caused a specific mode of changes similar to myogenic differentiation. Treatment of RNA with pancreatic or T_1 ribonuclease abolished the activity,

Table 1
RNA-DEPENDENT DIFFERENTIATION IN PNP

Additions	A_{260}[a]	No. of PNP used	No. of PNP differentiated[b]
Chick embryonic heart RNA			
Total RNA	15.00	94[c]	60[c]
Unbound RNA	5.00	10	0
	15.00	51	0
Bound I RNA	0.25	8	0
	0.50	9	0
Bound II RNA	0.07	9	3
	0.14	9	5
	0.28	42[c]	30[c]
Bound II, fraction 1	0.01	19	16
	0.02	6	5
	0.05	9	6
Bound II, fraction 2	0.03	6	0
	0.07	6	0
Bound II, fraction 3	0.02	6	1
	0.08	8	3
Bound II, fraction 4	0.05	4	9
Bound II, fraction 1'a'	0.005	20[c]	15[c]
	0.01	9	6
Bound II, fraction 1'a' (heat treatment)	0.01	9	6
Bound II, fraction 1'a' (pan. RNase treated)	0.01	9	0
Bound II, fraction 1'a' (RNase treated)	0.01	8	0
Chick embryonic brain RNA			
Unbound RNA	11.00	8	0
Bound RNA	0.25	8	0
	0.50	8	0
Rat liver RNA			
Unbound RNA	10.00	8	0
Bound RNA	0.03	8	0
	0.60	8	0
Calf kidney RNA			
Unbound RNA	12.000	8	0
Bound RNA	0.25	8	0
	0.50	8	0
δ-Crystallin mRNA	0.34	6	0
Rous sarcoma viral RNA	0.90	6	0
Rabbit macrophage RNA	3.00	6	0
Bombyx mori rRNA	15.00	6	0
Bombyx mori tRNA	12.00	16[c]	0
E. coli rRNA	12.00	9	0
E. coli tRNA	12.00	21[c]	0
Total chick embryonic RNA (DNase treated)	15.00	3	2
Poly(A)	0.005	9	0
	0.01	9	0
	0.05	6	0
	0.10	6	0

Table 1 (continued)
RNA-DEPENDENT DIFFERENTIATION IN PNP

Additions	A_{260}[a]	No. of PNP used	No. of PNP differentiated[b]
Poly(C)	0.09	3	0
Poly(G)	0.09	3	0
Poly(I)	0.10	3	0
Control (without RNA)	—	57[c]	0
Control (with carrier RNA)	15.00	90[c]	0

Note: Culture technique, addition of RNA, incubation conditions, enzymic diges-
tions, and heat treatment are all described in detail elsewhere.[66]

[a] The final concentration was adjusted to 15 A_{260}/mℓ by addition of carrier RNA.
E. coli or *Bombyx mori* rRNA and tRNA, in a ratio of 101, was used as carrier
RNA.

[b] Initial scoring of the differentiated explants was based on the beating property
of the PNP cells. The beating and the nonbeating explants were pooled separately and then divided randomly into batches which were then subjected to
ultrastructural histo- and biochemical examinations as described earlier.[66]

[c] The number of explants represents a cumulative total of several experiments.

whereas pronase or DNase had no effect. Partial alkali hydrolysis also caused loss of activity,
whereas the activity was resistant to heat treatment. The RNA prepared from chick embryonic
brain, kidney, and liver, and RNA from calf kidney and rat liver did not induce the differentiation as above, nor did they support any other morphological change attributable to a
specific mode of differentiation. Total poly(A) RNA from nonmuscle tissues or purified
mRNA such as δ-crystallin mRNA had no effect. The addition of δ-crystallin mRNA did
not lead to δ-crystallin synthesis in PNP cells when examined by an immunoprecipitation
assay. Synthetic polynucleotide used in a variety of concentrations had no effect.

Thus, it appears that the induction of myogenic-like changes in PNPs is indeed dependent
upon the presence of a specific RNA in the culture medium.

C. Do All PNP Cells Respond to Inducer RNA?

It is believed that during the early stages of development, i.e., stage 3+ to 4−,[53] the
precardiogenic cells lie in the bilateral regions of the epiblast adjacent and approximately
halfway between the anterior and posterior ends of the growing primitive streak.[55] Precardiac
mesoderm begins to invaginate at stage 4 moving anteriorly along the growing primitive
streak to form well-defined bilateral regions at stage 5. These regions are centered at the
level of Hensen's node, halfway between the node and the lateral edges of area pellucida.
When tissue fragments, isolated by sequential cuts at intervals of 0.1 mm beginning with
0.4 mm and up to 1.2 mm posterior to Hensen's node, were grown in presence of inducer
RNA, there was a gradient in potential to develop heart-like tissue (Figure 11). The fragments
obtained by cutting below 0.8 mm posterior to Hensen's node failed to develop into structures
similar to fragments containing the anterior portions of PNP cells. This led us to speculate
that heart-forming cells are located in area extending up to 0.8 mm behind the Hensen's
node, which is in agreement with the heart fate mapping experiments[54] using [³H] thymidine-
labeled grafts taken from various regions of the blastoderm.

D. Is the RNA Taken Up by the PNP Cells?

Substantial evidence does exist on the uptake and transfer of RNA by eukaryotic cells.
Mildly disaggregated PNP cells were incubated with [¹²⁵I]-labeled 7S RNA for 4 hr in

STAGE 4

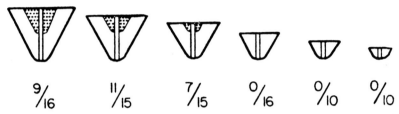

$$\frac{9}{16} \qquad \frac{11}{15} \qquad \frac{7}{15} \qquad \frac{0}{16} \qquad \frac{0}{10} \qquad \frac{0}{10}$$

FIGURE 11. Sequential transection of chick stage 4 blastoderm was done as indicated by dotted lines. The resulting postnodal explants (second row) were then grown in culture as usual.[66] The number of beating explants out of the total explants used is shown on the bottom. Shaded area represents an approximation of precardiac region in stage 4 blastoderm.

medium-199 and the cells were washed free of RNA by repeated washings till no more radioactivity was released. RNA was extracted from the cells and examined by electrophoresis on polyacrylamide gel in formamide. The RNA isolated from the PNP migrated identically to the RNA added (unpublished results). A fingerprint analysis of the two RNA showed identical profiles. The autoradiographic examination of the sectioned PNP revealed a random distribution of grains in the cells. One can deduce from these experiments that 7S RNA might enter the PNP cells in an intact form, although one cannot exclude the possibility that the RNA visualized as grains had undergone modifications prior to or after its uptake by the cells. Further experiments are needed to determine the intracellular fate of 7S RNA and to ascertain whether the uptake occurred in macromolecular form or after specific modifications.

E. What Are the Structural and Functional Properties of Inducer RNA?

The 7S CEH RNA purified by successive electrophoresis on polyacrylamide gel containing formamide migrates with a mobility of 250 ± 20 nucleosides. However, when the total poly(A) RNA preparation from which 7S RNA is routinely isolated was subjected to an extensive denaturation by incubation with Me_2SO followed by fractionation on an 80% formamide-containing sucrose gradient, it yielded an RNA fraction which moved faster than 7S CEH-RNA on polyacrylamide gel with the mobility of about 200 nucleosides. This RNA is referred here as 7S RNA. Based on end group analysis, base composition, poly (A) length measurements, and T_1 oligonucleotide fingerprint profiles, 7S RNA appeared to be identical to 7S CEH-RNA, but was shorter than 7S CEH-RNA by 50 ± 20 adenosine residues removed from the 3'-end. Since the RNA is not found when Me_2SO-formamide treatment is avoided, it appeared that 7S RNA is not a metabolic entity per se, but rather a product of the removal of a portion of poly(A) tail from the 7S CEH-RNA, presumably due to Me_2SO-formamide treatment. More important is the fact that it possessed the activity to induce the specific mode of changes in the chick embryonic cells as well as 7S CEH-RNA. Because of the ease with which 7S RNA can be purified relative to 7S CEH-RNA (short-poly(A) tail facilitates the resolution on polyacrylamide gel), we have studied the structural properties of 7S RNA, although 7S CEH-RNA was also subjected to similar characterization in parallel.

In preliminary experiments, it was found that although alkaline phosphatase treatment was necessary for 5'-end labeling of 7S RNA, tobacco acid pyrophosphatase treatment was not. This and the previous results on analysis of [³H]-labeled 7S CEH-RNA by reduction with borotritide suggested that 7S RNA does not contain a capped 5'-terminus. Separation of labeled 7S RNA on a high resolution two-dimensional polyacrylamide gel, which produces a remarkably high resolution of tRNA isoacceptor species yields a single spot for 7S RNA. Complete digestion of 5'-end labeled 7S RNA with P-1 or snake venom phosphodiesterase revealed that more than 95% of the label resides in single nucleotide 5'-pA. Likewise, when the RNA was labeled with [³²P]pCp and hydrolyzed to completion with T₂, more than 90% of label was transferred to A. Identical results were obtained when 7S CEH-RNA was used. The base ratio analysis done by postlabeling of T₂ digests of 7S RNA with [γ-³²P]ATP clearly showed the presence of four major nucleosides with no detectable minor nucleoside.

7S RNAs also served as efficient templates for oligo(dT) primed cDNA synthesis using AMV reverse transcriptase, and the average size of the product was a full-size cDNA transcript. We also undertook the partial base sequencing analysis to ascertain whether the initiation signal, AUG, could be located near the 5'-end of the molecule. Based on the combined results of sequencing gels and mobility shift analyses, the 5'-terminal sequence

$$5'\text{AAGAGCGGAGGGXGGAACPyGGCCGXUCGUGCGPyCGAAGAAGAUUXPy}_{46}$$

(X represents an ambiguous base cut) was deduced. The sequence shows that the RNA does not contain the initiator codon AUG in the 5'-region, at least up to 46 nucleotides, however, GUG usually found in prokaryotic viral RNAs was present. Furthermore, the alignment of the bases into any reading frame results in a large number of nonsense codons.

F. Is the Inducer RNA a Physiological Entity?

Several lines of evidence suggest that the RNA is not produced via a nonspecific degradation of a large molecular weight RNA(s) during extraction procedures. The relative amount of 7S CEH-RNA obtained at a particular stage of chick development is constant regardless of the procedure employed for RNA extraction. Preparation of RNA under conditions optimized for isolation of functionally intact mRNAs yields 7S CEH-RNA in amounts comparable with those utilizing phenol extraction of the total chick heart tissue homogenate. The specificity with which the RNA influences the biological transition in the presumptive heart-forming cell of the early chick blastoderm suggests that the RNA is a physiologically relevant entity with a possible function in regulatory processes during development.

VI. ON POSSIBLE ROLE OF 7S RNA IN CARDIAC MUSCLE CELL DIFFERENTIATION

We have previously observed that 7S RNA inhibits both homo- and heterologous poly(A) RNA translation in cell-free systems.[99] This suggested to us the observed activity of 7S RNA in inducing a specific mode of differentiation in PNP might be related in some way to an effect in translation process in vivo. The RNA also shares some characteristics previously described for the low molecular RNA species isolated from chick muscle,[100] *A. salina*,[101] rabbit reticulocytes,[102] and Novikoff hepatoma cells.[103] These RNAs have been shown to participate as modulators of proteins synthesis in cell-free translation assays and, therefore, are implicated to play a role in regulation of cell development and differentiation. Although 7S RNA is capable of inhibiting effectively the translation of both homo- and heterologous mRNA without any apparent selection of specific mRNA or group of mRNAs, the inhibition appears to be effected by the adenosine-rich segment of 7S RNA, since the removal of nonpoly(A) fragment of 7S RNA produced the same kinetics of inhibition as the intact molecule.[99] These results would, therefore, suggest that a control at the level of translation is possibly not one of the mechanism(s) of actions of 7S RNA in causing heart-like differentiation in the early embryonic cells.

As described in Section IV, we have recently constructed recombinant plasmids containing gene sequences for MLC (pML10) and MHC (pMHC8). During the course of characterization of these plasmids, we made an unexpected observation. The cDNA for 7S RNA hybridized effectively with both pML10 and pMHC8 plasmids DNA (unpublished observations). The possibility of nonspecific hybridization to DNA and adventitious binding to nitrocellulose membrane was discounted by the use of appropriate controls. Several explanations seemed compatible with this observation.

7S RNA might have sequences complementary to chick heart MHC and MLC mRNAs. This is being tested at present by direct hybridization of [^{32}P] 7S RNA with purified MHC and MLC mRNAs. Distinct low molecular weight RNA species are known to exist in eukaryotic cells which specifically bind to poly(A)-containing mRNAs.[21,22,24] The eukaryotic mRNAs with known base sequences contain pyrimidine-rich sequences in the 3'-noncoding region which would serve as the potential binding site for the purine-rich sequences found in many of the low-molecular weight RNAs reported.[24] The 4.5S RNA isolated from the Chinese hamster ovary cells (CHO) was recently shown to hybridize specifically to the 3'-noncoding segment of mouse liver and CHO mRNAs.[24] Preliminary experiments utilizing [^{32}P] cDNA for 7S RNA indicated that the complementarity between the cDNA and the cloned DNA in plasmid pMHC8 and pML10 resides in a small fragment obtained by restriction digestion of the plasmid DNA. The orientation of these fragments with respect to the corresponding mRNA sequences is currently under investigation. Partial base sequence analysis of 7S RNA revealed that the RNA lacks coding sequence, in at least the first 60 5'-terminal nucleotides. If this is true for the entire molecule, the 7S RNA would apparently be complementary to the 3'-noncoding sequence of the mRNAs.

The other possible site of hybridization of 7S RNA to myosin mRNAs is the sequence at the 3'- and 5'-ends of splice junctions. The consensus sequence at splice junctions of pre-mRNAs is complementary to SnRNAs (see Section II). However, the number of complementary bases located in the exon region of the splice junctions, which ranges from one to only a few, cannot account for the strong hybridization signal observed with [^{32}P] 7S cDNA and pMHC8 (or pML10) DNA under high stringency conditions employed. It would thus appear that 7S RNA might not be involved in the splicing mechanism.

It was proposed[20] that functional expression of a particular set of genes is regulated by specific interactions between the repetitive DNA transcripts and the single copy structural gene transcripts (see Section II). The repetitive DNA is found interspersed with single copy DNA in many eukaryotes and the poly(A) RNA transcripts of single copy DNA are known to contain short repetitive sequences covalently linked to the RNA.[104,105] The precise role of repetitive sequences in gene regulation is yet to be elucidated. It would be of interest to know whether 7S RNA interacts with the repetitive DNA (Alu family) and/or the RNA transcripts of the DNA. Since 7S RNA was isolated from the predominantly cytoplasmic fraction, its interaction with DNA would appear less probable. On the other hand, cytoplasmic 7S RNA can still perform a nuclear function, since SnRNA is known to exist in cytoplasm and shuttle between the cytoplasm and nuclei.[106] The finding that 7S cDNA hybridizes to DNA of clones pMHC8 and pMLC10 suggests that 7S RNA may interact with specific sequences on mRNAs or their corresponding gene segments. Whether the site of interaction is the repetitive DNA or its short transcript contained in the myosin mRNA molecule is yet to be ascertained.

We have recently constructed recombinant plasmids using the cDNA for 7S RNA and are currently isolating genomic DNA clones carrying 7S MHC and MLC specific sequences.

These DNAs would be used to characterize the exact relationship between 7S and the myosin-specific sequences in genomic DNA. The localization of 7S sequences in the chick genome and its orientation with respect to myosin and other muscle-specific genes would provide insights into the mechanism of action of 7S RNA during muscle gene expression.

VII. SUMMARY

The understanding of events that control early gene expression would require a comprehensive analysis of transcriptional activity of a specific population of cells. For a particular developmental pathway, the gene activity can be studied by identifying and isolating the genes that code for the proteins involved. However, the complexity of eukaryotic genome precludes a direct isolation of structural genes and elucidation of the regulating elements. One way to circumvent this problem is to isolate the mRNAs coding for the developmentally regulated marker proteins and use the DNA probe to monitor the transcriptional activity throughout the early stages of development. To this end, we initiated our studies on cardiac muscle development in the chick embryo, since it offers a unique advantage in the availability of the predetermined population of embryonic cells which are highly primitive and can be grown successfully in the laboratory to differentiate into well-defined cardiac tissue. The cloned DNA probes isolated in our laboratory will afford a reliable detection and quantitation of myosin-specific mRNA levels in developing blastodermal cells. Apart from the usefulness as probes, the DNAs would be used for analysis of structure and organization of the muscle genes within the chick genome.

We have also demonstrated unequivocally the existence of a low molecular weight RNA, 7S RNA, that is capable of inducing a specific mode of changes in the embryonic cells of myogenic origin. The mode of differentiation was indistinguishable from that of myogenesis. 7S RNA appeared to be an authentic component of chick heart cells. Whether it is present in the early blastodermal cells and serves as an inducing factor in vivo for expression of muscle specific genes is yet to be ascertained. With the help of recombinant plasmids containing 7S RNA sequences the biogenesis of 7S RNA and its localization can be determined. These cloned DNAs will also facilitate establishing the relationship between the muscle-specific mRNA and 7S RNA. The knowledge of the orientation of muscle genes with respect to 7S genes within the physical map of chick genome would provide leads to the physiological relevance of 7S RNA. The specific interaction of 7S RNA with myosin gene sequences is an important observation. Although it does not elucidate the function of 7S RNA, it does focus attention on the potential role of the molecule in synthesis, utilization, and/or transport of myosin mRNAs. It is, nonetheless, conjectural at present to relate the presence of homology between the 7S RNA and myosin mRNAs to the inducing activity of 7S RNA in the PNP cells of chick blastoderm.

ACKNOWLEDGMENTS

I am thankful to my associates, A. Ahmed, H.-H. Arnold, K. Datta, A. K. Deshpande, H. J. Drabkin, S. B. Jakowlew, M. Krauskopf, W.-L. Lin, C. Mendola, S. K. Narula, and A. M. Zarraga for their valuable contributions and highly qualified assistance during the conduct of this work. The excellent secretarial help of Ms. B. Kerr is also gratefully acknowledged.

REFERENCES

1. **Child, C. M.**, in *Patterns and Problems of Development*, Chicago University Press, Chicago, 1941.
2. **Wolpert, L.**, *J. Theor. Biol.*, 25, 1, 1969.
3. **Goldacre, R. J. and Bean, D.**, *Nature (London)*, 186, 294, 1960.
4. **Goodwin, B. C. and Cohen, M. H.**, *J. Theor. Biol.*, 25, 49, 1969.
5. **Riley, P. A.**, *Differentiation*, 1, 183, 1973.
6. **Davidson, G. H.**, in *Gene Activity in Early Development*, Academic Press, New York, 1968.
7. **Raff, R. A.**, *BioScience*, 27, 394, 1977.
8. **Davidson, E. H. and Britten, R. J.**, *J. Theor. Biol.*, 32, 123, 1971.
9. **DeRobertis, E. M. and Gordon, J. B.**, *Proc. Natl. Acad. Sci. U.S.A.*, 74, 2470, 1977.
10. **Getz, M. I., Birnie, G. D., Young, B. D., MacPhail, E., and Paul, J.**, *Cell*, 4, 121, 1975.
11. **Davidson, E. H., Hough, B. R., Amerson, C. S., and Britten, E. H.**, *J. Mol. Biol.*, 77, 1, 1973.
12. **Deininger, P. H. and Schmid, C. W.**, *J. Mol. Biol.*, 106, 773, 1976.
13. **Davidson, E. H., Hough, B. K., Klein, W. H., and Britten, R. J.**, *Cell*, 4, 217, 1975.
14. **Scheller, R. H., Constantine, F. D., Kozlowski, M. R., Britten, R. J., and Davidson, E. H.**, *Cell*, 15, 189, 1982.
15. **Davidson, E. H., Galan, G. A., Angerer, R. C., and Poritten, R. I.**, *Chromosomes*, 51, 253, 1975.
16. **Manning, J. E., Schmid, C. W., and Davidson, N.**, *Cell*, 4, 141, 1975.
17. **Levy, J. and McCarthy, B. J.**, *Biochemistry*, 14, 2440, 1975.
18. **Britten, R. J. and Davidson, E. H.**, *Science*, 165, 349, 1969.
19. **Davidson, E. H. and Britten, R. J.**, *Q. Dev. Biol.*, 46, 111, 1973.
20. **Davidson, E. H. and Britten, R. J.**, *Science*, 204, 1052, 1979.
21. **Harada, F. and Ikawa, Y.**, *Nucleic Acids Res.*, 7, 895, 1979.
22. **Jelinck, W. and Linwand, L.**, *Cell*, 15, 205, 1978.
23. **Weiner, A. M.**, *Cell*, 22, 209, 1980.
24. **Harada, F. and Kato, N.**, *Nucleic Acids Res.*, 8, 1273, 1980.
25. **Jelinek, W., Toomey, T., Linwant, L., Duncan, C., Bio, P., Chondary, P., Weissman, S., Rubin, C., Honk, C., Deninger, P., and Schmid, L.**, *Proc. Natl. Acad. Sci. U.S.A.*, 77, 1398, 1980.
26. **Shibata, H., Ro-Chai, T. S., Reddy, R., Choi, Y.C., Henning, D., and Busch, H.**, *J. Biol. Chem.*, 250, 3909, 1975.
27. **Reddy, R., Henning, D., and Busch, H.**, *J. Biol. Chem.*, 254, 11097, 1979.
28. **Rogers, J. and Wall, R.**, *Proc. Natl. Acad. Sci. U.S.A.*, 77, 1877, 1980.
29. **Roop, D. R., Kristo, P., Stumph, W. E., Tsai, M. J., and O'Mally, B. W.**, *Cell*, 23, 671, 1981.
30. **Lerner, M. R. and Steitz, J. A.**, *Cell*, 25, 298, 1981.
31. **Lerner, M. R. and Steitz, J. A.**, *Proc. Natl. Acad. Sci. U.S.A.*, 76, 5495, 1979.
32. **Yang, V. W., Lerner, M. R., Steitz, J. A., and Flint, J. A.**, *Proc. Natl. Acad. Sci. U.S.A.*, 78, 1371, 1981.
33. **Denison, R. A., Van Arsdell, S. W., Bernstein, C. B., and Weimer, A. M.**, *Proc. Natl. Acad. Sci. U.S.A.*, 78, 810, 1981.
34. **Diener, J. O.**, *Science*, 205, 859, 1979.
35. **Owens, R. A. and Diener, J. O.**, *Proc. Natl. Acad. Sci. U.S.A.*, 79, 113, 1982.
36. **Pinch, Y. H., Fritze, D., Waldman, S. R., and Kern, D. H.**, in *Curr. Top. Microbiol. Immunol.*, 72, 187, 1975.
37. **Meiss, H. K. and Fishman, M.**, *J. Immunol.*, 108, 1172, 1971.
38. **Billelo, P., Fishman, M., and Koch, G.**, *Cell. Immunol.*, 23, 309, 1976.
39. **Bester, A. J., Kennedy, D. S., and Heywood, S. M.**, *Proc. Natl. Acad. Sci. U.S.A.*, 72, 1523, 1975.
40. **Zieve, G. and Penmann, S.**, *Cell*, 8, 19, 1976.
41. **Zeichner, M. and Breitkrentz, M.**, *Arch. Biochem. Biophys.*, 188, 410, 1978.
42. **Plenskal, M. E. and Sarkar, S.**, *Biochemistry*, 20, 2048, 1981.
43. **Holmes, D. S., Mayfield, J. E., Sanders, G., and Bonner, J.**, *Science*, 177, 72, 1972.
44. **Kanechisa, T., Oki, Y., and Ikata, K.**, *Arch. Biochem. Biophys.*, 165, 146, 1974.
45. **Bogdanasky, D., Herman, W., and Schapira, E.**, *Biochem. Biophys. Res. Commun.*, 54, 25, 1973.
46. **Lee-Huang, S., Sierra, J. M., Navajo, R., Filipowicz, W., and Ochoa, S.**, *Arch. Biochem. Biophys.*, 180, 276, 1977.
47. **Bhargava, P. and Shamurgam, G.**, *Progress in Nucleic Acid Research and Molecular Biology*, Vol. 11, Davidson, J. N. and Colon, W. E., Eds., Academic Press, New York, 103.
48. **Slavkin, H. and Crossant, R.**, in *The Role of RNA in Reproduction and Development*, Niu, M. C. and Segal, S. J., Eds., North-Holland, Amsterdam, 1973, 247.
49. **Bishop, J. M.**, *Annu. Rev. Biochem.*, 67, 35, 1978.
50. **Erickson, E. and Robertson, H. D.**, *Cancer Res.*, 36, 3387, 1976.

51. **Siddiqui, M. A. Q., Deshpande, A. K., Arnold, H.-H., Jakowlew, S. B., and Crawford, P. A.,** *Brookhaven Symp. Biol.,* 29, 62, 1977.
52. **Siddiqui, M. A. Q.,** in *Genes and Protein Structure,* Weissbach, H., Siddiqui, M. A. Q., and Krauskopf, M., Eds., Academic Press, New York, 1981, 109.
53. **Hamburger, V. and Hamilton, H. L.,** *J. Morphol.,* 88, 49, 1951.
54. **Rosenquist, G. C.,** *Dev. Biol.,* 22, 461, 1970.
55. **DeHaan, R. L., Dunning, J. O., Hirakow, R., McDonald, T. F., Rash, J., Sachs, H., Wildes, J., and Wolf, K. M.,** in *Annu. Rep. Dir. Carnegie Ins. Washington Yearbook,* 70, 66, 1970.
56. **Hunt, T. E.,** *J. Exp. Zool.,* 59, 395, 1931.
57. **Rudnick, D.,** *Anat. Rec.,* 70, 351, 1938.
58. **Rawles, M. E.,** *Physiol. Zool.,* 16, 22, 1943.
59. **Spratt, N. T., Jr.,** *J. Exp. Zool.,* 120, 109, 1952.
60. **Mulherkar, L.,** *J. Embryol. Exp. Morphol.,* 6, 1, 1958.
61. **Butros, J.,** *J. Embryol. Exp. Morphol.,* 17, 119, 1967.
62. **Chauhan, S. D. S. and Rao, K. V.,** *J. Embryol. Exp. Morphol.,* 23, 71, 1970.
63. **Butros, J.,** *J. Embryol. Exp. Morphol.,* 13, 119, 1965.
64. **Sanyal, S. and Niu, M. C.,** *Proc. Natl. Acad. Sci. U.S.A.,* 55, 743, 1966.
65. **Niu, M. C. and Deshpande, A. K.,** *J. Embryol. Exp. Morphol.,* 29, 455, 1973.
66. **Deshpande, A. K. and Siddiqui, M. A. Q.,** *Dev. Biol.,* 58, 230, 1977.
67. **Deshpande, A. K., Jakowlew, S. B., Arnold, H.-H., Crawford, P. A., and Siddiqui, M. A. Q.,** *J. Biol. Chem.,* 18, 6521, 1977.
68. **Deshpande, A. K. and Siddiqui, M. A. Q.,** *Differentiation,* 10, 1, 1978.
69. **Siddiqui, M. A. Q., Arnold, H.-H., Deshpande, A. K., Jakowlew, S. B., and Crawford, P. A.,** *Arch. Biol. Med. Exp.,* 12, 331, 1979.
70. **Romanoff, A. L.,** *The Avian Embryo,* Macmillan, New York, 1960.
71. **Paterson, B. and Strohman, R. C.,** *Dev. Biol.,* 29, 113, 1972.
72. **Emerson, C. P., Jr. and Beckner, S. K.,** *J. Mol. Biol.,* 93, 431, 1975.
73. **Garrels, J. I. and Gibson, W.,** *Cell,* 9, 793, 1976.
74. **Yablonka, Z. and Yaffe, D.,** *Differentiation,* 8, 133, 1977.
75. **Devlin, R. B. and Emerson, C. P., Jr.,** *Cell,* 13, 599, 1978.
76. **Roy, R. K., Streter, F. A., and Sarkar, S.,** *Dev. Biol.,* 69, 15, 1979.
77. **Keller, L. R. and Emerson, C. P.,** *Proc. Natl. Acad. Sci. U.S.A.,* 77, 1020, 1980.
78. **Strohman, R. C., Moss, P. S., Micou-Eastwood, J., Spector, D., Przybyla, A., and Paterson, B.,** *Cell,* 10, 265, 1977.
79. **Paterson, B. M. and Bishop, J. O.,** *Cell,* 12, 751, 1977.
80. **Devlin, R. B. and Emerson, C. P., Jr.,** *Dev. Biol.,* 69, 202, 1979.
81. **Heywood, S. M., Kennedy, D. S., and Bester, A. J.,** *Proc. Natl. Acad. Sci. U.S.A.,* 71, 2428, 1974.
82. **Heywood, S. M. and Kennedy, D. S.,** *Biochemistry,* 15, 3314, 1976.
83. **Masaki, T. and Yoshizaki, C.,** *J. Biochem.,* 76, 123, 1974.
84. **Streter, F. A., Balint, M., and Gergely, J.,** *Dev. Biol.,* 46, 317, 1975.
85. **Obinata, T., Hasegawa, T., Masaki, T., and Hayashi, T.,** *J. Biochem.,* 79, 521, 1976.
86. **Arnold, H.-H. and Siddiqui, M. A. Q.,** *Biochemistry,* 18, 647, 1979.
87. **Arnold, H-H. and Siddiqui, M. A. Q.,** *Biochemistry,* 18, 5641, 1979.
88. **Jakowlew, S. B. and Siddiqui, M. A. Q.,** *Eur. J. Biochem.,* in press.
89. **Benoff, S. and Nadal-Ginard, B.,** *Proc. Natl. Acad. Sci. U.S.A.,* 76, 1853, 1979.
90. **Benoff, S. and Nadal-Ginard, B.,** *J. Mol. Biol.,* 140, 283, 1980.
91. **Wetmur, J. G. and Davidson, N.,** *J. Mol. Biol.,* 31, 349, 1968.
92. **Jakowlew, S. B., Khandekar, P., Datta, K., Arnold, H.-H., Narula, S. K., and Siddiqui, M. A. Q.,** *J. Mol. Biol.,* 156, 673, 1982.
93. **Arnold, H.-H., Domdey, H., Weibauer, K., Datta, K., and Siddiqui, M. A. Q.,** *J. Biol. Chem.,* in press.
94. **Drabkin, H. J. and Siddiqui, M. A. Q.,** *Biochem. Int.,* 3, 533, 1981.
95. **Manasek, F. J.,** *J. Morphol.,* 125, 329, 1968.
96. **Pollinger, I. S.,** *Exp. Cell. Res.,* 76, 243, 1972.
97. **Hauschka, S. D.,** in *The Stability of the Differentiated State,* Vol. 1, Urspring, H., Ed., Springer-Verlag, Berlin, 1968, 32.
98. **Karnovsky, M. J. and Roots, L. J.,** *Histochem. Cytochem.,* 12, 219, 1964.
99. **Arnold, H.-H., Innis, M. A., and Siddiqui, M. A. Q.,** *Biochemistry,* 17, 2050, 1978.
100. **Bester, A. J., Kennedy, D. S., and Heywood, S. M.,** *Proc. Natl. Acad. Sci. U.S.A.,* 72, 1523, 1975.
101. **Bogdavsky, D., Herman, W., and Schapira, E.,** *Biochem. Biophys. Res. Commun.,* 54, 2532, 1973.
102. **Lee-Huang, S., Sierra, J. M., Navanjo, R., Filipowicz, W., and Ochoa, S.,** *Arch. Biochem. Biophys.,* 180, 276, 1977.

103. **Rao, N. J., Blackstone, M., and Busch, H.,** *Biochemistry,* 16, 2756, 1977.
104. **Kimmel, A. R. and Firtel, R. A.,** *Cell,* 16, 787, 1979.
105. **Constantini, F. D., Britten, R. J., and Davidson, E. H.,** *Nature (London),* 287, 111, 1980.
106. **Goldstein, L. and Ko, C.,** *Cell,* 2, 259, 1974.

INDEX

C

Calcium, 8
  acrosome reaction and, 35—37
  role in fertilization, 53—56
Calcium ions, 172
Calmodulin, 35, 54
5′ Capping, 100
Cardiac muscle development
  chick heart development (model system), 258—259
  control of gene activity in, 255—279
  gene-controlling RNA, 256—258
  myosin mRNAs, biogenesis of, 259—268
  7S RNA, identification and characterization of, 269—274
  7S RNA, role in cell differentiation, 274—276
Cardiac transformation, 246
CD blastomeres, 86, 95, 97, 175
CD cell, 97
CD lineages, 97
cDNA probe, 96—97, 106, 215
cDNA synthesis, 262—263
Cell contacts, 175
Cell determination, 106
Cell differentiation, 190
Cell lineages, 80—82, 85, 92, 96—98, 105, 107
  morphogenetic fate of, 74
Cell membrane, interaction between genome and, 2
Cell surface, 7—19
  cortical cytoplasm, 7—9
  external surface, 15—19
  plasma membrane, 9—11
  surface architecture, 11—15
Cell-to-cell communication, 189—190
Cell-to-cell contact, see also Direct cell-to-cell contacts, 189—190
Cell-to-cell interactions, 180—190, 230, 232—233
  molluscs, 178—179, 181
Cellular differentiation, 188
Cellular interactions, see Cell-to-cell interactions
Cenogenetic characters, 160
Centrifugation, 100, 103, 169—170, 172, 176
  experiments, 2—4
Cephalopoda, 146
Chaetognaths, 75, 77, 79, 81
Chaetopleura apiculata, 84
Chaetopterus, 93—94, 96
Chick, 124—125, 127—128, 139—145, 149, 151, 189, 194—195
  protein denaturation and stabilization, 154—155
Chick heart cells, myosin gene transcription in, 265—269
Chick heart development (model system), 258—259
Chick/mouse tissue recombinations, 205—207
Chironomid midge, 103
Chorda-mesoderm, 191
Chorio-allanotic membrane (CAM), 194—195
Chorio-allantois, 139—140
Chorion, 33, 103

Chromatin, 244, 247—248
Chromomere, 247—248, 250
Chromosome diminution, 103
Chromosome elimination, 103
Ciona, 106, 138
  intestinalis, 84, 90, 97
Cleavage, 74, 85, 88
  egg surface during, 19
Cleavage chronology in molluscs, 170—172
Cleavage furrows, 14, 95, 172
Cleavage pattern, 2, 4, 19, 171—172, 178
Cleavage planes in molluscs, 170—172
Cleavage products, 97
Clock mechanism, 170—171
Cloned DNA probes, see cDNA probe
Clones of spermatozoa, 35
C macromeres, 175
Colcemid, 169
Colchicine, 8, 10, 169
Collagen, 223—224
Compartmentalization, 2, 56
Compression, 170
Concanavalin A, 9, 11, 19, 34
Copper, 123
Cordulia aenea, 84
Corneal development, 19
Cortex, 90—91, 93, 100, 154, 168—170, 172, 175—176
Cortical contractions, 8
Cortical cytoplasm, 3, 7—9, 19, 85, 93
Cortical cytoskeleton, 6
Cortical determinants, 172
Cortical granules, 3, 6, 8, 10, 39, 45—46, 49, 51, 54
Cortical pattern, 8
Cortical prepattern, 4—5
Cortical protein, 8
Cortical reaction, 51—52
Cortical scaffold, 3—4
Cranial neural crest cells, 192
Crayfish, 116, 123
Cross-furrow macromeres, 179
Crustaceans, 81, 84
Crystallin genes, 242
Cycloheximide, 211
Cyclopic embryos, 145
Cyclops strenuus, 81
Cytochalasin, 5, 8—12, 97, 169, 171—172
Cytochalasin D, 11
Cytodifferentiation, 211
Cytokinesis, 14
Cytoplasm, 96, 170—171, 176—177
Cytoplasmic determinants, 3
Cytoplasmic factors, 74
Cytoplasmic inclusions, 3
Cytoplasmic localizations, 2, 10—11, 19, 168, 174, 181
Cytoplasmic maturation, 37—38
Cytoplasmic organelles, 81—89
  mRNAs, 88—89
  nuage, 82—85